272
Topics in Current Chemistry

AF786005

Topics in Current Chemistry
Recently Published and Forthcoming Volumes

Bioactive Conformation II
Volume Editor: Peters, T.
Vol. 273, 2007

Bioactive Conformation I
Volume Editor: Peters, T.
Vol. 272, 2007

Biomineralization II
Mineralization Using Synthetic Polymers and Templates
Volume Editor: Naka, K.
Vol. 271, 2007

Biomineralization I
Crystallization and Self-Organization Process
Volume Editor: Naka, K.
Vol. 270, 2007

Novel Optical Resolution Technologies
Volume Editors:
Sakai, K., Hirayama, N., Tamura, R.
Vol. 269, 2007

Atomistic Approaches in Modern Biology
From Quantum Chemistry
to Molecular Simulations
Volume Editor: Reiher, M.
Vol. 268, 2006

Glycopeptides and Glycoproteins
Synthesis, Structure, and Application
Volume Editor: Wittmann, V.
Vol. 267, 2006

Microwave Methods in Organic Synthesis
Volume Editors: Larhed, M., Olofsson, K.
Vol. 266, 2006

Supramolecular Chirality
Volume Editors: Crego-Calama, M.,
Reinhoudt, D. N.
Vol. 265, 2006

Radicals in Synthesis II
Complex Molecules
Volume Editor: Gansäuer, A.
Vol. 264, 2006

Radicals in Synthesis I
Methods and Mechanisms
Volume Editor: Gansäuer, A.
Vol. 263, 2006

Molecular Machines
Volume Editor: Kelly, T. R.
Vol. 262, 2006

Immobilisation of DNA on Chips II
Volume Editor: Wittmann, C.
Vol. 261, 2005

Immobilisation of DNA on Chips I
Volume Editor: Wittmann, C.
Vol. 260, 2005

Prebiotic Chemistry
From Simple Amphiphiles to Protocell Models
Volume Editor: Walde, P.
Vol. 259, 2005

Supramolecular Dye Chemistry
Volume Editor: Würthner, F.
Vol. 258, 2005

Molecular Wires
From Design to Properties
Volume Editor: De Cola, L.
Vol. 257, 2005

Low Molecular Mass Gelators
Design, Self-Assembly, Function
Volume Editor: Fages, F.
Vol. 256, 2005

Anion Sensing
Volume Editor: Stibor, I.
Vol. 255, 2005

Bioactive Conformation I

Volume Editor: Thomas Peters

With contributions by

S. V. Evans · J. A. Alfaro · D. Fischer · A. Geyer
N. J. Harmer · S. W. Homans · H. Kessler · B. W. Koenig
S. R. LaPlante · A. Modlinger · S. Narayanan · B. Reif
B. Schuman · T. Weide

Springer

The series *Topics in Current Chemistry* presents critical reviews of the present and future trends in modern chemical research. The scope of coverage includes all areas of chemical science including the interfaces with related disciplines such as biology, medicine and materials science. The goal of each thematic volume is to give the nonspecialist reader, whether at the university or in industry, a comprehensive overview of an area where new insights are emerging that are of interest to a larger scientific audience.
As a rule, contributions are specially commissioned. The editors and publishers will, however, always be pleased to receive suggestions and supplementary information. Papers are accepted for *Topics in Current Chemistry* in English.
In references *Topics in Current Chemistry* is abbreviated Top Curr Chem and is cited as a journal.
Visit the TCC content at springerlink.com

ISSN 0340-1022

ISBN 978-3-642-08035-7 e-ISBN 978-3-540-49078-4

Springer is a part of Springer Science+Business Media

springer.com

Cover design: WMXDesign GmbH, Heidelberg

Volume Editor

Prof. Dr. Thomas Peters

Institute of Chemistry
University of Lübeck
Ratzeburger Allee 160
23538 Lübeck, Germany
Thomas.peters@chemie.uni-luebeck.de

Editorial Board

Topics in Current Chemistry Also Available Electronically

For all customers who have a standing order to Topics in Current Chemistry, we offer the electronic version via SpringerLink free of charge. Please contact your librarian who can receive a password or free access to the full articles by registering at:

springerlink.com

If you do not have a subscription, you can still view the tables of contents of the volumes and the abstract of each article by going to the SpringerLink Homepage, clicking on "Browse by Online Libraries", then "Chemical Sciences", and finally choose Topics in Current Chemistry.

You will find information about the

- Editorial Board
- Aims and Scope
- Instructions for Authors
- Sample Contribution

at springer.com using the search function.

Preface

Specific binding of a ligand to a receptor is a key step in a variety of biological processes, such as immune reactions, enzyme cascades, or intracellular transport processes. The ligand–receptor terminology implies that the receptor molecule is significantly larger than the ligand, and the term "bioactive conformation" usually characterizes the conformation of a ligand when it is bound to a receptor. In a more general sense, bioactive conformation applies to any molecule in a biologically relevant bound state regardless of size considerations. Most of the contributions to this book address ligands that are much smaller than their receptors.

X-ray crystallography and high resolution NMR spectroscopy are the two main experimental techniques used to study bioactive conformations. Therefore, the two volumes of this book cover approaches that use either of the two techniques, or a combination thereof. The combination of X-ray crystallography and NMR spectroscopy is particularly useful when a crystal structure of a receptor protein, but not of the receptor protein–ligand complex, is available. A number of experimental techniques to analyze the bioactive conformation of a ligand with NMR are based on the observation of the resonance signals of the free ligand that is in exchange with the bound ligand. Several chapters focus on such approaches that range from "classical" transferred NOE experiments, to transferred dipolar couplings, to STD (Saturation Transfer Difference) NMR techniques. In cases where tight binding in the sub-nanomolar range prevents the analysis of the bioactive conformation via free ligand signals, the ligand–protein complex has to be analyzed with protein NMR-based techniques or by crystallography. Since this area has been the subject of many reviews and monographs it will not be covered here in particular detail. As a unifying theme, all contributions target the question of how molecular recognition of biologically active molecules is achieved on the atomic scale. Depending on the research topic the results from these studies have a strong impact not only in basic research but also in several fields of application ranging from pharmaceutical applications to the use of biomolecules as, for example, cryoprotectants.

Almost all contributions to the two volumes highlight the fact that ligand–protein complexes cannot be treated as static ensembles. On both sides, the ligand and the receptor side, dynamic processes contribute to the molecular recognition. In this sense it is hoped that these two volumes of *Bioactive Conformation* will sensitize us for the need to invent and develop more experimental techniques to study the dynamic aspects of bioactive conformations.

Lübeck, October 2006 Thomas Peters

Contents

Contents of Volume 273

Bioactive Conformation II

Volume Editor: Thomas Peters
ISBN: 978-3-540-49079-1

Top Curr Chem (2007) 272: 1–50
DOI 10.1007/128_052

Published online: 22 July 2006

Spatial Screening for the Identification of the Bioactive Conformation of Integrin Ligands

Timo Weide · Armin Modlinger · Horst Kessler (✉)

Department Chemie, Technische Universität München, Lichtenbergstr. 4, 85747 Garching, Germany
horst.kessler@ch.tum.de

Abstract The development of low-molecular-weight, non-peptidic, and orally available drugs starting from a biologically active peptide is one of the great challenges in medicinal chemistry. In the absence of a crystal structure of the receptor, success in mimicking natural peptide ligands with potent non-peptides has been elusive. A systematic stepwise strategy has been developed to accomplish these goals. These include determining the primary amino acid side chain residues required for molecular recognition via an "Ala scan" and the preferred backbone conformation, which can serve as a template for the bioactive conformation. By "spatial screening" the recognition motif is embedded in cyclic hexa- or pentapeptides, whose conformation is controlled by distinct incorporation of turn-inducing D-amino acid. This procedure was demonstrated on the integrin-binding tripeptide motifs RGD and LDT. The bioactive conformation was derived from conformation–activity relations and used for the design of non-peptidic biased libraries. The recently available crystal structure of the $\alpha_V\beta_3$-integrin head groups with the highly active cyclic peptide developed in our group confirmed the indirectly derived receptor bound conformation and now allows a structure-based design of new integrin ligands. More recently, the head group of the integrin $\alpha_{IIb}\beta_3$ in complex with antibodies was also structurally solved. Homology modeling of other related integrins is used to understand and design integrin subtype specificity.

Keywords Bioactive conformation · Cyclic peptides · LDT and RGD mimetics · Ligand-based drug design · Spatial screening

Abbreviations

CPP	Cyclic pentapeptide
CS-1	Peptide from epitope mapping of the IIICS domain from fibronectin
ECM	Extracellular matrix
IgSF	Immunoglobulin superfamily
LPAM	Lymphocyte peyer's patch specific adhesion molecule-1
MAdCAM-1	Mucosal addressin cell adhesion molecule-1
SAR	Structure–activity relationship
VCAM-1	Vascular cell adhesion molecule-1
VLA	Very late activation antigen
VLA-4	$\alpha_4\beta_1$-integrin

1 Introduction

Many important biological and pathological processes at a molecular level are triggered by protein–protein or by peptide–protein interactions. Interference with these interactions remains one of the great challenges in the development of new therapeutics. The use of therapeutic antibodies for this purpose represents the fastest growing segment of the prescription-drug market [1]. They are highly specific for their molecular targets and tend to be very stable in human serum. But antibodies lack oral bioavailability and cell permeability. They also suffer from difficulties in manufacture and are very expensive. Hence, the search for orally available small-molecule modulators for protein–protein interactions is an ongoing challenge [2–5]. However, several problems exist in the application of small molecules for inhibiting protein–protein interactions. The size of a typical protein–protein interface is approximately 750–1500 Å^2 and examinations of the shape usually do not reveal small and deep cavities that look like small-molecule-binding sites [6].

Nevertheless, peptides and small molecules may modulate protein–protein interactions when binding to the "hot spots" (subset of the interface that contributes to high-affinity binding) of the protein. However, there are already several examples demonstrating the possibility of interfering with protein–protein interactions: e.g. taxol and epothilone derivatives stimulate tubulin polymerization [7,8]. On the other hand, integrins that bind to large extracellular matrix proteins are bidirectional cellular receptors which recognize only a small peptide sequence, typically a tripeptide such as RGD or LDT [9]. In this field the main challenge is to find ligands which are specific, selective, and superactive to inhibit subtypes of these receptors [10]. The general procedure in the search for ligands to modulate protein–protein interactions in cases with unknown three-dimensional (3D) structures of the receptor begins with the identification of binding peptides (Fig. 1).

One may start from sequence comparison of already existing natural ligands or by epitope mapping of the interacting proteins. Another approach is phage display [11,12]. Then, small peptides having similar sequences compared to binding sites of natural ligands are synthesized and tested for their binding affinity. The usual procedure for examining the importance of amino acid side chains is an "Ala scan", in which each amino acid is successively substituted by an alanine and the influence of this exchange on biological activity is investigated [13]. A first hint for identification of turn structures in the peptide is obtained from successive substitution of each amino acid by its D-enantiomer (D-scan). Also, double D-scans have been used [14]. The next

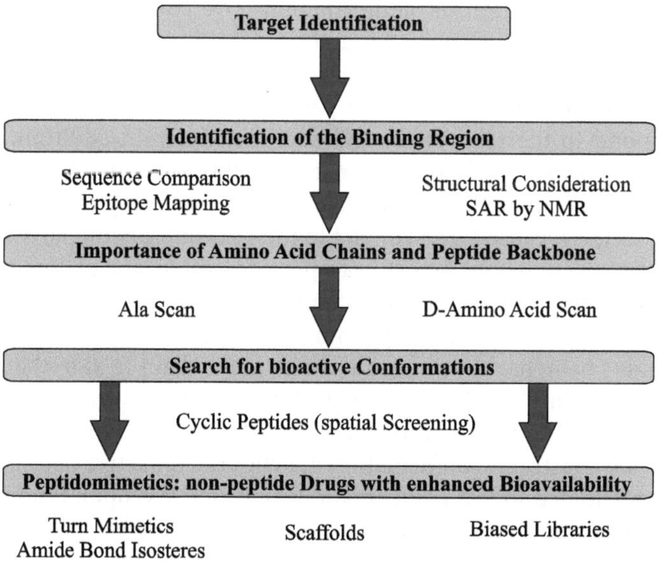

Fig. 1 *De novo* approach for peptidomimetic design. For further explanation see text

step is the introduction of further restriction of the conformational space by peptide cyclization [15–17]. A procedure to screen the space has been introduced by embedding of the binding motif in cyclic penta- and hexapeptides which constrains the conformation. Shifting of the binding motif around the cyclic peptide by systematic use of D- or L-amino acids results in different spatial orientations of the pharmacophoric groups aimed to elucidate the bioactive conformation [18, 19].

Side chains of peptides also have intrinsic flexibility. With increasing preference of backbone conformations (usually called: with increasing "rigidity") often also distinct side chain conformations are more preferred. This is especially true for rotamers about the $C_\alpha - C_\beta$ bond (χ_1) (see Fig. 4). Extensive synthetic efforts have been made to restrict or to prefer different side chain conformations and to correlate them with biological activity [20]. Today, the concept of "rigid" complementarity has led to attempts to develop conformationally restricted analogues of inherently flexible substances in order to delineate the "bioactive conformation" [15]. Conformational restriction may result in the discovery of specificity among the members of multireceptor families, as receptor subtypes have often evolved to recognize different (bioactive) conformations. However, even highly conformational restricted molecules usually can adopt multiple, closely related conformations so that the term "bioactive conformation" describes an average of a limited population of closely related conformers.

Peptides possess a large number of flexible torsion angles which produce an enormous number of conformations in solution. A tripeptide with six flexible bonds could adopt over 65 000 possible conformations. The number depends on the definition of "conformation". We consider here a conformation as an energy minimum characterized by barriers over kT between the next conformation [21]. However, one always has to be aware that the preferred conformations of (flexible) peptides in solution may not necessarily correspond to the receptor-bound conformations. In addition, even the bound conformation could be different from the "bioactive" one, when the structure of the complex is not identical with the one responsible for the biological effect, e.g. caused by a contact to the next interaction partner. Hence, conformational studies of free linear peptides often are of low relevance. A common approach to reduce the conformational space of linear peptides has been achieved by cyclizations (amino terminal to carboxy terminal, side chain to side chain, backbone to side chain, main chain to side chain, and so on) [22], often using amide, alkyl, alkenyl, thioether, etc., connectors [23].

Such a procedure is the more successful the smaller the ring size is and the better the constraint conformation matches the bioactive one. The reason for this is attributed to less conformational entropy lost upon binding and to induction of strains necessary to adopt the binding conformation [21, 24].

A short sequence consisting of three or four amino acids in loop regions of proteins often serves as the basic recognition element for ligand binding [25–27]. In these cases, the entire molecule provides a scaffold for fixing the sequence in its active conformation. In various binding studies, small peptides containing a short peptide sequence have been shown to successfully compete with the larger proteins as their recognition residues are associated with a β-turn or β-sheet, which can serve as a template for "bioactive conformation" [17, 28, 29]. From the viewpoint of structural chemistry, β-turns have been attributed a special importance as carriers of molecular recognition since, with the context of pharmacophore arrangement, they allow a sterically controlled presentation of two to four amino acid side chains. The replacement of a peptide with retention of the pharmacophoric groups by non-peptide, drug-like structures, so-called "β-turn mimetics", has progressed to become a topic in the structural design of peptide mimetics (Fig. 2). There are numerous examples of β-turn mimetics and this topic has been extensively reviewed [30–35].

In many cases, the secondary structure of a peptide or protein serves primarily as a template or scaffold for presentation of the specific side chain groups in three-dimensional space (shape) for molecular recognition and biological activity [36]. The first successful application of conformational restriction to peptide chemistry was carried out by Veber et al., who tried to simplify and reduce the structure of somatostatin in the attempt to produce selective derivatives for the glucagon or insulin receptor [37]. Replacement of L-tryptophan in position 8 of somatostatin by D-tryptophan produced an analogue that retained and even slightly increased biological activity by stabilizing a βII′-turn about Phe-Trp-Lys-Thr in the putative recognition site. After several iterations, a biologically active cyclic hexapeptide was discovered in which only six of the original 14 amino acids were needed to produce a fully active derivative [17]. Other prominent examples, where small cyclic peptides derived from natural ligands with retained or improved biological activities could be obtained, are the development of enkephalin analogues [38], cholecystokinin-like peptides, or melanotropin peptides [20, 39], or, more recently, in p53/HDM2 interaction [40].

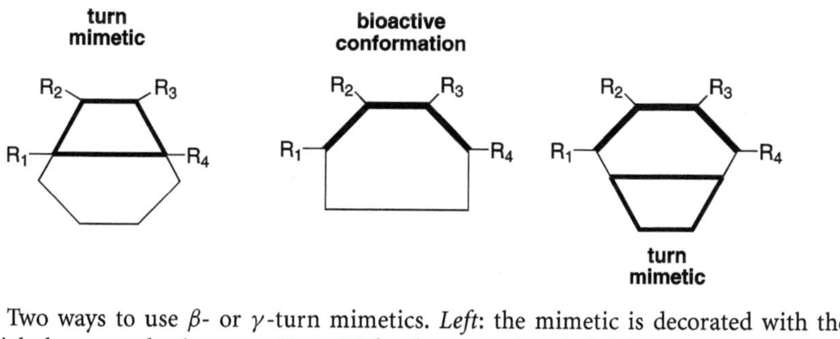

Fig. 2 Two ways to use β- or γ-turn mimetics. *Left*: the mimetic is decorated with the essential pharmacophoric groups R_{1-4}. *Right*: the turn mimetic is incorporated into small cyclic peptides and forces the (otherwise flexible) structure carrying the pharmacophores into the bioactive conformation

The concept of cyclization is also adopted by microorganisms. For defense mechanisms, nature favors cyclopeptides which are less prone to enzymatic degradation and which have a higher bioavailability and reduced conformational flexibility [41].

In this review we will describe the application of the concept using conformationally restricted compounds for the development of highly active and selective integrin ligands as modulators of protein–protein interactions. We will present the investigations associated with RGD- and LDT-derived compounds whose tripeptide sequences are known as common integrin recognizing motifs. In the next sections we focus mainly on the influence of the conformation of the binding profile.

Some members of the integrin family recognize and bind the RGD sequence [10,42,43]. Despite this common motif, a high degree of specificity of different RGD containing extracellular matrix (ECM) proteins for distinct integrin subtypes is observed. This could be explained by assuming that the RGD sequence shared binding sites are responsible for specificity. However, our work has demonstrated that the conformation controls most of the specificity at least between α_V- and α_{IIb}-integrin subtypes [44,45]. Another motif in endogenous ligands for $\alpha 4$ integrins is the tripeptide LDT, which is present in the CD loop of domain I of MAdCAM-1 and the QIDSP sequence in VCAM-1, respectively [46,47].

First we will discuss general aspects of conformations of cyclic peptides and the benefits that may arise from introducing conformational restraints. In the following section we will review the approach based on the design of preferred conformations of peptides and peptidomimetics with reference to RGD- and LDT-derived compounds.

2 Relationship Between Ligand Flexibility and Binding Affinity

It is generally assumed that the more the structure of the free ligand in solution already resembles the structure of the ligand in the complex, the stronger is the binding to a given receptor. A powerful tool for obtaining information about the conformation at the receptor site is by introducing conformational restraints. If the constrained compounds are active, the preferred structure of the binding motif should be more closely related to the receptor-bound peptide conformation than the corresponding linear sequences (matched case). *Vice versa*, if conformational restriction forces the pharmacophoric motif in a conformation not related to the active structure, reduced activity or inactivity results (mismatched case) [15,28].

For a qualitative rationalization of activity and selectivity for different ligands for various receptors, we represent the multidimensional conformational space in a two-dimensional ellipse (Fig. 3). The grey fields correspond to conformations of the receptor-bound ligand. Different receptor subtypes often recognize different "bioactive conformations" of the ligand. The empty circles represent the conformational space of the different conformationally restricted ligands in solution. Strongly constrained ligands (such as ligands a and b) have a smaller conformational space than more flexible ones (ligand d). Flexible ligands in general should have a higher probability to bind to given receptors, but they are less active (see ligand d). Constrained ligands with a small conformational space would in many cases exhibit no activity, but when the allowed conformational space strongly overlaps with the required space for a receptor, high activity and selectivity will result (matched case, ligand a).

It is obvious from these considerations that cyclization in most cases will not result in a matched case. The question of how to screen a large area of the conformational space is reviewed below.

Often cyclization via side chains, e.g. via disulfide bridged Cys residues, is discussed to obtain "rigid" conformations due to the easy introduction without special chemical steps. However, although forming a disulfide bridge restricts the conformational space to some extent, this does not necessarily introduce strong constraints. Even a cyclic CXYC peptide is not forcing a β-turn but it still allows forming it. Larger disulfide-cyclic peptides C–$(X)_n$–C with $n \geq 6$ certainly still allow the adoption of an enormous number of conformations [48]. Cyclization is also often assumed to be a particularly meritorious strategy for the design of ligands or drugs with higher affinity and selectivity [15, 49]. Reduced molecular flexibility by means of fewer rotatable bonds is known to increase the probability of oral bioavailability [50].

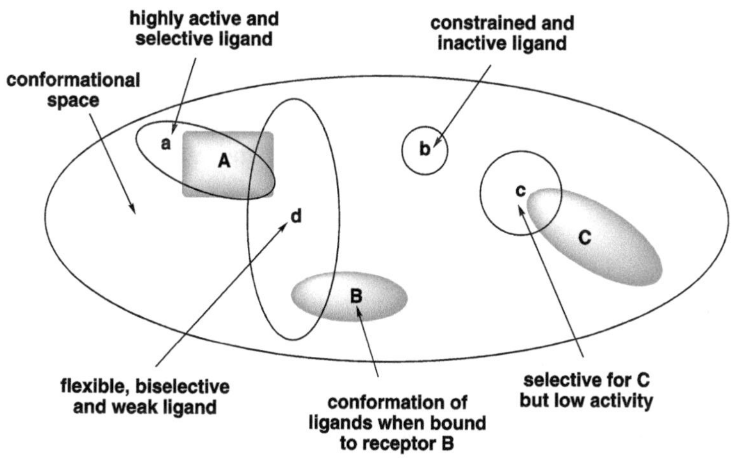

Fig. 3 The conformational space for flexible and rigid ligands. A, B, and C represent bioactive conformations of different receptor subtypes. Several possibilities of biological activities from constraint ligands are shown

In the following section we will briefly discuss conformations of cyclopeptides in solution and the procedure of "spatial screening" in order to find preferred ligand conformations.

2.1
Understanding Preferred Conformations of Cyclic Peptides

In proteins, secondary structure elements such as α- or 3_{10}-helices and β-sheets adopt rather distinct areas in their ϕ- and ψ-space (for the definition of bond angles, see Fig. 4). Plotting the values of ϕ- and ψ-angles for each amino acid in the Ramachandran diagram is also used to validate the quality of a protein structure [51]. Usually, loop regions of proteins are structurally less well determined and more flexible. Those loops often include β- or γ-turns for which several subtypes exist [25, 52, 53]. These turns are exposed and are therefore important convex recognition elements for binding cavities of interacting proteins.

Small linear peptides are usually very flexible and a large number of conformations with similar energies are in rapid exchange with each other [15, 21]. Some restraints could be introduced by the cyclic amino acid proline and/or internal hydrogen bonds, especially in non-aqueous solutions. Also, hydrophobic clusters, e.g. by interaction of aromatic residues with an adjacent proline, may lead to conformational preference. A successful *de novo* design of linear peptides which formed a β-hairpin structure was presented [54]. However, such preferences do not confirm a similar bioactive conformation [55, 56]. A recent example is the interaction of p53 with MDM2 which is shown to adopt a helical structure of the p53 site [57, 58] but also β-sheet peptides allow strong binding to this E3-type inhibition site in MDM2 [40].

More constraints can be introduced by cyclization [17]. We prefer head-to-tail cyclic penta- and hexapeptides because they provide enough restrictions in their backbone conformation without forcing *cis*-peptide bonds (such as in cyclic tetrapeptides) or having too much flexibility (such as in larger rings). Often ligands of receptors of hormones require *C*-terminal free carboxamides

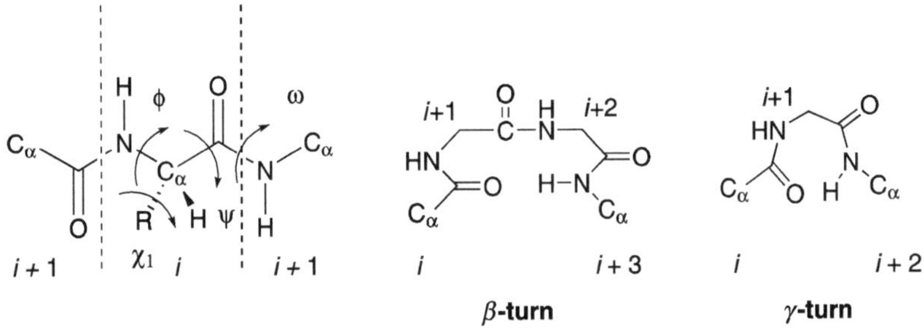

Fig. 4 Definition of the angles in peptides to describe the backbone conformation and schematic depiction of turns

which contradicts head-to-tail cyclization. Then, procedures such as the so-called "backbone cyclization" are used [22].

In linear peptides usually all secondary amide bonds are *trans* ($\omega = 180°$). The population of the *cis* isomer ranges from 0.1 to 1.0% [59–61]. This is also valid for cyclic penta- or hexapeptides. As a rare case, an unexpected *cis* peptide bond population of about 6% was observed in the somatostatin analogue cyclo(-D-Ala-Phe-Val-Lys(Z)-Trp-Phe-) containing only secondary amide bonds [62]. If the hydrogen of the secondary amide bond is substituted by an alkyl residue, such as in proline or other *N*-alkylated peptides, the similar size of alkyl and the substituted α-carbon leads to comparable steric hindrance in both *cis* and *trans* amide bonds. Therefore, both rotamers are observed [15, 53, 63, 64]. It is assumed but not yet generally proven that *N*-methylation of an amino acid restricts the ϕ- and ψ-angles of this amino acid as well as the angles of the residue preceding it in sequence.

Apart from *cis*/*trans* isomerization, the exchange between different conformations in cyclic penta- and hexapeptides is fast on the NMR time scale. Therefore, symmetric peptides such as cyclo[Gly_5], cyclo[Phe_5], and cyclo[Gly_6] exhibit only one single signal set for all amino acids in their NMR spectra [15, 65]. The explanation for this may either be an inherent C_5 symmetry of the molecule or a five-fold degenerate, rapidly established equilibrium of one or more conformations [15, 66, 67]. Evidence for a rapid equilibrium was provided from the following experiment: changing one Gly into Ala or Phe exhibited gradual differences in the NMR parameters of the glycines due to increased conformational preferences. Yet, fast introversions of conformations cannot be excluded. In fact, it is very difficult to decide whether a molecule is conformationally homogeneous or not [15].

The following NMR parameters serve as criteria for the presence of conformational preferences of the peptide backbone:

- Strong chemical shift differences of NH protons.
- Strong chemical shift differences of α-protons of chemically identical amino acids.
- Strong differences of temperature gradients of the NH chemical shifts. In DMSO, a value of > -2.5 ppb/K (e.g. a value of -1.0 ppb/K) indicates protection from the solvent and often involvement of intramolecular hydrogen bonds, whereas values < -5 ppb/K (e.g. a value of -6 ppb/K) are typical for solvent-exposed NH protons.
- Backbone conformational restriction is often accompanied by conformational preferences of side chains reflected by differentiation of the diastereotopic β-protons, especially in their scalar coupling to the α-proton ($^3J_{H_\alpha H_\beta}$, $^3J_{H_\alpha H'_\beta}$).

In contrast to general perception, the formation of intramolecular hydrogen bonds has only a small influence on the conformation of cyclic peptides [68]. However, we and others have found that the configuration of an amino acid has a significant influence on the conformation (steric hindrance by β-carbons) [36, 45].

In the search for preferred conformations of cyclic penta- and hexapeptides we synthesized all possible diastereomeric alanine peptides (Fig. 5) and investigated their NMR spectra according to the criteria given above for the presence of conformational preferences [69]. In addition, we also studied some of those peptides containing Gly, D-Pro, or L-Pro, because due to steric reasons Gly can occupy the position of a D- or L-amino acid. Proline has a stronger effect on conformational preferences than D- or L-alanine.

One of the strongest preferences in pentapeptides was found for cyclo(-D-Ala-Ala$_4$-) and cyclo(-D-Pro-Ala$_4$-). Hence, we consider this molecule as a conformational template, i.e. a substitution of any of the alanine by any other amino acid (except Gly or Pro) will exhibit a similar conformation [70]. The D-amino acid is found in the position $i+1$ of a βII$'$-turn in such peptides allowing the formation of a γ-turn around Ala^3. The latter position is less well defined and often shows flexibility [70–74].

In cyclic hexapeptides conformations with two β-turns are often observed [75]. If the ring contains only one D-Ala or one D-Pro, this residue again prefers the position $i+1$ of a βII$'$-turn. The second β-turn is found on the opposite side, often in fast equilibrium between βI and βII [36]. Again, the position of the D-amino acid "dictates" the conformational preference.

To explore the bioactive conformations of the RGD sequence these two templates (penta- and hexapeptides with one D- and four or five L-amino acids, respectively) were selected. The gene data bank revealed that the RGD sequence in proteins of the extracellular matrix (ECM) is often flanked by an N-terminal valine and by a C-terminal phenylalanine. Hence, we took the sequences RGDFV and RGDFXV and synthesized five cyclopentapeptides (CPPs) and six cyclohexapeptides in which each amino acid was successively chosen as its D-enantiomer (Gly was taken as D-amino acid equivalent). Assuming a similar influence of the D-residues in each of these peptides such as

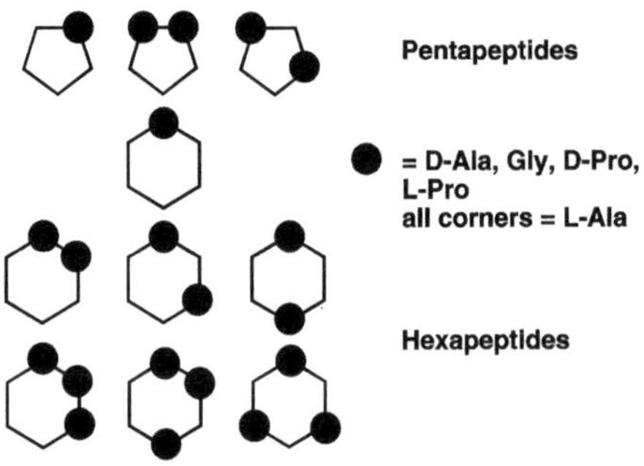

Fig. 5 Search for a preferred (not necessarily rigid!) conformation. The configuration of the amino acids accounts for the spatial arrangement

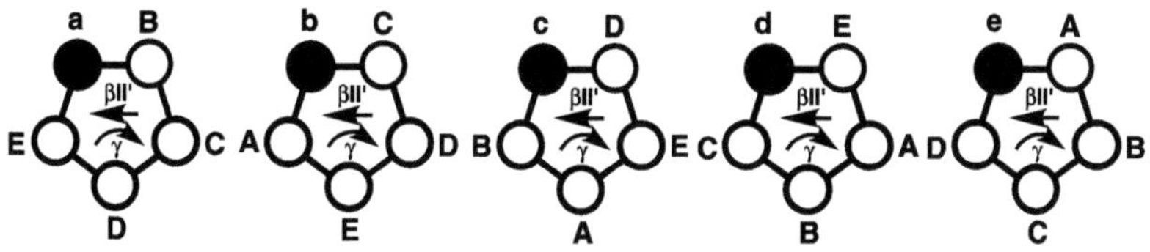

Fig. 6 Cyclopentapeptides containing one D-amino acid, which preferably is located at the position $i + 1$ of the βII′-turn [19]

D-Ala in the alanine peptide, we can shift the RGD sequence around the scaffold conformation as shown for the case of pentapeptides in Fig. 6 [44, 76]. NMR investigation proved this hypothesis. As the peptides only differ in their stereochemistry, we dubbed this procedure spatial screening [19].

As anticipated, the biological activities and selectivities for $\alpha_V\beta_3$ and $\alpha_{IIb}\beta_3$ were totally different. The cyclic pentapeptide cyclo(-RGDfV-) exhibited a 10^3 higher inhibition of $\alpha_V\beta_3$/vitronectin binding than the linear reference peptide GRGDSP but at the same time an about 10 fold lower activity against $\alpha_{IIb}\beta_3$/fibrinogen. *Vice versa*, the cyclic hexapeptide cyclo(-GRGDfV-) exhibited high activity for $\alpha_{IIb}\beta_3$/fibrinogen but very low activity for the $\alpha_V\beta_3$/vitronectin system [77].

The "spatial screening", originally applied by us for exploring the bioactive conformation of thymopentin [78, 79], is now also used successfully for other systems. There are numerous studies of the utility of CPP-based libraries which consist of both "conformational-based" and "sequence-based" libraries of lead discovery in drug research. For example, Fujii et al. generated libraries of CPPs of which three compounds exhibited bioactivities comparable to a 14 amino acid containing antagonist (T140) of CXCR4 [80, 81]. Libraries of CPPs were also investigated in the search for active endothelin antagonists [82, 83], LHRH antagonists [84], and antagonists for the B_2 kinin receptor [85, 86]. For hexapeptides, the induction of distinct conformations was also used to embed the active sequence of tendamistat, an α-amylase inhibitor, into a cyclic hexapeptide template maintaining the spatial arrangement of the residues in their native, biologically active conformation [36, 87]. More recently, spatial screening has also been applied to the LDT sequence [88]. As long as the three-dimensional structure of the protein/protein or protein/ligand complexes are not available, cyclic peptides are optimally suited to explore the structural requirements with respect to the arrangement of the pharmacophoric groups on a semirational basis.

3 RGD-derived Peptides

3.1 Introduction

The ability of various cell types to adhere to and interact with other cells or with components of the extracellular matrix is essential for maintaining cellular function and tissue integrity via signaling between and within the communicating cells. The interaction of various cells with the ECM is *inter alia* carried out via a family of cell surface heterodimeric receptors called integrins that are present on most cell types, including lymphocytes and platelets [89]. It is important to note that the expression and activation of different integrin subtypes strongly depends on cell types and the differentiation level of cells. Integrin levels are regulated. These receptors are composed of one α- and one β-subunit, both characterized by a large *N*-terminal extracellular domain, a transmembrane domain, and a short *C*-terminal intracellular tail (Fig. 7). Eighteen α- and eight β-mammalian subunits are known, which assemble non-covalently into more than 24 different heterodimers (Table 1) [9, 90–92].

The number of integrins and the remarkable breadth of their cellular distribution support the statement that the phenotype of virtually every cell is uniquely influenced by its display of integrins [93–96]. The most common integrin binding sequence is the RGD [97] motif found within many

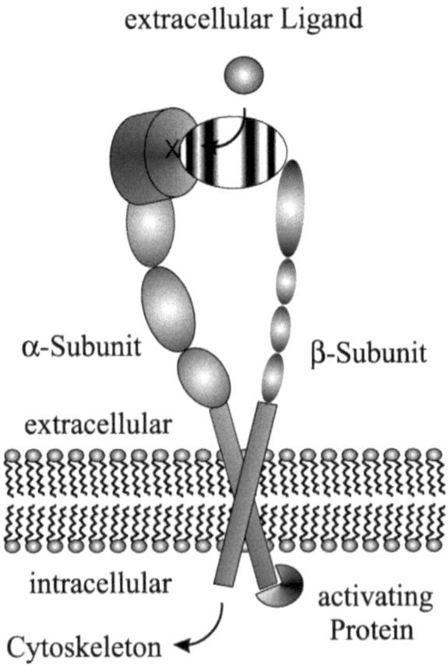

Fig. 7 Schematic representation of a typical integrin consisting of an α- and a β-subunit

Table 1 Selected recognition sites of integrins

Integrin	Ligand	Sequence
$\alpha 4\beta 7$	MadCAM-1, fibronectin, VCAM-1	LDT
$\alpha 4\beta 1$	Fibronectin, invasin, VCAM-1	IDS
$\alpha 5\beta 1$	Fibronectin	RGD
$\alpha 6\beta 1$	Laminin	RGD
$\alpha IIb\beta 3$	Fibrinogen, fibronectin, vitronectin, von Willebrand factor, thrombospondin	RGD
$\alpha v\beta 3$	Vitronectin, laminin, fibronectin, von Willebrand factor, thrombospondin, tenascin, osteopontin, fibrillin, fibulin	RGD
$\alpha 6\beta 1$	Laminin	RGD
$\alpha L\beta 2$	ICAM-1, ICAM-2	RGD
$\alpha 9\beta 1$	Collagen, laminin, tenascin	RGD
$\alpha 7\beta 1$	Laminin	RGD

extracellular matrix proteins and disintegrins [98, 99]. Disintegrins are low molecular mass (5–9 kDa), cystein-rich peptides isolated from the venom of viper snake species which exert their biological activities by preventing the binding of adhesive ligands to integrin receptors. Disintegrins such as echistatin [26, 27], kistin [97], flavoriden [97], and albolabrin [100, 101] carry the (R/K)GD tripeptide integrin-binding motif present at the tip of a highly mobile loop. The activity of disintegrins depends on the appropriate pairing of eight to 14 cysteines by disulfide bridges that maintain the RGD-containing loop in its active conformation. It was shown that extracellular matrix proteins including vitronectin, von Willebrand factor, fibrinogen, thrombospondin, fibronectin, and laminin are RGD-dependent adhesion proteins which bind to a variety of either specific or promiscuous integrin receptors [90]. Integrins are known to play significant roles in diseases and have been extensively studied in areas such as thrombosis, inflammation, angiogenesis, osteoporosis, and cancer. Of pioneering importance in this respect is the platelet- and megakaryocyte-specific integrin $\alpha_{IIb}\beta_3$, which has received considerable attention as a drug target due to its requisite role in platelet aggregation, a significant mechanism in mediating arterial thrombosis [102]. The $\alpha_v\beta_3$-integrin, also called the vitronectin receptor, is expressed on the surface of a variety of migrating cell types. This integrin plays a key role in numerous physiological processes such as angiogenesis [10, 45, 103, 104], apoptosis, and bone formation and resorption [105]. Both of these integrins bind to RGD.

These investigations led to the concept that the shape and size of the RGD loop are important structural features that affect fitting of the ligand to the binding pocket of integrin receptors. For instance, wider RGD loops result in an increase of anti-$\alpha_{IIb}\beta_3$ and a decrease of anti-$\alpha_v\beta_3$-activity [48, 76, 106, 107]. Extensive screening studies with RGD peptides and mimetics without the knowledge of the protein structure have led to the development of peptidic and non-peptidic superactive and selective RGD-based compounds. In 2002 the structure of the head group of $\alpha_V\beta_3$-integrin complexed with Cilengitide® [cyclo(-RGDf(NMe)Val-)] [108] was solved by X-ray diffraction [109]. The crystal structure of the extracellular domains of the $\alpha_v\beta_3$- [109, 110] and $\alpha_{IIb}\beta_3$-integrins [111] allows us to explain numerous results of structure–activity-relationship (SAR) studies in the RGD field (see below) at atomic level [109–114].

3.2
Cyclic RGD Peptides

Cyclic RGD peptides containing cystein disulfide bridges have been studied very early on [100, 115, 116]. Such peptides may also achieve receptor selectivity. Nevertheless, we do not treat this field here.

Intensive studies in pharmaceutical industries have already led to highly active compounds for the inhibition of platelet aggregation mediated by the $\alpha_{IIb}\beta_3$-integrin [102, 117]. As mentioned above, spatial screening of the sequence VRGDFX (X = Ala or nil) yielded the cyclic pentapeptide cyclo(-RGDfV-) which exhibited an about 1000-fold increase of activity to inhibit binding of $\alpha_v\beta_3$ to vitronectin than the reference compound GRGDSPK and more than a 10-fold reduced activity for the $\alpha_{IIb}\beta_3$-integrin [44, 76, 77]. Hence, the spatial presentation of the pharmacophoric groups obviously matches the receptor bound conformation for $\alpha_v\beta_3$. The conformation of cyclo(-RGDfV-) was later studied in detail [70]. The βII′-turn about D-Phe-Val is well defined but the γ-turn about glycine is less rigid (Fig. 8). As found in other cyclic pentapeptides, the RG- and GD-peptide bonds can flip to a conformation in which two adjacent peptide bonds orient their NH protons

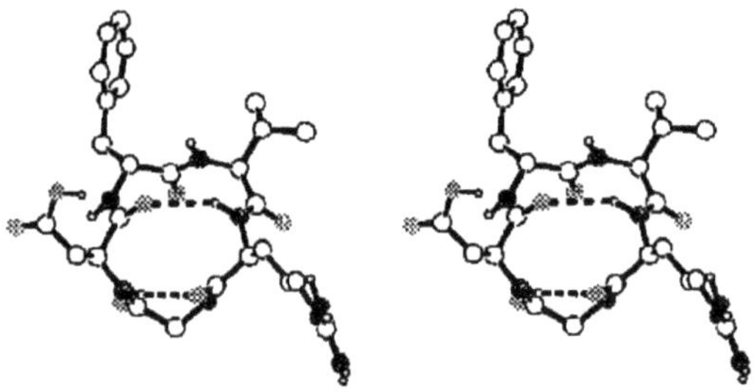

cyclo(-RGDfV-)
$IC_{50}/IC_{50}(GRGDSPK) = 0.006$

Fig. 8 Stereoplot (relaxed view) of cyclo(-RGDfV-) (**1**). The oxygen atoms are *stippled* and the nitrogen atoms are *filled*. The hydrogen bonds are indicated as *dotted lines* [44]

to the same side of the molecule. Although our conformational analysis was criticized [74], a careful analysis with modern NMR techniques confirmed our postulated structure of pentapeptides with one D- and four L-amino acids [118].

The influence of different amino acids in positions X and Y in cyclo (-RGDXY-) on inhibition of vitronectin and fibronectin binding was investigated and an analysis of side chain and backbone conformation by NMR and molecular dynamics (MD) was used to derive SARs [71]. Generally, all these peptides reveal similar backbone structures and are more or less flexible due to motions of the amide bond about its adjacent ϕ and ψ angles (see Fig. 4). However, this flexibility does not influence the relative positions of the α-carbon atoms to each other or the spatial arrangement of the pharmacophoric groups. Hydrophobic aromatic amino acids in position X increase the activity for $\alpha_V\beta_3$-integrins. In contrast, neither hydrophobic nor hydrophilic substitutions in position Y influence the activity (Fig. 9).

A number of different cyclic peptides have been studied. It turned out that there is a strong correlation between the opening of the kink and the reduced α_V-selectivity [119]. Cyclo(-PRGDpG-) displays the common two β-turn cyclic hexapeptide conformation. While residues Gly-Asp-D-Pro-Gly are claimed to occupy a type βI′-turn, residues Gly-Pro-Arg-Gly are assembled in a type βI-turn [120, 121]. The RGD motif is suggested to adopt a turn–extended-turn conformation that maintains a 7 Å separation between the α-carbons of Arg and Asp. Cyclic peptides which present the RGD sequence in either a turn–extended-turn conformation or in a Gly-Asp β-turn display good activity for the $\alpha_{IIb}\beta_3$-integrin, with examples of the former being better than those of the latter conformational type. Overlays of conformations of molecules representing each conformational type reveal a remarkable simi-

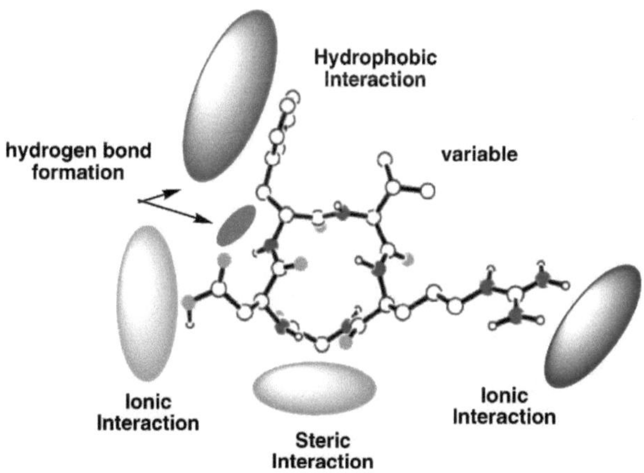

Fig. 9 Receptor model summarizing the results of substrate optimization of cyclo (-RGDXY-) as a ligand for $\alpha_V\beta_3$-integrin [71]

larity of the conformations of the Gly-Asp region but a distinct difference in the placement of the arginine residue [121].

All the results show that RGD is recognized by $\alpha_{IIb}\beta_3$ in a more extended conformation whereas α_V-integrins recognize a kinked arrangement [108, 119, 122]. It seems that the higher conformational flexibility induced by the central glycine residue is used by nature to develop receptor subtypes which recognize different conformations. It is well known that glycine can adopt a considerably larger area in the Ramachandran energy diagram than any other amino acid due to the lacking side chain.

β-peptides composed of β-amino acids display new types of conformations such as helical structures [123, 124]. Incorporation of β-amino acids in cyclic peptides can result in the stabilization of the overall secondary structure while β-amino acid residues preferably occupy the central sequence position of a modified γ-turn conformation. β-amino acid residues may even outplay the preference of D-amino acid residues, commonly found in position $i+1$ of βII′-turns. Generally, the β-amino acid is considered as a γ-turn mimetic ($\Psi\gamma$-turn). However, when glycine is substituted with β-alanine in cyclo(-RGDfV-), due to the increasing distance of the R- and D-side chains, activity is destroyed. Hence, the peptide cyclo(-RβADfV-) is often used as a negative control peptide. Also, substitution of glycine by alanine destroys activity due to steric reasons [125], which can now be understood by docking studies based on the X-ray structure of Xiong et al. [109] after removing Cilengitide from the binding pocket [113]. A systematic β-amino acid scan on the cyclic model peptide cyclo(-RGDfV-) with permutational replacement of one α-amino acid by the corresponding β-analogue led to the cyclic pentapeptide cyclo(-RGD-D-β-Phe-V-) (Fig. 10), which inhibits fibrinogen-mediated cell adhesion to $\alpha_{IIb}\beta_3$-integrin with an activity comparable to that of the known hexapeptide cyclo(-RGDfVG) [126, 127]. The RGD sequence is now found in positions $i+1$ to $i+3$ of a β-turn while the β-amino acid

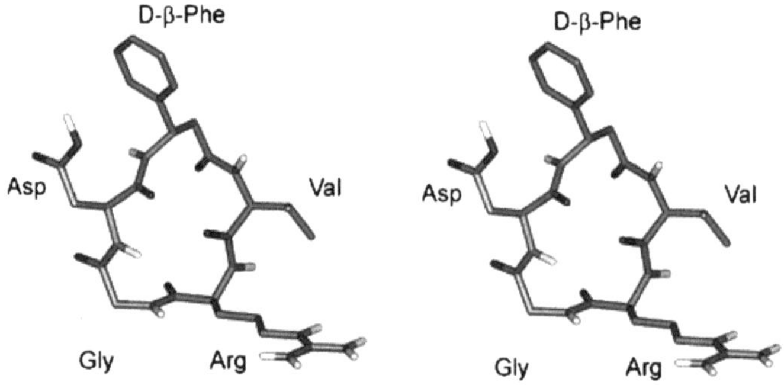

Fig. 10 Solution conformation of cyclo(-RGD-D-β-Phe-V-). D-β-Phe induces a $\Psi\gamma$-turn (upper part of the molecule) [126]

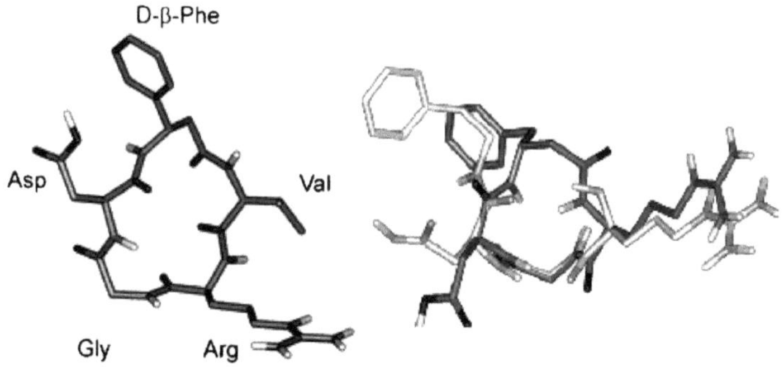

Fig. 11 Solution conformation of cyclo(-RGD-D-β-HPhe-). D-β-HPhe induces a $\Psi\gamma$-turn (*left*). Overlay of the solution conformation of the cyclic pentapeptide cyclo(-RGDfV-) and the cyclic tetrapeptide cyclo(-RGD-D-β-HPhe-) (*right*) [126]

residue occupies the central position of a tight reverse turn closely related to a γ-turn conformation.

Consequently, the structural replacement of the dipeptide sequence D-Phe-Val in cyclo(-RGDfV-) by the β-amino acid D-β-Phe yielded the cyclic homotetrapeptide cyclo(-RGD-D-β-Phe), which indeed reveals nanomolar affinity to $\alpha_V\beta_3$-integrin as it induces the required γ-turn in the RGD motif for the opposite peptide moiety [127] (Fig. 11).

3.3 Modification of Peptide Bonds

3.3.1 *N*-Methylation

Structural modifications of peptide bonds provide a powerful approach to modify profiles of biological activity [22, 128]. *N*-methylation of a specific peptide bond in a bioactive peptide is often used to achieve protection against proteolytic degradation. But, in addition, incorporation of *N*-methylated amino acids into different biologically important peptides has led to analogues with enhanced potency or to conversion of an agonist into an antagonist [129, 130]. With the synthesis of a series of five *N*-methylated cyclic peptides it was investigated whether *N*-methylation of a single peptide bond in the lead structure cyclo(-RGDfV-) has any effects on activity and selectivity [108]. This *N*-methylation scan has provided the drug candidate cyclo (-RGDf-N(Me)V-) (Cilengitide) with enhanced biological activity and 1500-fold-higher selectivity for the vitronectin-binding $\alpha_V\beta_3$-integrin while revealing only low activities in inhibiting fibrinogen binding to $\alpha_{IIb}\beta_3$-integrin (see below, Sect. 5.1). In contrast to its non-methylated parent compound, Cilengitide has also considerable activity for $\alpha_V\beta_5$. The structure differs from cyclo

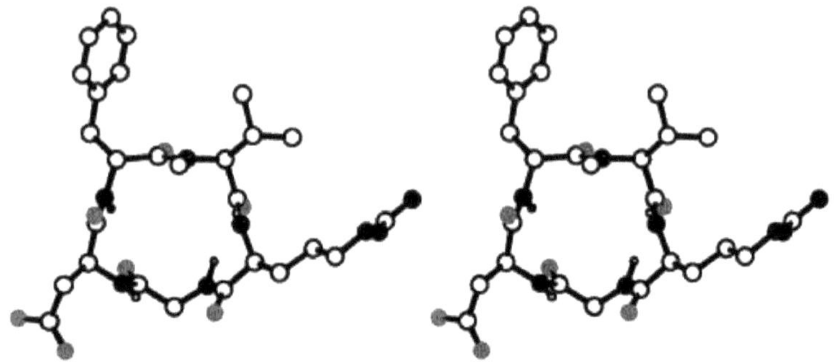

cyclo(-RGDf-N(Me)V-), Cilengitide
IC_{50}/IC_{50}(GRGDSPK) = 0.0028

Fig. 12 Stereoplot (relaxed view) of cyclo(-RGDf-N(Me)V-) from the 100-ps restrained MD simulation [108]

(-RGDfV-): it can be characterized by three turns, two γ_i-turns with Arg and Asp at position $i+1$ and one γ-turn with Gly at position $i+1$ (Fig. 12). Due to steric repulsion the amide groups Asp-D-Phe and Val-Arg turn into a more perpendicular orientation with respect to the plane of the peptide ring. The conformational difference between cyclo(-RGDfV-) and cyclo(-RGDf-N(Me)V-) is correlated with the effect on the $\alpha_v\beta_3$-activity. While adopting a pseudoaxial position in cyclo(-RGDfV-) the side chains turned towards a more pseudoequatorial conformation in the *N*-methylated peptide. These studies reveal an almost complete freezing of the peptide backbone structure including the C_β atoms. The arginine side chain with its C_γ and δ atoms, the guanidine group, and the aspartate carboxylate group are, however, flexible. Changes in the binding affinity are therefore mainly caused by modification of the fixed backbone conformation.

3.3.2
Retro–inverso Peptides and Other Stereoisomers

The advantage of retro–inverso peptides, in which the peptide sequence is reversed and the chirality of each amino acid is inverted, is attributed to their high metabolic stability, because peptide bonds adjacent to D-amino acids normally are stable to enzymatic cleavage. However, with the development of highly active and selective cyclic penta- and hexapeptides there is no need to modify the sequence into retro–inverso structures as cyclic pentapeptides such as Cilengitide are not enzymatically cleaved [131]. In fact, Cilengitide is excreted in animal and man without detectable metabolization.

It is postulated that the total inversion of chirality should be combined with the inversion of the sequence to achieve an orientation of the side chains similar to the parent peptide [28, 32, 132]. However, in reality the biological activity of retro–inverso peptides mostly is modified compared to that of their parent compounds because this oversimplified concept does not involve conformational aspects. Due to the shifted hydrogen bond pattern and the different spatial structure compared to its parent compound, these conformational changes often result in a dramatic decrease of activity. The replacement of CONH with NHCO resulted in different bond angles and bond lengths [133]. To delineate the different hydrogen bonding pattern restrictions a comparison of the retro–inverso peptide analogue to cyclo(-RGDfV-) is suggested to be most illustrative [134]. The retro–inverso isomer cyclo(-vFdGr) contains only one L-amino acid (besides Gly) and we expected a βII-conformation about Phe-D-Asp. Indeed, this conformation was adopted but, due to the reversed sequence, this cyclic retro–inverso peptide did not mimic the side chain topology of its parent peptide and very low activity was observed (Fig. 13).

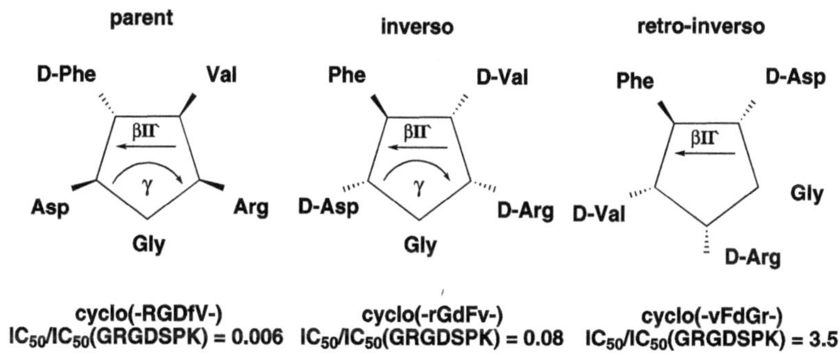

Fig. 13 A schematic illustration of a cyclic parent pentapeptide with one D-amino acid and its inverso and retro–inverso isomers: the case of cyclo(-RGDfV-). The peptide sequence is shown clockwise from *N*- to *C*-terminus [134]

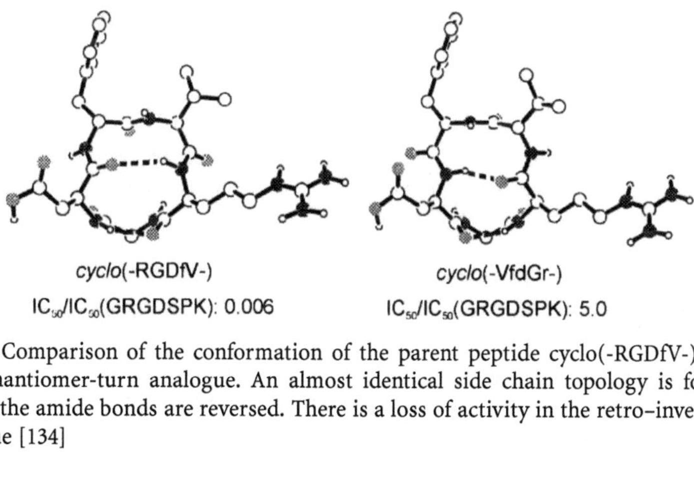

Fig. 14 Comparison of the conformation of the parent peptide cyclo(-RGDfV-) and its retro-enantiomer-turn analogue. An almost identical side chain topology is found although the amide bonds are reversed. There is a loss of activity in the retro–inverso-turn analogue [134]

We synthesized all 16 stereoisomers of the parent peptide cyclo(-RGDfV-). Cyclo(-VfdGr-) exhibited an identical orientation of all side chains although the amide bonds were inverted (Fig. 14). The dramatically lower activity of cyclo(-VfdGr-) to $\alpha_V\beta_3$ by three orders of magnitude proved that the peptide backbone of the RGD sequence is involved in the binding. This observation – derived here from indirect SAR studies – was later nicely confirmed by the X-ray structure of the $\alpha_V\beta_3$-Cilengitide complex that revealed the involvement of the Asp-NH in a hydrogen bridge to the amide-CO of Arg^{216} of the $\alpha_V\beta$-receptor [109, 113].

3.3.3 Reduced Amide Bond

To elucidate whether amide bonds contribute to integrin binding, attempts were made to remove one amide bond by introducing reduced peptides $\Psi[CH_2NH]$ or thioamides $\Psi[CSNH]$. However, these modifications were accompanied by a dramatic change in backbone conformation (Fig. 15) [135]. Amide bond modifications additionally alter torsional freedom of the backbone and modify hydrogen acceptor and donor properties. In conclusion, the altered hydrogen bond forming properties can be of major influence on the entire peptide backbone conformation [136–138]. Therefore, it was not clear whether the reduced activity is due to the lack of the amide bond or to the overall change in the conformation.

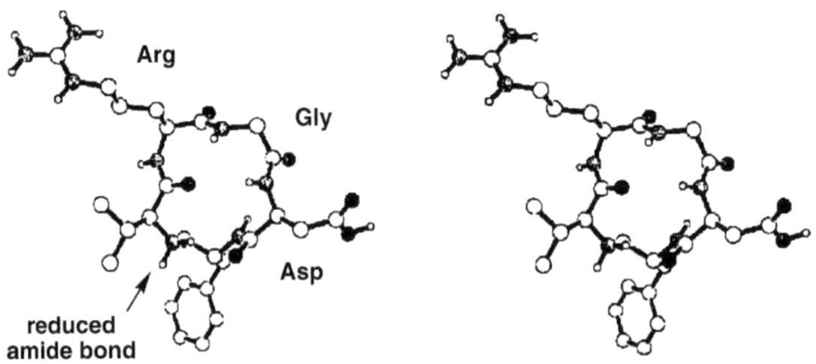

Fig. 15 Stereoview (relaxed view) of the conformation of cyclo(-RGD-D-Phe$\psi[CH_2-NH_2^+]$V-). Its activity lies between that of the linear reference peptide GRGDS and that of the cyclic parent peptide cyclo(-RGDfV-) [135]

3.3.4 Turn Mimetics

Spatial screening led to the highly active first-generation peptide cyclo (-RGDfV-). Further reduction of the flexibility should be achieved by incorporating different rigid building blocks, so-called β-turn mimetics [28, 32, 33], like the (*S*)- and (*R*)-Gly[ANC-2]Leu dipeptide, the β-turn dipeptide (BTD), and the (*S*,*S*)-spiro-Pro-Leu moiety. These distinct β-turn mimetics are introduced by replacing the D-Phe-Val dipeptide in the lead structure cyclo (-RGDfV-) (Fig. 16) [139].

In the peptide analogues cyclo(-RGD″*S*-ANC″-) (**T1**), cyclo(-RGD″*R*-ANC″-) (**T2**), and cyclo(-RGD″BTD″-) (**T3**) the turn mimetic does not adopt the desired position in the β-turn; instead, Gly occupies the position $i+1$ of a βII′-turn while the turn mimetic is located in the positions $i+3$ and $i+4$. Only cyclo(-RGD″Spiro″-) (**T4**) led to the desired βII′/γ-turn arrangement (Fig. 17). In general, some of the "β-turn mimetics" do not exactly correspond to the geometry of a β-turn in terms of their bond angles and distances.

Peptide analogues **T1**, **T2**, and **T3** reveal an increased activity compared to the linear peptide GRGDSPK with respect to the inhibition of vitronectin binding to the $\alpha_V\beta_3$-receptor while **T2** shows an even higher inhibitory activity for vitronectin binding than the lead structure cyclo(-RGDfV-). In contrast, **T2** reveals an increased activity, while **T1** and **T3** display a decreased activity for fibrinogen-mediated cell adhesion to $\alpha_{IIb}\beta_3$-integrins. Thus, among the "β-turn mimetics", **T2** is the most active compound, but possesses only little selectivity (Table 2).

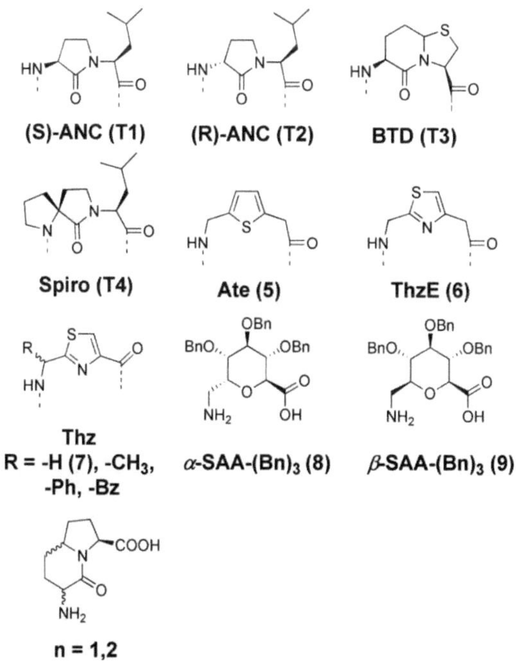

Fig. 16 Various β-turn mimetics used to replace the -D-Phe-Val-dipeptide unit of cyclo (-RGDfV-). For the activities of the turn mimetics, see Table 2

T1

T2

T3

T4

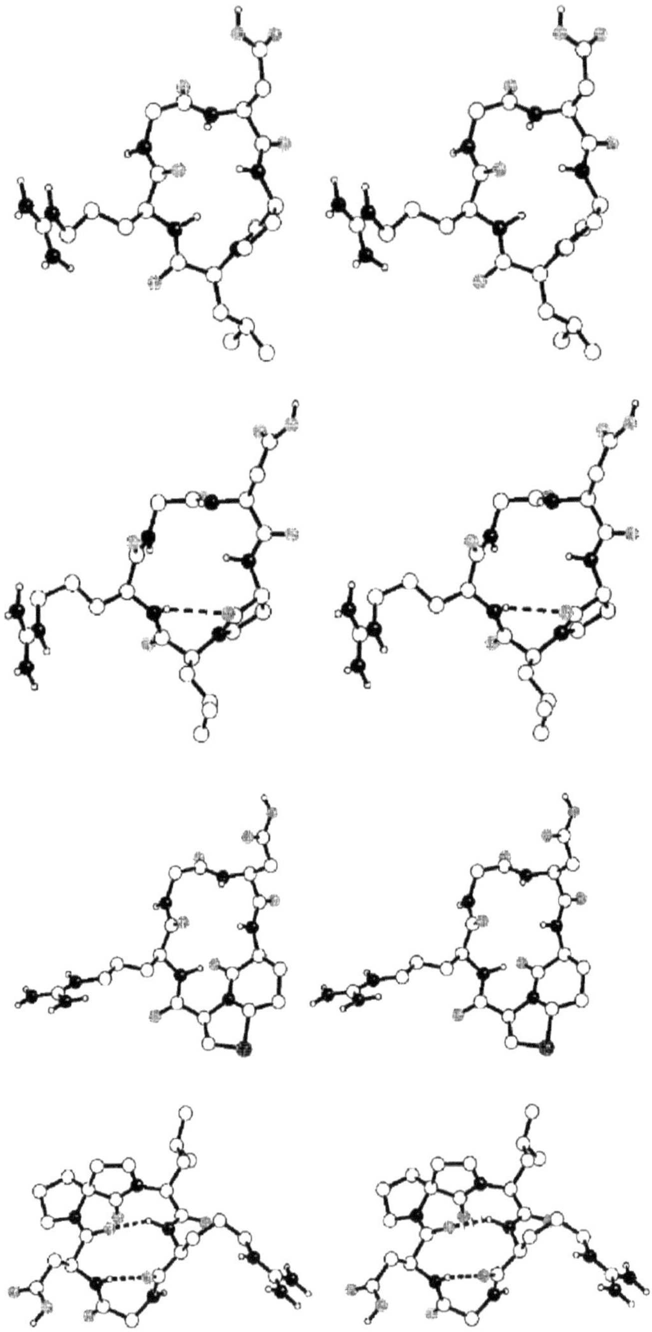

Fig. 17 Stereoplots (relaxed view) of conformations of cyclo(-RGD″S-ANC″-) (**T1**), cyclo(-RGD″R-ANC″-) (**T2**), cyclo(-RGD″BTD″-) (**T3**), and cyclo(-RGD″spiro″-) (**T4**). Carbon and hydrogen atoms are *white*; oxygen atoms are *grey* [139]

Table 2 Inhibition of fibrinogen and vitronectin binding. Values are given as IC_{50} and as quotients Q = IC_{50}[peptide]/IC_{50}[GRGDSPK]. This procedure was used because of the necessity of a reference value. Here, the well-known GRGDSPK peptide was set as standard. Its activity for $\alpha_{IIb}\beta_3$ vs. fibrinogen and $\alpha_v\beta_3$ vs. vitronectin is about 1.7 μM and 0.21 μM, respectively

#	Peptide analogue	$\alpha_{IIb}\beta_3$ IC_{50} (μM)	Q	$\alpha_v\beta_3$ IC_{50} (μM)	Q
	GRGDSPK		1.0		1.0
	c(RGDfV)	8.3×10^{-1}	5.0	2.2×10^{-3}	6.1×10^{-3}
T1	c(RGD″S-ANC″)	2.8	1.6	4.0×10^{-2}	1.1 – 2
T2	c(RGD″R-ANC″)	8.5×10^{-3}	4.9×10^{-3}	8.5×10^{-4}	2.3×10^{-4}
T3	c(RGD″BDT″)	4.8	2.7	2.8×10^{-1}	7.3×10^{-2}
T4	c(RGD″Spiro″)		Not active		Not active
5	c(RGD″Ate″)		Not active	2.4×10^{-3}	2.0×10^{-3}
6	c(RGD″ThzE″)	3.5×10^{-2}		3.2×10^{-3}	1.8×10^{-2}
7	c(RGD″GlyThz″)		Not active	2.5	2.1
8	c(RGD″α-SAA(Bn)″$_3$)	0.72	0.42	0.15	0.71
9	c(RGD″β-SAA(Bn)″$_3$)	1.3×10^{-2}	7.9×10^{-3}	2.5×10^{-2}	0.12

Incorporation of both 2-(2-(aminomethyl)thiophen-5-yl)- and 2-(2-(aminomethyl)thiazol-4-yl) acetic acid as a β-turn mimetic yielded highly active (low-nanomolar range) compounds in which the D-Phe-Val dipeptide in the lead structure cyclo(-RGDfV-) was replaced. Embedding substituted (2-aminomethyl)thiazol-4-carboxylic acid into the cyclic parent peptide resulted in a significant loss of activity (Table 2; for structures see Fig. 16). This finding can be explained by the methylene (– CH_2) contraction and the consequential effect on the conformation [140].

Another turn mimetic consisting of a series of 1-aza-2-oxobicyclo[X.3.0] alkane amino acids (Fig. 16) was introduced by replacing the D-Phe-Val dipeptide, which should adopt a βII′-turn and hence force the RGD sequence into a kinked, $\alpha_V\beta_3$-selective conformation [141, 142].

In the last decade modified carbohydrates were used as peptide mimetics. Whereas Hirschmann et al. used the pyranoid form of a sugar to mimic a cyclic hexapeptide and derived a sugar-based somatostatin analogue [143], we applied sugar amino acids as turn mimetics or dipeptide surrogates [144, 145]. These sugar amino acids were also incorporated into the cyclic pentapeptide cyclo(-RGDfV-) to mimic the D-Phe-Val dipeptide [119, 146]. However, the missing lipophilic Phe residue caused only low $\alpha_V\beta_3$ affinity. In contrast, the high activity of the β-SAA$(Bn)_3$-derivative against the $\alpha_{IIb}\beta_3$-receptor was unexpected (Table 2; for structures see Fig. 16).

3.4 Peptide Analogues for $\alpha_{IIb}\beta_3$-Integrin Binding

The arginine's guanidine group binds "side-on" to Asp^{280} of the $\alpha_V\beta_3$-receptor [109, 113]. Hence, a substitution of the guanidinium group with an amino group destroys α_V-activity whereas it is allowed for binding to $\alpha_{IIb}\beta_3$. The disintegrin "barbourin" which contains amino acid substitution of lysine for arginine in the RGD motif reveals high specificity for $\alpha_{IIb}\beta_3$. Incorporation of this motif in conformational constrained, disulfide-bridged, cyclic peptide templates, followed by systematic optimization of the cyclic ring size, has resulted in peptide analogues which displayed potency and selectivity comparable to that of barbourin. The cyclic heptapeptide mimic of "barbourin" is called "eptifibatide" which reveals a βII′-turn displaying arginine and aspartic acid side chains roughly in the same plane and diagonally disposed across a slightly cupped Gly-Asp β-turn (Fig. 18) [147–149].

Only modest improvements in potency were achieved with linear analogues of Ac-RGD-NH_2, while increased affinity for $\alpha_{IIb}\beta_3$ was achieved with cyclic disulfide-bridged analogues [115], culminating in the cyclic disulfide Ac-cyclo-*S*,*S*-[Cys-(NMe)Arg-Gly-Asp-Pen]–NH_2 with improved bioavailability and comparable affinity to fibrinogen [150]. The activities of analogues delineate that the arginine side chain cannot be modified without a decrease in potency; the glycine and aspartic acid are essential. The addition of a methyl group to the α-amino group of Arg and replacement of cysteine Cys-5 with the more lipophilic Pen are both called to account for restriction of the flexibility. Since the discovery of the analogue Ac-cyclo-*S*,*S*-[Cys-(NMe)Arg-Gly-Asp-Pen]–NH_2 several modifications to the Ac-Cys-, -(NMe)Arg-, and -Pen–NH_2 groups have been explored, which are described in detail in the literature [102, 151].

Further modification by replacing the disulfide tether *N*-acetylcysteine/penicillamineamide with the constrained and more lipophilic tether 2-mer-

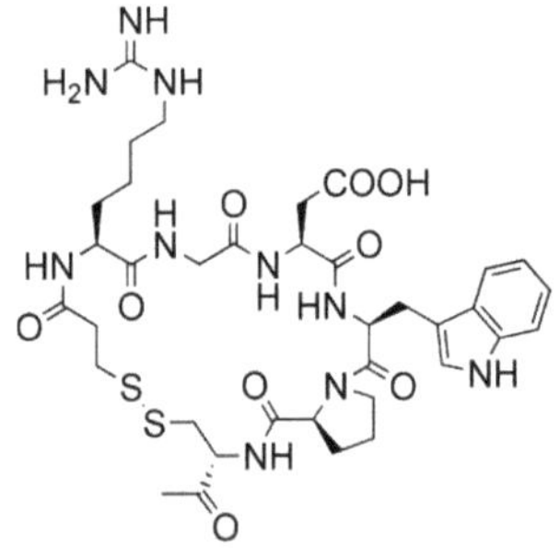

Eptifibatide
$IC_{50}(\alpha_{IIb}\beta_3)$ = 120 nM

Fig. 18 Eptifibatide, a disulfide-bridged cyclic peptide derived from a disintegrin

captobenzoyl/2-mercaptoaniline (Mba/Man) afforded the semipeptide cyclo-*S*,*S*-[Mba-(NMe)Arg-Gly-Asp-Man] (Fig. 19), which exhibited significant enhancement in both affinity and potency for $\alpha_{IIb}\beta_3$. It is capable of introducing a greater degree of conformational constraint upon the macrocyclic ring as the χ-torsion angle is constrained to 0° [152].

The RGD motif in the *S*,*S*-bridged cyclic peptides is to a certain extent conformationally restricted. The crystal structures of both peptides display *all-trans* peptide bonds and exhibit an extended Gly residue flanked by an (NMe)-Arg residue in a conformation roughly equivalent to the position $i+2$ of a β-turn and the Asp residue in the position $i+1$ of a βI-turn. The Asp residue can also approximate the position $i+1$ of a γ-turn-like structure. This RGD-sequence arrangement can be regarded as a "turn–extended-turn" conformation [151].

The advantage of incorporating a rigid non-peptide template into a macrocycle can be rationalized by limitations of the rotatable bonds which should confer the greatest degree of conformational restriction to the RGD motif. The amino acid *m*-(aminomethyl)benzoic acid "Mamb" was incorporated into an initial cyclic peptide, cyclo(-Gly-Arg-Gly-Asp-Mamb-), in which the Cys-Pen disulfide has been replaced by Gly-Mamb [153, 154]. The most active compounds were cyclo(-D-Abu-(NMe)Arg-Gly-Asp-Mamb), also termed DMP, and cyclo(-D-Val-(NMe)Arg-Gly-Asp-Mamb). The D-configuration of the amino acid at position 1 of the sequence and the *N*-methylated arginine were found to be important for obtaining high binding affinity to the $\alpha_{IIb}\beta_3$-receptor [129]. The predominant solution structure (Fig. 20) is similar to that of classical cyclic hexapeptides with two connected β-turns. The first turn is an almost ideal type βII′-turn centered at the D-Abu-(NMe)Arg residue. The Gly residue is generally extended, with the molecule presenting an elongated face between the (NMe)Arg and the Asp residue.

Replacement of the D-Xaa-(NMe)Arg dipeptide with L-Ala-Arg resulted in a striking reversal of selectivity to favor binding to $\alpha_v\beta_3$-integrin [130]. In cyclic peptides, a dipeptide consisting of a D-amino acid followed by an *N*-methylamino acid induces a type βII′-turn, while an L,L-dipeptide in the

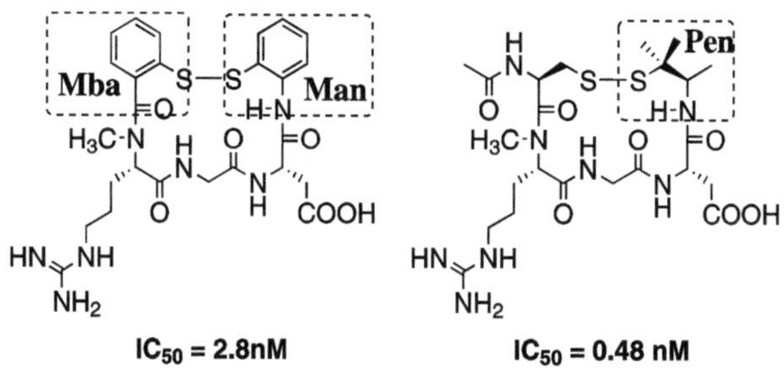

Fig. 19 Chemical structures of lead compounds. Activities for $\alpha_{IIb}\beta_3$ are shown

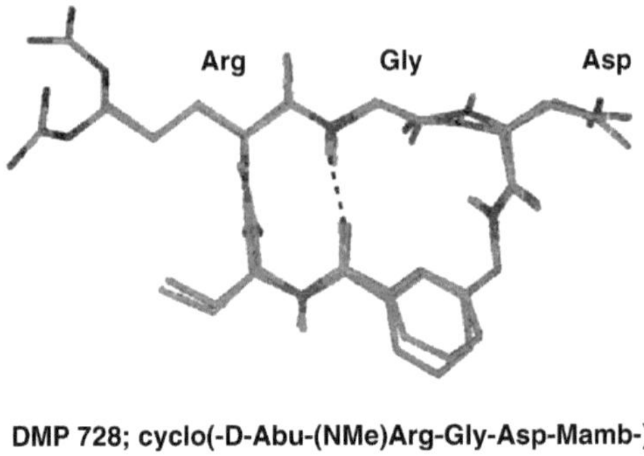

Fig. 20 Overlay of the centroid conformer calculated for DMP728 and its X-ray crystal structure. The two conformations are very similar [129]

same position favors a type βI-turn. Thus, a change in turn type might modulate receptor specificity. Except for the opposite orientation of the hydrogen bond relative to the plane formed by the backbone, the turns of both β-types vary in their relative orientation of the incoming and waning peptide strands. Considering the virtual dihedral angle formed by the four α-carbons of the turn, also termed "twist", the βI-turn prevails in a roughly gauche arrangement with an angle of 49° while the βII′-turn adopts an eclipsed orientation [155]. This difference in the turns interferes with the relative arrangement of the Asp and Arg residues and, therefore, with the specificity

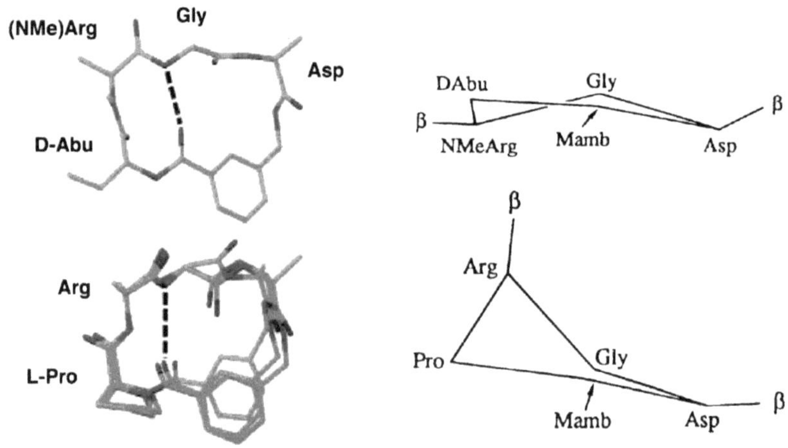

Fig. 21 Conformation of both compounds, featuring a βII′-turn centered at D-Abu-(NMe)Arg and a βI-turn centered at Pro-Arg (*left*). The side chains have been truncated in both panels. Cα backbone traces of both conformers showing their different twists. The Mamb *C*-terminal aromatic carbon is used as the Cα [130]

and affinity to $\alpha_{IIb}\beta_3$- and $\alpha_V\beta_3$-integrins. The arginine side chain in the $\alpha_V\beta_3$-selective peptide is raised above the plane formed by the backbone into a pseudoaxial position, while the (NMe)Arg side chain in the $\alpha_{IIb}\beta_3$-selective peptide adopts a pseudoequatorial conformation (Fig. 21).

3.5 RGD Mimetics

The receptor-bound conformation of RGD ligands was derived indirectly by performing SAR of small linear peptides as well as of small and large cyclic peptides containing the RGD motif [26, 27, 97, 147]. To investigate the structural requirements for RGD recognition by cognate integrins, various structural mimetics of the tripeptide sequence were designed. Structures containing guanidinium and carboxylic groups separated by manifold scaffolds are ascribed to mimic the distal configuration of the functional RGD sequence [106]. These findings suggest that the role of the backbone of the RGD sequence is mainly to control the correct spacing between the two ionic groups involved in interaction with the receptor while hydrophobic interactions can almost be excluded [156]. However, it was shown later by SAR and finally by the structure of the Cilengitide/$\alpha_V\beta_3$-complex that the peptide backbone participates in the binding (see below, Sect. 3.3.2). In this article we will only shortly review a few out of a large set of compounds.

Almost every member of the non-peptide ligands contains a β-alanine group in which the carboxylate mimics the aspartic acid in RGD, a linking group which provides a variety of scaffolds delivering the acid and basic pharmacophore groups in the spatial geometry, and an amine or guanidine to mimic the arginine group [102, 157–164]. It became evident early on that the distance between the guanidinium group and the carboxyl group has to be shorter for $\alpha_V\beta_3$-integrin activity than for $\alpha_{IIb}\beta_3$ activity [76, 77, 106, 163, 165, 166]. Furthermore, the selectivity depends on the effects of structural changes in both the guanidine mimetic and the substituent α to the carboxylate. It appeared that a less basic guanidine mimetic imparted greater selectivity for $\alpha_V\beta_3$-mediated adhesion vs. $\alpha_{IIb}\beta_3$-mediated platelet aggregation [157, 160, 162–164, 166–173]. Activity for $\alpha_V\beta_3$-integrin can also be ascribed to the α substituent culminating in 2,6-disubstituted arylsulfonamides. Different studies pointed out that aromatic β-amino acids are suitable building blocks in potent RGD mimetics [164, 172, 174, 175]. These findings illustrate the versatility of different scaffolds and highlight the structural requirements for the design of selective compounds in terms of the arginine mimetic, the distance between the *N*- and *C*-terminus, and the lipophilic side chain.

A number of heterocycles were taken as core structures. The 3,5-disubstituted isoxazoline was selected resulting in isoxazolinylacetamides of β-alanine as effective frameworks for the display of the basic and acidic functionality [176]. Modification of the α-substituent of the $\alpha_{IIb}\beta_3$-

isoxazoline ligand eventually led to a series of most potent α-sulfonamide diaminopropionate-isoxazolinylacetamides which were found to be inhibitors of in vitro platelet aggregation exemplified by roxifiban (DMP 754) and DMP 802 (Fig. 22) [158, 161].

SmithKline Beecham scientists focused their work on the examination of the crystal structure of the potent cyclic peptide cyclo-*S*,*S*-[Mba-(NMe)Arg-Gly-Asp-Man] which contains a turn at the arginine residue, an extended conformation of the glycine, and a C_7-turn at the aspartic acid residue. This structure affirms the "turn–extended-turn" conformation about the RGD sequence which is responsible for $\alpha_{IIb}\beta_3$-activity [152]. On the basis of these observations, the 3-oxo-1,4-benzodiazepine scaffold was chosen to mimic the C_7-turn accommodated by $\alpha_{IIb}\beta_3$-ligands and the extended glycine conformation and resulted in several potent ligands exemplified by SB-208651 [102, 177, 178]. Attempts to improve the limited oral bioavailability of this series focused on replacing the benzamidine function with a less basic piperidine functionality in analogues such as lotrafiban (SB-214857) (Fig. 23) [179–181].

In analogy to previous approaches, the preferred conformation of a RGD-containing cyclic peptide analogue of eptifibatide (see Fig. 18) serves as a template to mimic a slightly cupped β-turn at the Gly-Asp position with a fused 6,6-ring system. The turn itself contains 10 atoms and resembles a decalin ring system substituted at positions 2 and 6 with the pharmacophoric groups. Initial analogues to verify this hypothesis utilized the 3,4-dihydro-1-

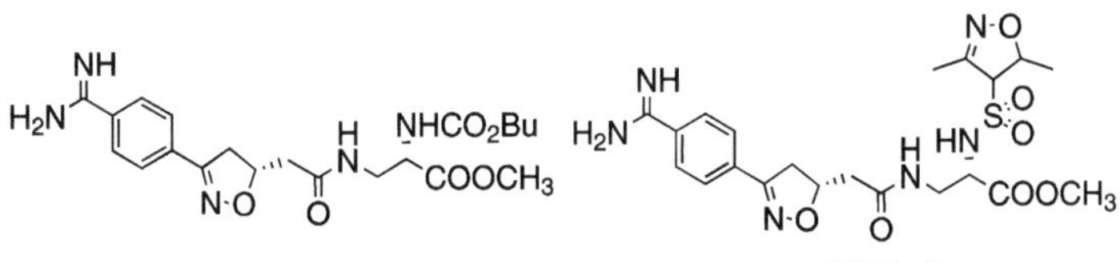

Fig. 22 Two potent $\alpha_{IIb}\beta_3$-integrin ligands. Roxifiban (*left*) and DMP 802 (*right*). Both are inhibitors of human platelet aggregation

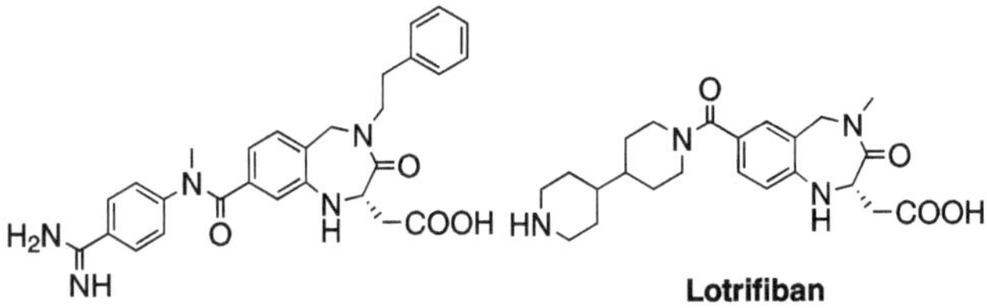

Fig. 23 The 1,4-benzodiazepine nucleus was chosen to mimic the C_7-turn and the extended glycine conformation resulted in several potent ligands for platelet aggregation

oxoisoquinolone ring system [102, 149] and several [4.4.0]-bicyclic templates (Fig. 24) [182–184].

Historically, $\alpha_{IIb}\beta_3$-integrin antagonists had attracted the interest of the pharmaceutical industry very early, whereas the $\alpha_V\beta_3$-integrin was discovered as a drug target later. Research into $\alpha_V\beta_3$-integrins has benefited from the discoveries made in $\alpha_{IIb}\beta_3$-research and several rigid scaffolds have been used to construct $\alpha_V\beta_3$-ligands mimicking the RGD motif [102, 185]. Generally, these studies support the observation that $\alpha_V\beta_3$-ligands recognize a shorter Arg-to-Asp distance than $\alpha_{IIb}\beta_3$-ligands, and demonstrate that a potent and selective $\alpha_{IIb}\beta_3$-ligand can be converted to a potent and selective $\alpha_V\beta_3$-ligand by changing the guanidine mimic. It turned out that a less basic guanidine mimic imparts greater selectivity for $\alpha_V\beta_3$-mediated adhesion, but it seems that an amidine-like or guanidine-like disposition of nitrogens is more important than basicity for recognition by $\alpha_V\beta_3$-integrins (side-on binding). Selectivity could be optimized through careful choice of pK_a, geometry, and hydrogen-bonding characteristics. Previous studies revealed that incorporation of a benzimidazole scaffold into the benzodiazepine Gly-Asp mimic afforded potent and selective non-peptidic $\alpha_V\beta_3$-ligands [167, 168, 171]. In these studies of benzimidazole-containing non-peptidic $\alpha_V\beta_3$-ligands, investigations delineated that a free NH, an amidine-like disposition of nitrogens, and a fused aromatic residue were important for optimal binding. Further studies reported investigations of 2-aminopyridines, a ring system with neutral pK_a value which presents the desired amidine-like disposition of nitrogen atoms favored by $\alpha_V\beta_3$-ligands [169, 171]. Generally, cyclic guanidine mimics are preferred over non-cyclic guanidine scaffolds [163, 186, 187]. These results already suggested that binding to $\alpha_{IIb}\beta_3$ occurs through

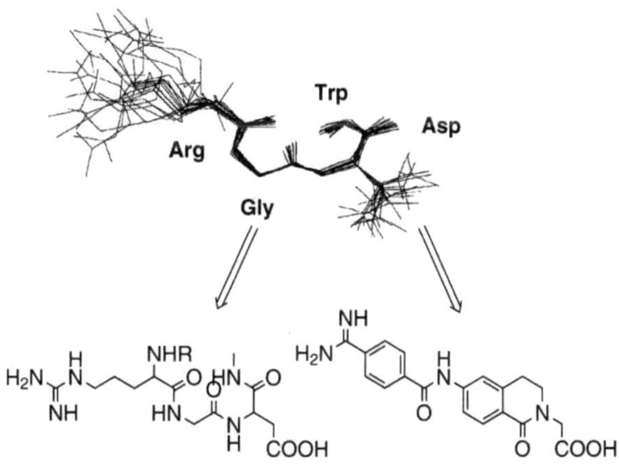

Fig. 24 View of the βII′-turn structure resulting from a Trp-Arg hydrogen bond. The figure displays the backbone of the Arg-Gly-Asp region along with the tryptophane amide nitrogen and proton (*above*). A two-dimensional representation of the Gly-Asp β-turn is shown along with its turn scaffold (*below*) [149]

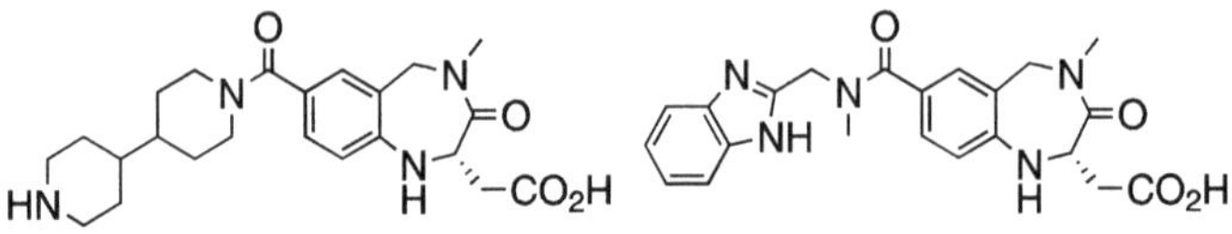

$\alpha_{IIb}\beta_3$: **IC$_{50}$ = 28 nM**
$\alpha_V\beta_3$: **IC$_{50}$ = 10340 nM**

$\alpha_{IIb}\beta_3$: **IC$_{50}$ = 30000 nM**
$\alpha_V\beta_3$: **IC$_{50}$ = 2 nM**

Fig. 25 The selectivity for $\alpha_{IIb}\beta_3$- or $\alpha_V\beta_3$-integrin is based on the guanidine mimetic

an "end-on" interaction, so that non-cyclic guanidines are preferred. Binding to $\alpha_V\beta_3$ occurs through a "side-on" interaction, so that cyclic guanidines are tolerated [109, 111, 113].

Now we know that the side-on binding of an amidine by Asp^{218} is crucial for binding to the α_V-subunit. Even though an 1,4-benzodiazepine scaffold acts as a Gly-Asp mimic which introduces a γ-turn about aspartic acid, it might afford a potent and selective $\alpha_V\beta_3$-ligand scaffold. The remarkable versatility of the benzodiazepine scaffold as a Gly-Asp mimetic illustrates that potent non-peptide ligands can be designed with selectivity for either $\alpha_{IIb}\beta_3$- or $\alpha_V\beta_3$-integrins simply by altering the length and hydrogen-bonding characteristics of the arginine mimetic (Fig. 25) [185, 188].

3.6 From Ligand-based to Structure-based Rational Drug Design

Although enormous approaches have been applied to the discovery of potent integrin ligands, all efforts had to be conducted in the absence of structural information until the crystal structures of the $\alpha_V\beta_3$- [109, 110] and $\alpha_{IIb}\beta_3$-integrin extracellular domains [111] were solved. Identification of the ligands' bioactive conformation by mimicking natural peptide ligands in a "ligand-based" design without a structure of the complex is quite elusive. The recently available crystal structures allowed the re-examination of structurally diverse ligands with regard to distinct interactions to afford selectivity and affinity for $\alpha_V\beta_3$- over $\alpha_{IIb}\beta_3$-ligands or *vice versa*. So far, only one crystal structure of an integrin head group with a small ligand (Cilengitide) is known. The usual, mutual double induced fit (flexible target, flexible ligand) may be misleading in the design of new ligands, as these incidents may also induce slight changes in the receptor conformation. On the other hand it was possible to use the above-mentioned $\alpha_V\beta_3$/Cilengitide structure to apply docking procedures by removing Cilengitide from the complex and to understand the structure–activity relationship on a detailed molecular level [112, 113]. A step forward is the use of homology modeling to construct other integrin subtypes such as $\alpha_V\beta_5$ [189], $\alpha_5\beta_1$ [190], and $\alpha_{IIb}\beta_3$ [112]. These receptor models are the better the stronger the sequence homologies are, but the results should be taken with care.

The X-ray-derived integrin–ligand complex of the $\alpha_V\beta_3$ head group with Cilengitide reveals that the ligands insert into a crevice between the β-propeller of the α-subunit and the βA domain on the integrin head group. The central glycine residue or mimic lies directly on the surface and might also add to the stability of the complex. Any substitution of the glycine unit decreases $\alpha_V\beta_3$-activity. The loss of activity upon Gly to Ala substitution proves that a close contact between these two polar amide groups adjacent to glycine and the receptor is essential. The main interactions are between the positively charged arginine and negatively charged side chains in the α-subunit and between the anionic aspartic acid and the metal cation in the MIDAS (metal-ion-dependent adhesion site) region of the β-subunit. Further stabilization occurs through hydrogen bonds between the backbone of the peptide and the ligand. Although the distance between the arginine and aspartic acid residues in $\alpha_V\beta_3$-selective RGD ligands is about 13 Å, the shortest distance between the Ca^{2+} bound to the ADMIDAS (adjacent to metal-ion-dependent adhesion site) region and the β-propeller of the α_V-subunit is in the order of 17 Å. Therefore, we can deduce that the integrin activation mechanism in structural terms requires conformational changes for ligand binding. They involve a shift in the position of the metal ion from the ADMIDAS region to the MIDAS region of the β_3-unit, which is surprisingly not populated by any metal ion in the crystal structure of free $\alpha_V\beta_3$-integrins. Movement of the Ca^{2+} ion towards the α-subunit allows ligand interactions with both subunits. The ion shift is accompanied by rearrangements of loops surrounding the MIDAS region which in turn trigger the dissociation of the integrin head groups [113, 114]. In addition, the interacting transmembrane helices of the α- and β-subunits rearrange by a 110° twist of the β_3-subunit [191] and can trimerize and reassociate with α-subunits to form the focal adhesion complex [192]. An intermediate glycophorin-like conformation of the transmembrane helices of the $\alpha\beta$-heterodimer has been proposed [193, 194]. However, the important outside-in signal transduction requires homooligomeric association and multivalent binding of RGDs in the focal adhesion [195].

The $\alpha_V\beta_3$-selective Cilengitide [108] is bound by the arginine and aspartic acid side chains of the ligand pointing in opposite directions [109]. The arginine side chain inserts into a narrow groove at the top of the propeller domain, and its guanidinium group binds over a bidentate salt bridge to (α)-Asp^{218} and an additional salt bridge to (α)-Asp^{150}. On the opposite side, one of the aspartic acid carboxylate oxygens contacts the metal ion at the MIDAS region, while the other carboxylate oxygen forms a hydrogen bond with the backbone amide hydrogen of (β)-Asn^{215}. Furthermore, the backbone NH of aspartic acid is involved in a hydrogen bond with the backbone carbonyl oxygen of (β)-Arg^{216}. Additionally, D-Phe is involved in a T-shaped interaction with (β)-Tyr^{122} (Fig. 26).

Sequence alignment shows that Phe^{231} of the α_{IIb}-subunit corresponds to Asp^{218} of the α_3-subunit. Phe^{231} cannot stabilize the arginine residue of the

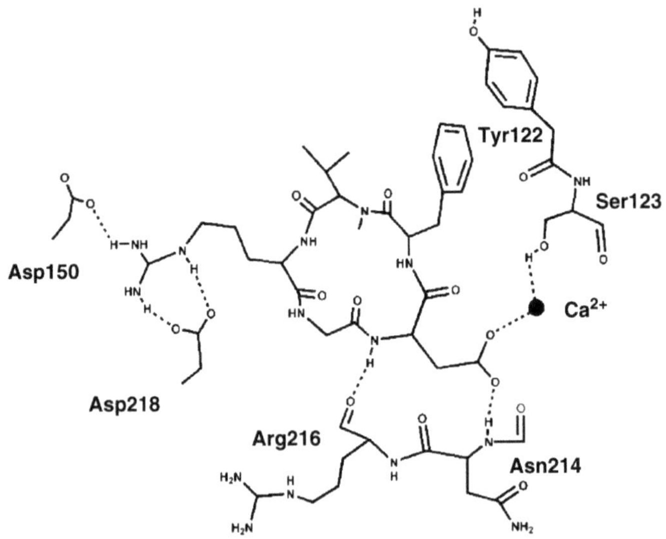

Fig. 26 Schematic representation of the X-ray structure of the cyclo(-RGDf-(NMe)V-)/$\alpha_V\beta_3$-integrin complex depicting the major protein–ligand interaction [112]

RGD ligand by "side-on" binding. The aspartic acid residue of Asp^{224} of the α_{IIb}-subunit might serve as an "end-on" binding partner for lysine and basic arginine mimics of $\alpha_{IIb}\beta_3$-integrin ligands. The residue that corresponds to α_V-Asp^{150} is α_{IIb}-Gln^{157}, which can also be ascribed to "end-on" binding. Superimposition of α_{IIb} on α_V reveals that α_V-specific ligands are not suited to bind to $\alpha_{IIb}\beta_3$-integrins as the loop at the *C*-terminus of α_V-Asp^{150} and α_{IIb}-Gln^{157} is not conserved between the two subtypes. The longer loop of α_{IIb}-integrin closes the binding pocket in which α_V-specific ligands bind and leads to steric clashes. The solved X-ray crystal structure contributed to the understanding of the different subtype specificities [111, 113]. Recently, the crystal structure of the $\alpha_V\beta_3$-integrin has also been used for homology modeling of $\alpha_V\beta_5$ [189] and $\alpha_5\beta_1$ [190] to understand selectivity within these RGD-recognizing integrins.

4 LDT-derived Peptides

4.1 Introduction

Inflammatory processes leading to tissue damage and disease are mediated in part by the α_4-integrins, $\alpha_4\beta_7$ (LPAM) and $\alpha_4\beta_1$ (VLA-4), which are both frequently expressed on leukocyte surfaces [196, 197]. The vascular cell adhesion molecule-1 (VCAM-1) and mucosal addressin cell adhesion molecule-1 (MAdCAM-1), which are both expressed on activated endothelium, are the most important endogenous ligands for α_4-integrins [198, 199]. Whereas VCAM-1 binds to both $\alpha_4\beta_7$- and $\alpha_4\beta_1$-integrins, MAdCAM-1 on mucosal endothelial cells is an exclusive ligand for $\alpha_4\beta_7$-integrins under physiological conditions [200, 201]. These molecular interactions are required for lymphocytes to enter the central nervous system and the mucosa-associated lymphoid tissue of the intestine, selectively mediated by $\alpha_4\beta_7$-integrin binding to its cognate ligand MAdCAM-1 [202, 203]. In contrast to other cell adhesion molecules, MAdCAM-1 is expressed only on a few cell types. Therefore, the $\alpha_4\beta_7$/MAdCAM-1 adhesion pathway represents a potent and organ-specific target for therapeutic modulation of inflammatory diseases of the gastrointestinal tract, including Crohn's disease, ulcerative colitis, and hepatitis C [204, 205]. In addition, VLA-4 is an attractive target for therapeutic intervention in several models of inflammatory and autoimmune pathologies such as asthma, diabetes, multiple sclerosis, and arthritis [197]. Impressive results have been reported for the treatment of multiple sclerosis and other inflammatory diseases on a humanized monoclonal antibody called "Antegren®" (natalizumab) which binds to the α_4-integrin subunit and thus reduces leukocyte migration [206, 207]. However, three months after its expedited approval by the FDA, natalizumab was removed from the market after two cases of deadly progressive multifocal leukoencephalopathy were reported among a few thousand patients [208].

In analogy to the previously described RGD motif, the critical residues for α_4-integrin-mediated cell attachment are contained in a short amino acid sequence. The critical epitope of VCAM-1 for binding to VLA-4 was identified as QID(40)SP in domain 1, of which Asp^{40} was found to be essential [209, 210].

The tripeptide LDT was found to be the recognition motif for MAdCAM binding to the $\alpha_4\beta_7$-integrin [211]. The Leu^{41}, Asp^{42}, and Thr^{43} residues are associated with a turn between strands C and D of the protruding CD loop of domain 1 [47, 212]. The most striking structural feature of MAdCAM-1 is a negatively charged DE β-ribbon which extends from the strands D and E of domain 2 and is predicted to orientate the molecule optimally for integrin binding [213, 214].

Both MAdCAM-1 and VCAM-1 share a similar immunoglobulin-like topology and they present their key integrin-binding residues at a protruding loop at the edge of their first IgSF domain. In either case an aspartic acid residue in domain 1 is essential for integrin binding.

4.2 LDT Motif in Peptides

There is no X-ray structure of a ligand-bound complex with an α_4-integrin head group. Hence, ligand-based drug design is restricted to X-ray structures of natural ligands (MAdCAM-1, VCAM-1) or active, conformational restricted small molecules [215, 216]. One always has to bear in mind that flexibility of proteins in their mutual interaction can drastically change their conformation. Thus, the "design" of smaller, more rigid (or more conformationally preferred) ligands should allow screening for a larger conformational space than given by the isolated natural ligand.

As the development of $\alpha_4\beta_1$-ligands as drug targets has progressed, the number of publications also increased [196, 197, 217–221]. Therefore, we will not treat this topic here but rather concentrate on $\alpha_4\beta_7$.

4.2.1
Spatial Screening Applied on LDT

Random peptidic phage libraries containing peptides of the general sequence $X_LCM_MCX_N$, where X represents any of the 20 naturally occurring amino acids, revealed a disulfide-bridged cyclic peptide containing a SDT-tripeptide sequence and demonstrating a 10-fold selectivity favoring $\alpha_4\beta_7$ over $\alpha_4\beta_1$. Truncation of the *C*- and *N*-terminal residues flanking the two cysteine residues did not result in a significant loss of potency or selectivity for $\alpha_4\beta_7$. Quite the contrary, the resulting peptide C*RSDTLC*–NH_2 (where * indicates disulfide-linked residues) showed enhanced potency and selectivity [12].

For the development of potent and selective $\alpha_4\beta_7$-integrin ligands, it is of crucial importance to determine the bioactive conformation of the binding motif in MAdCAM-1. Therefore, a library of constrained homodetic cyclic peptides was synthesized to elucidate the bioactive conformation [88]. As described for RGD peptides, five cyclic pentapeptides and six hexapeptides of the sequence LDTAAX (X = Ala or nil) in which each amino acid is once in the D-configuration were investigated. However, D-alanine was replaced with D-proline because of its strong structurally inducing effect. Only the hexapeptide cyclo(-LDTApA-) inhibited binding of $\alpha_4\beta_7$-integrin-containing cells to MAdCAM-1 effectively (7.0% adhesion at 1 mg/mL) while the linear precursor showed no activity. This peptide did not affect binding of cells with the closely related $\alpha_4\beta_1$-integrin to VCAM-1. In the bioactive peptide the LDT motif approximates a βI-turn with aspartic acid in position $i+1$; D-proline is at position $i+1$ of a βII′-turn (Fig. 27). Structure–activity approaches elucidate the essential role of the LDT sequence for binding to the $\alpha_4\beta_7$-integrin. Further approaches, centered upon substitution of alanine in position 6 with hydrophobic amino acids, indicate that hydrophobic interactions strongly enhance binding activity culminating in the most active peptide cyclo(-LDTDpF-).

For the further development of non-peptidic and highly selective drugs, we scrutinized the structural and functional requirements of the LDT recognition sequence by using several peptidomimetic variations such as peptoids, azapeptides, and reduced amide bonds [222]. Structure–activity approaches

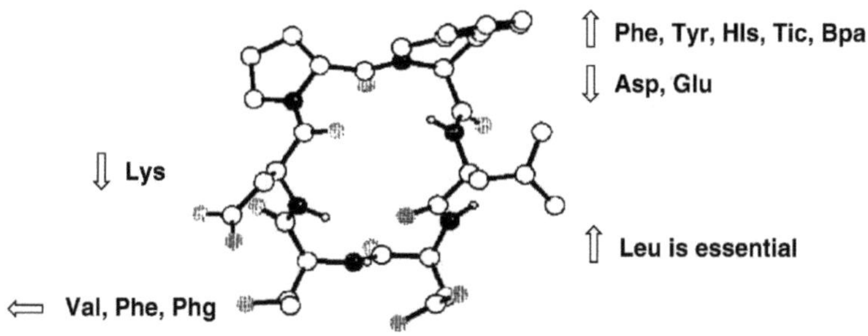

Fig. 27 Effect of substituents on $\alpha_4\beta_7$/MAdCAM-1 activity. The homodetic cyclic hexapeptides normally prefer a conformation with two facing β-turns. The figure shows the plot of the solution structure of cyclo(-Tic-LDTDp-)

revealed that only *N*-terminal modifications to the LDT recognition sequence resulted in biologically active and selective compounds. Systematic reduction of the amide bonds in the $\alpha_4\beta_7$-selective peptide cyclo(-LDTDpF-) is suitable to elucidate the importance of the peptide backbone. The successive reduction of the amide bonds led to inactive peptides. Only the amide bond between Thr[4] and Asp[3] in cyclo(-FLD-ThrΨ(CH_2NH)-Dp-) could be reduced without loss of activity. This finding is not only a result of the interaction of the backbone with the integrin but, also, or even exclusively, implies an involvement of amide bonds in stabilizing the bioactive conformation.

N-terminal acylation of the LDT motif and modification of the *C*-terminal carboxamide with amines led to low micromolar $\alpha_4\beta_7$-integrin ligands [211, 223]. All *C*-terminal modifications in the linear peptide yielded completely inactive compounds, except for substitution of threonine with β-threonine, D-threonine, or valine. Furthermore, the most active tripeptide Iquin-L-D-T–NH_2 (Iquin = isoquinoline-3-carboxylic acid) (Fig. 28) does not show any selectivity for $\alpha_4\beta_1$-binding to VCAM-1 (entry **a** in Table 3). The amide bond between threonine and aspartic acid is not essential. Often it is possible to substitute an amide bond by a phenyl moiety without loss of activity [224]. We also found that 3-amino-3-aryl-propionic acid can mimic the Asp-Thr-residue. Based on both these results and on the results illustrated above, we developed a biased library culminating in a peptidomimetic with a reduced amide bond (Fig. 28) [225, 226].

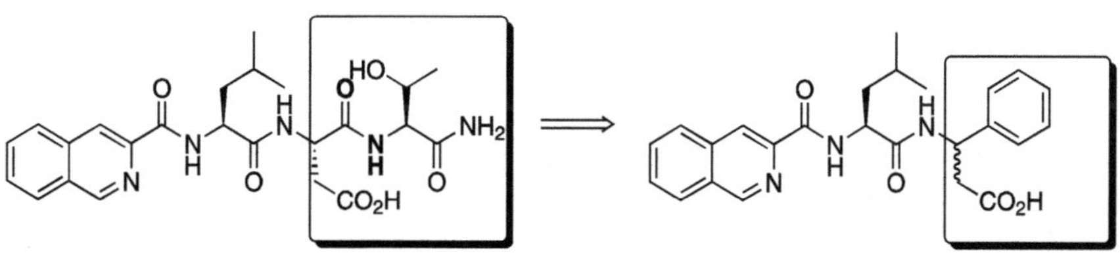

Fig. 28 Amide bond replacement by a phenyl ring

Only a few successful approaches to the synthesis of peptidomimetics based on rigid scaffolds such as cyclohexane [227, 228] and pyranose sugars have been reported [143, 229–233]. We used the information of the $\alpha_4\beta_7$-selective cyclic hexapeptide described above (Fig. 29) [88] to develop β-mannose-based derivatives [234]. An overlay of the conformations exhibited a relatively good match of the essential pharmacophoric groups.

In contrast to the expectations derived from the superposition of the potent and selective $\alpha_4\beta_7$-integrin ligand, the compounds based on the β-D-mannopyranose scaffold showed no activity for this integrin. However, a mannose derivative with side chains that mimic the LDS recognition motif led to an enhanced activity towards the cognate $\alpha_4\beta_1$-integrin recognizing the QIDS(P) sequence, which represents the key motif in VCAM-1 (entry a in Table 3).

Following the approaches of several research groups who discovered phenylalanine-based α_4-ligands, the serine side chain known to be irrelevant for activity to the $\alpha_4\beta_7$-integrin can be replaced by a hydrophobic side chain [196, 197, 235–239]. Therefore, another small library of β-mannose

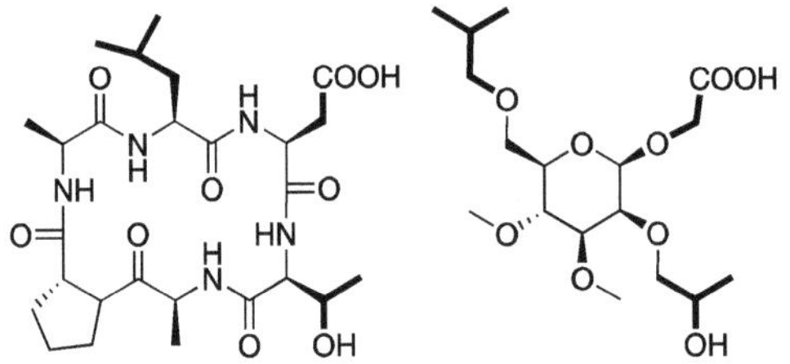

Fig. 29 Cyclic peptide (*left*) and carbohydrate-based peptidomimetic (*right*). The carbohydrate scaffold presents the essential pharmacophoric groups in the same relative orientation as the lead peptide [234]

Table 3 Effect of mannose-based peptidomimetics on VCAM-1 and MAdCAM-1/integrin interaction

#	*n*	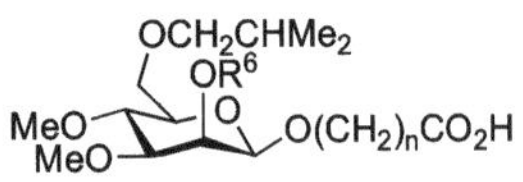 R^6	[%] Adhesion $\alpha_4\beta_1$/VCAM-1	[%] Adhesion $\alpha_4\beta_7$/MAdCAM
a	1	CH_2CH_2OH	30	104
b	1	$CH_2CH_2CH_2OH$	75	108
c	1	$CH_2CH(OH)CH_3$	79	122
d	2	CH_2CH_2OH	93	130
e	2	$CH_2CH_2CH_2OH$	81	132
f	2	$CH_2CH(OH)CH_3$	68	126

Table 4 Cell adhesion is presented as the percentage of medium control in the presence of 1 mg/mL antagonist; [a] Shroff HN et al. (1998) Bioorg Med Chem Lett 8:1601; [b] Boer J et al. (2001) Angew Chem Int Ed 40:3870

#	Structure	[%] Adhesion $\alpha_4\beta_7$/MAdCAM-1	[%] Adhesion $\alpha_4\beta_1$/VCAM-1
Ref. [a]	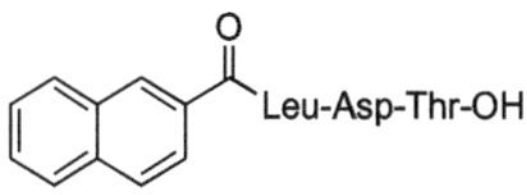	10	10
1-β [b]	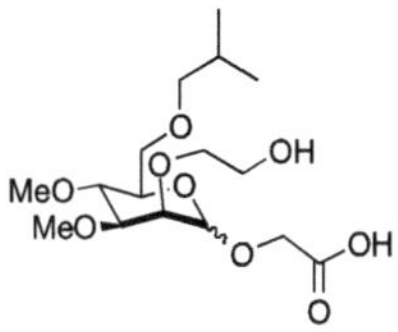	100	60
1-α		100	100
2-β	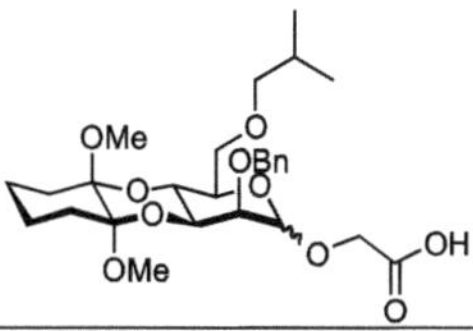	15	77
2-α		90	89

derivatives was synthesized; among them, one compound had about the same activity for inhibition of the $\alpha_4\beta_7$-integrin/MAdCAM-1 interaction as the reference peptide (Table 4) [240]. SAR studies confirmed the β-orientation of the aspartic acid side chain to be necessary for activity. In contrast, the $\alpha_4\beta_1$/VCAM-1 interaction was not affected significantly by the β-anomer, demonstrating its high selectivity for the $\alpha_4\beta_7$-integrin.

5 Medical Application

5.1 RGD Peptides and Peptidomimetics as Drugs

There are numerous publications about using integrins as drug targets. We will here shortly point out only a few applications exemplified by our own work also discussing Cilengitide as a major drug derivative.

The $\alpha_V\beta_3$-integrin is expressed on the surface of a variety of cell types, including osteoclasts, vascular smooth muscle cells, endothelial cells, and various tumor cells and mediates several biologically relevant processes such as adhesion of osteoclasts to bone, vascular smooth muscle cell migration, and angiogenesis. As a result, $\alpha_V\beta_3$-integrin ligands are drug candidates for the treatment of cancer [10, 45, 241], osteoporosis [105, 188, 242, 243], proliferative diabetic retinopathy [244], restenosis [245, 246], and acute renal injury [247–249]. As the $\alpha_V\beta_3$-integrin is overexpressed on the endothelium of tumor tissue, it may contribute to a malign phenotype by supporting angiogenesis. Angiogenesis is an anchorage-dependent process that can be inhibited by interfering with the attachment of endothelial cells to the extracellular matrix [103, 104]. Inhibition of the $\alpha_V\beta_3$-mediated cell–matrix interaction leads to apoptosis of activated endothelial cells and suppresses blood-vessel formation. Thus, RGD ligands can inhibit tumor-induced angiogenesis [45, 250] and can be used as cancer therapeutics exemplified by Cilengitide, cyclo(-RGDf-(NMe)V-), which is in clinical phase II for the treatment of glioblastoma (a brain tumor); for the results of clinical phase I, see Eskens et al. [251].

5.2 Biomaterials

The coating of biomaterials with cyclic RGD peptides offers an attractive strategy for controlling the material interface and achieving a bioactive implant. The boundary between implant and tissue can be strengthened by coating such implants with integrin-specific and cell-selective cyclo(-RGDfK-), which binds and activates the $\alpha_V\beta_3$-integrin expressing osteoblasts [252–255]. The amino acid in position X in cyclo(-RGDfX-) has no significant influence on selectivity and activity towards the $\alpha_V\beta_3$-integrin. Therefore, incorporating side chain functionalized amino acids in this position gives rise to anchoring various linker molecules that are used to bind the peptide covalently to the surface [256, 257]. Coating of implant surfaces with cell-adhesive molecules provides a strong mechanical contact between cells and the graft surface. Recent approaches also involve the application of non-peptidic ligands [258].

5.3 RGD for Tumor Imaging

Another application is in imaging of cancer metastasis. As $\alpha_V\beta_3$-integrins are highly expressed on both tumor and endothelial cells, radiolabeled $\alpha_V\beta_3$-ligands are promising tumor-imaging agents that allow non-invasive visualization of tumors and monitoring of cancer therapy [259, 260]. Several ways to introduce the radiolabel inside cyclic peptides are illustrated in Fig. 30.

The rapid rearrangement of integrins in forming adhesion complexes leads to a rapid dissociation of integrin/RGD peptides [192, 261, 262]. Hence, the signal is washed out easily. Introduction of a sugar moiety improves the pharmacokinetic behavior of a hydrophilic peptide-based tracer [263, 264]. Recently, we could show that radiolabeled multimeric RGD peptides are

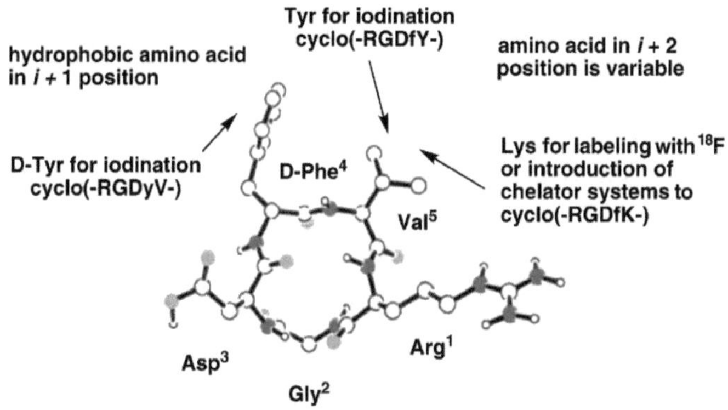

Fig. 30 Structure of cyclo(-RGDfV-). Possible targets for modification are the amino acids in positions 4 and 5 as they do not induce structural transformation [259]

especially suitable to achieve very good contrast between $\alpha_V\beta_3$-expressing cells such as growing endothelial cells in angiogenesis and $\alpha_V\beta_3$-negative tissue [195]. The compounds are efficiently synthesized by linking a *p*-^{18}F-benzaldehyde to an aminooxy-functionalized RGD dendrimer (oxime ligation) [265, 266]. Also, radiolabeled dimeric RGD peptides using ^{64}Cu as positron-emitting radionuclide have been presented recently [267, 268]. The concept of polyvalency was further extended to polymers. Line et al. investigated (99m)Tc(CO)(3)-labeled HPMA copolymer-RGD4C and RGD4C and found significantly higher tumor uptake of HPMA-RGD4C conjugate relative to monomeric RGD4C [269].

5.4 RGD for Drug Targeting

Low water solubility and renal clearance and negative side effects represent the major barriers that limit the therapeutic use of many hydrophobic tumor agents. Therefore, drug-delivery systems such as polymeric micelles have emerged as very important carriers. The utilization of unique molecular markers that are specifically overexpressed seems to be a convenient strategy to achieve cancer-targeted drug delivery. To this end, cyclo(-RGDfK-) as an $\alpha_V\beta_3$-ligand was attached to drug-loaded micelles with the functionalized lysine side chain. Cyclo(-RGDfK-) was selected as the targeting ligand since it can selectively bind to $\alpha_V\beta_3$-integrins, which are over-expressed on tumor cells [270].

Cilengitide has also been used for drug delivery of highly cytotoxic agents like doxorubicin [271] or paclitaxel [272] to tumor tissue. Another example of selective drug delivery using RGD peptides was cancer siRNA. It was embedded in self-assembling nanoparticles consisting of siRNA and polyethyleneimine (PEI) which were PEGylated with an RGD peptide ligand. The ligand was attached at the distal end of the polyethylene glycol [273].

5.5 Biological Role of the $\alpha_{IIb}\beta_3$-Integrin

The platelet fibrinogen receptor $\alpha_{IIb}\beta_3$, exclusively expressed on platelets, has received an enormous amount of attention over the past decade, since ligands have utility in thrombotic disorders such as angina, myocardial infarction, and perhaps atherosclerosis and stroke. The binding of fibrinogen to the activated form of $\alpha_{IIb}\beta_3$-integrin is both a necessary and sufficient event that mediates the process of platelet aggregation which may lead to the occlusion of blood vessels. The obligatory step in the formation of platelet aggregates involves the cross linking of platelets by plasma fibrinogen or other matrix proteins [274]. Platelets are activated by a wide variety of agonists including adenosine diphosphate, serotonin, arachidonic acid, thrombin, and collagen. Arterial occlusive disorders still remain a major cause of morbidity and mortality despite advances in their treatment and diagnosis. As antiplatelet agents, such as aspirin, can only block one of the pathways, prevention of platelet aggregation by inhibition of the $\alpha_{IIb}\beta_3$/fibrinogen interaction represents a more efficacious antiplatelet therapy [102, 117, 274–277]. Clinical trials with several compounds are under way exemplified by early reports of efficacy with both peptide mimetics and peptides derived from cyclic disulfide-bridged peptides, such as eptifibatide [148], roxifiban [158], and lotrafiban [180, 181].

6 Conclusion

Much effort in the pharmaceutical research area is devoted to the tuning of protein–protein interactions by small molecules. Determining the primary amino acid side chain residues required for molecular recognition and the preferred backbone conformation can be regarded as the first step to accomplish this goal. The preferred backbone conformation serves as a template for the bioactive recognition motif. For the identification of the ligands' bioactive conformation, the procedure of "spatial screening" was introduced in the early 1990s. The recognition motif often resides in a turn of an exposed loop. Therefore, application of cyclopenta- and hexapeptides as receptor probes with known 3D structures was initiated. This approach was applied to RGD and LDT peptides and resulted in highly active and selective peptides. Meanwhile, highly active and selective integrin ligands are used for various medical applications such as drug targeting, coating of biomaterials, and development of diagnostic imaging reagents based on radiolabeled compounds.

References

1. Arkin MR, Wells JA (2004) Nat Rev Drug Discov 3:301
2. Ball LJ, Kühne R, Schneider-Mergener J, Oschkinat H (2005) Angew Chem Int Ed 44:2852
3. Berg T (2003) Angew Chem Int Ed 42:2462
4. Gadek TR, Nicholas JB (2003) Biochem Pharmacol 65:1
5. Toogood PL (2002) J Med Chem 45:1543
6. Lo Conte L, Chothia C, Janin J (1999) J Mol Biol 285:2177
7. Heinz DW, Schubert WD, Höfle G (2005) Angew Chem Int Ed 44:1298
8. Bollag DM, McQueney PA, Zhu J, Hensens O, Koupal L, Liesch J, Goetz M, Lazarides E, Woods CM (1995) Cancer Res 55:2325
9. Hynes RO (2002) Cell 110:673
10. Arndt T, Arndt U, Reuning U, Kessler H (2004) In: Weber GF (ed) Cancer Therapy: Molecular Tumor. Horizon Bioscience, Wymondham, pp 93–141
11. Li RH, Hoess RH, Bennett JS, DeGrado WF (2003) Protein Eng 16:65
12. Dubree NJP, Artis DR, Castanedo G, Marsters J, Sutherlin D, Caris L, Clark K, Keating SM, Beresini MH, Chiu H, Fong S, Lowman HB, Skelton NJ, Jackson DY (2002) J Med Chem 45:3451
13. DuVigneaud V, Denning GS Jr, Drabarek S, Chan WY (1964) J Biol Chem 239:472
14. Rothemund S, Beyermann M, Krause E, Krause G, Bienert M, Hodges RS, Sykes BD, Sonnichsen FD (1995) Biochemistry 34:12 954
15. Kessler H (1982) Angew Chem Int Ed 21:512
16. Hruby VJ (1982) Life Sci 31:189
17. Freidinger RM (2003) J Med Chem 46:5553
18. Kessler H, Kutscher B, Klein A (1986) Liebigs Ann Chem, p 893
19. Kessler H, Gratias R, Hessler G, Gurrath M, Müller G (1996) Pure Appl Chem 68:1201
20. Hruby VJ, Balse PM (2000) Curr Med Chem 7:945
21. Payne JW, Marshall NJ, Grail BM, Gupta S (2002) Curr Org Chem 6:1221
22. Gilon C, Mang C, Lohof E, Friedler A, Kessler H (2003) In: Goodman M, Felix L, Moroder L, Toniolo C (eds) Houben-Weyl: Methods of Organic Chemistry, vol E 22b. Thieme, Stuttgart, p 461
23. Li P, Roller PP (2002) Curr Top Med Chem 2:325
24. Davidson JP, Lubman O, Rose T, Waksman G, Martin SF (2002) J Am Chem Soc 124:205
25. Rose GD, Gierasch LM, Smith JA (1985) Adv Protein Chem 37:1
26. Saudek V, Atkinson RA, Pelton JT (1991) Biochemistry 30:7369
27. Chorev M (2002) Chim Oggi 20:81
28. Giannis A (1993) Angew Chem Int Ed 32:1244
29. Bisang C, Jiang L, Freund E, Emery F, Bauch C, Matile H, Pluschke G, Robinson JA (1998) J Am Chem Soc 120:7439
30. Hölzemann G (1991) Kontakte (Darmstadt), p 55
31. Hölzemann G (1991) Kontakte (Darmstadt), p 3
32. Gante J (1994) Angew Chem Int Ed 33:1699
33. Souers AJ, Ellman JA (2001) Tetrahedron 57:7431
34. Kee KS, Jois SD (2003) Curr Pharm Des 9:1209
35. Cluzeau J, Lubell WD (2005) Biopolymers 80:98
36. Matter H, Kessler H (1995) J Am Chem Soc 117:3347
37. Veber DF, Holly FW, Nutt RF, Bergstrand SJ, Brady SF, Hirschmann R, Glitzer MS, Saperstein R (1979) Nature 280:512

38. Malicka J, Czaplewski C, Groth M, Wiczk W, Oldziej S, Lankiewicz L, Ciarkowski J, Liwo A (2004) Curr Top Med Chem 4:123
39. Rew Y, Malkmus S, Svensson C, Yaksh TL, Chung NN, Schiller PW, Cassel JA, DeHaven RN, Taulane JP, Goodman M (2002) J Med Chem 45:5414
40. Fasan R, Dias RLA, Moehle K, Zerbe O, Vrijbloed JW, Obrecht D, Robinson JA (2004) Angew Chem Int Ed 43:2109
41. Wipf P (1995) Chem Rev 95:2115
42. Pierschbacher MD, Ruoslahti E (1984) Nature 309:30
43. Ruoslahti E, Pierschbacher MD (1987) Science 238:491
44. Aumailley M, Gurrath M, Müller G, Calvete J, Timpl R, Kessler H (1991) FEBS Lett 291:50
45. Haubner R, Finsinger D, Kessler H (1997) Angew Chem Int Ed 36:1375
46. Komoriya A, Green LJ, Mervio MM, Yamada SS, Yamada KM, Humphries MJ (1991) Faseb J 5:A1617
47. Viney JL, Jones S, Chiu HH, Lagrimas B, Renz ME, Presta LG, Jackson D, Hillan KJ, Lew S, Fong S (1996) J Immunol 157:2488
48. Weisshoff H, Frost K, Brandt W, Henklein P, Mugge C, Frommel C (1995) FEBS Lett 372:203
49. Hruby VJ (2002) Nat Rev Drug Discov 1:847
50. Veber DF, Johnson SR, Cheng HY, Smith BR, Ward KW, Kopple KD (2002) J Med Chem 45:2615
51. Laskowski RA, MacArthur MW, Moss DS, Thornton JM (1993) J Appl Crystallogr 26:283
52. Balaram H, Uma K, Balaram P (1990) Int J Peptide Protein Res 35:495
53. Hutchinson EG, Thornton JM (1994) Protein Sci 3:2207
54. Ramirez-Alvarado M, Kortemme T, Blanco FJ, Serrano L (1999) Bioorg Med Chem 7:93
55. Dyson HJ, Wright PE (1991) Annu Rev Biophys Biophys Chem 20:519
56. Yao J, Bruschweiler R, Dyson HJ, Wright PE (1994) J Am Chem Soc 116:12 051
57. Kussie PH, Gorina S, Marechal V, Elenbaas B, Moreau J, Levine AJ, Pavletich NP (1996) Science 274:948
58. Kritzer JA, Hodsdon ME, Schepartz A (2005) J Am Chem Soc 127:4118
59. Scherer G, Kramer ML, Schutkowski M, Reimer U, Fischer G (1998) J Am Chem Soc 120:5568
60. Fischer G (2000) Chem Soc Rev 29:119
61. Schiene-Fischer C, Fischer G (2001) J Am Chem Soc 123:6227
62. Kessler H, Anders U, Schudok M (1990) J Am Chem Soc 112:5908
63. Gierasch LM, Deber CM, Madison V, Niu CH, Blout ER (1981) Biochemistry 20:4730
64. Kofron JL, Kuzmic P, Kishore V, Gemmecker G, Fesik SW, Rich DH (1992) J Am Chem Soc 114:2670
65. Bara YA, Friedrich A, Hehlein W, Kessler H, Kondor P, Molter M, Veith HJ (1978) Chem Ber 111:1029
66. Manjula G, Ramakrishnan C (1979) Int J Peptide Protein Res 13:353
67. Ramakrishnan C, Rao BNN (1980) Int J Peptide Protein Res 15:81
68. Bewley CA, He H, Williams DH, Faulkner DJ (1996) J Am Chem Soc 118:4314
69. Kurz M (1991) Cyclische Modellpeptide als Template für ein konformationell orientiertes Peptiddesign. Dissertation, Technische Universität München
70. Mierke DF, Kurz M, Kessler H (1994) J Am Chem Soc 116:1042
71. Haubner R, Gratias R, Diefenbach B, Goodman SL, Jonczyk A, Kessler H (1996) J Am Chem Soc 118:7461

72. Rodriguez AM, Baldoni HA, Suvire F, Vazquez RN, Zamarbide G, Enriz RD, Farkas O, Perczel A, McAllister MA, Torday LL, Papp JG, Csizmadia IG (1998) J Mol Struct Theochem 455:275
73. Weisshoff H, Prasang C, Henklein P, Frommel C, Zschunke A, Mügge C (1999) Eur J Biochem 259:776
74. Nikiforovich GV, Kover KE, Zhang WJ, Marshall GR (2000) J Am Chem Soc 122:3262
75. Schwyzer R, Ludescher U (1969) Helv Chim Acta 52:2033
76. Gurrath M, Müller G, Kessler H, Aumailley M, Timpl R (1992) Eur J Biochem 210:911
77. Pfaff M, Tangemann K, Müller B, Gurrath M, Müller G, Kessler H, Timpl R, Engel J (1994) J Biol Chem 269:20 233
78. Kessler H, Kutscher B, Kerssebaum R, Klein A, Lautz J, Obermeier R, Muellner H, Van Gunsteren WF, Boelens R, Kaptein R (1985) In: Peptides: Structure and Function, Proc 9th Am Pept Symp, p 83
79. Kessler H, Kutscher B (1985) Tetrahedron Lett 26:177
80. Fujii N, Oishi S, Hiramatsu K, Araki T, Ueda S, Tamamura H, Otaka A, Kusano S, Terakubo S, Nakashima H, Broach JA, Trent JO, Wang ZX, Peiper SC (2003) Angew Chem Int Ed 42:3251
81. Tamamura H, Mizumoto M, Hiramatsu K, Kusano S, Terakubo S, Yamamoto N, Trent JO, Wang ZX, Peiper SC, Nakashima H, Otaka A, Fujii N (2004) Org Biomol Chem 2:1255
82. Fukami T, Nagase T, Fujita K, Hayama T, Niiyama K, Mase T, Nakajima S, Fukuroda T, Saeki T, Nishikibe M, Ihara M, Yano M, Ishikawa K (1995) J Med Chem 38:4309
83. Spatola AF, Crozet Y, DeWit D, Yanagisawa M (1996) J Med Chem 39:3842
84. Gao XM, Ye YH, Bernd M, Kutscher B (2002) J Peptide Sci 8:418
85. Thurieau C, Feletou M, Hennig P, Raimbaud E, Canet E, Fauchere JL (1996) J Med Chem 39:2095
86. Monteagudo ES, Calvani F, Catrambone F, Fincham CI, Madami A, Meini S, Terracciano R (2001) J Peptide Sci 7:270
87. Etzkorn FA, Guo T, Lipton MA, Goldberg SD, Bartlett PA (1994) J Am Chem Soc 116:10 412
88. Boer J, Gottschling D, Schuster A, Semmrich M, Holzmann B, Kessler H (2001) J Med Chem 44:2586
89. Elbe JA, Kühn K (1997) Integrin–Ligand Interaction. Springer, Heidelberg
90. Plow EF, Haas TK, Zhang L, Loftus J, Smith JW (2000) J Biol Chem 275:21785
91. Arnaout MA, Goodman SL, Xiong JP (2002) Curr Opin Cell Biol 14:641
92. Shimaoka M, Springer TA (2003) Nat Rev Drug Discov 2:703
93. Albelda SM, Buck CA (1990) Faseb J 4:2868
94. Hynes RO, Lander AD (1992) Cell 68:303
95. Collins TL, Kassner PD, Bierer BE, Burakoff SJ (1994) Curr Opin Immunol 6:385
96. Humphries MJ (2000) Biochem Soc Trans 28:311
97. Adler M, Lazarus RA, Dennis MS, Wagner G (1991) Science 253:445
98. Ruoslahti E, Pierschbacher MD (1986) Cell 44:517
99. D'Souza SE, Ginsberg MH, Plow EF (1991) Trends Biochem Sci 16:246
100. Calvete JJ, Schafer W, Soszka T, Lu WQ, Cook JJ, Jameson BA, Niewiarowski S (1991) Biochemistry 30:5225
101. Smith KJ, Jaseja M, Lu XJ, Williams JA, Hyde EI, Trayer IP (1996) Int J Peptide Protein Res 48:220
102. Scarborough RM, Gretler DD (2000) J Med Chem 43:3453
103. Strömblad S, Cheresh DA (1996) Chem Biol 3:881

104. Strömblad S, Cheresh DA (1996) Trends Cell Biol 6:462
105. Schaffner P, Dard MM (2003) Cell Mol Life Sci 60:119
106. Müller G, Gurrath M, Kessler H (1994) J Comput Aided Mol Des 8:709
107. Calvete JJ, Moreno-Murciano MP, Theakston RDG, Kisiel DG, Marcinkiewicz C (2003) Biochem J 372:725
108. Dechantsreiter MA, Planker E, Mathä B, Lohof E, Hölzemann G, Jonczyk A, Goodman SL, Kessler H (1999) J Med Chem 42:3033
109. Xiong JP, Stehle T, Zhang RG, Joachimiak A, Frech M, Goodman SL, Arnaout MA (2002) Science 296:151
110. Xiong JP, Stehle T, Diefenbach B, Zhang RG, Dunker R, Scott DL, Joachimiak A, Goodman SL, Arnaout MA (2001) Science 294:339
111. Xiao T, Takagi J, Coller BS, Wang JH, Springer TA (2004) Nature 432:59
112. Marinelli L, Lavecchia A, Gottschalk KE, Novellino E, Kessler H (2003) J Med Chem 46:4393
113. Gottschalk KE, Kessler H (2002) Angew Chem Int Ed 41:3767
114. Gottschalk KE, Günther R, Kessler H (2002) ChemBioChem 3:470
115. Cheng S, Craig WS, Mullen D, Tschopp JF, Dixon D, Pierschbacher MD (1994) J Med Chem 37:1
116. Craig WS, Cheng S, Mullen DG, Blevitt J, Pierschbacher MD (1995) Biopolymers 37:157
117. Scarborough RM, Kleiman NS, Phillips DR (1999) Circulation 100:437
118. Heller M, Mierke DF, Reif B, Kessler H (2006) J Am Chem Soc, in press
119. Lohof E, Planker E, Mang C, Burkhart F, Dechantsreiter MA, Haubner R, Wester HJ, Schwaiger M, Hölzemann G, Goodman SL, Kessler H (2000) Angew Chem Int Ed 39:2761
120. Bean JW, Kopple KD, Peishoff CE (1992) J Am Chem Soc 114:5328
121. Peishoff CE, Ali FE, Bean JW, Calvo R, Dambrosio CA, Eggleston DS, Hwang SM, Kline TP, Koster PF, Nichols A, Powers D, Romoff T, Samanen JM, Stadel J, Vasko JA, Kopple KD (1992) J Med Chem 35:3962
122. Goodman SL, Hölzemann G, Sulyok GAG, Kessler H (2002) J Med Chem 45:1045
123. Seebach D, Kimmerlin T, Sebesta R, Campo MA, Beck AK (2004) Tetrahedron 60:7455
124. Seebach D, Beck AK, Bierbaum DJ (2004) Chem Biodiversity 1:1111
125. Pierschbacher MD, Ruoslahti E (1987) J Biol Chem 262:17 294
126. Schumann F, Müller A, Koksch M, Müller G, Sewald N (2000) J Am Chem Soc 122:12 009
127. Zimmermann D, Guthohrlein EW, Malesevic M, Sewald K, Wobbe L, Heggemann C, Sewald N (2005) ChemBioChem 6:272
128. Ron D, Gilon C, Hanani M, Vromen A, Selinger Z, Chorev M (1992) J Med Chem 35:2806
129. Bach AC, Eyermann CJ, Gross JD, Bower MJ, Harlow RL, Weber PC, Degrado WF (1994) J Am Chem Soc 116:3207
130. Bach AC, Espina JR, Jackson SA, Stouten PFW, Duke JL, Mousa SA, DeGrado WF (1996) J Am Chem Soc 118:293
131. Tyndall JDA, Nall T, Fairlie DP (2005) Chem Rev 105:973
132. Chorev M, Goodman M (1993) Acc Chem Res 26:266
133. Freidinger RM, Veber DF (1979) J Am Chem Soc 101:6129
134. Wermuth J, Goodman SL, Jonczyk A, Kessler H (1997) J Am Chem Soc 119:1328
135. Geyer A, Müller G, Kessler H (1994) J Am Chem Soc 116:7735
136. Suguna K, Padlan EA, Smith CW, Carlson WD, Davies DR (1987) Proc Natl Acad Sci 84:7009

137. Miller M, Schneider J, Sathyanarayana BK, Toth MV, Marshall GR, Clawson L, Selk L, Kent SBH, Wlodawer A (1989) Science 246:1149
138. Ma SG, Spatola AF (1993) Int J Peptide Protein Res 41:204
139. Haubner R, Schmitt W, Hölzemann G, Goodman SL, Jonczyk A, Kessler H (1996) J Am Chem Soc 118:7881
140. Finsinger D (1997) Peptidische und peptidanaloge Antagonisten des alphaVbeta3-integrins. Dissertation, Technische Universität München
141. Davies DE, Doyle PM, Farrant RD, Hill RD, Hitchcock PB, Sanderson PN, Young DW (2003) Tetrahedron Lett 44:8887
142. Belvisi L, Bernardi A, Checchia A, Manzoni L, Potenza D, Scolastico C, Castorina M, Cupelli A, Giannini G, Carminati P, Pisano C (2001) Org Lett 3:1001
143. Hirschmann R, Nicolaou KC, Pietranico S, Leahy EM, Salvino J, Arison B, Cichy MA, Spoors PG, Shakespeare WC, Sprengeler PA, Hamley P, Smith AB, Reisine T, Raynor K, Maechler L, Donaldson C, Vale W, Freidinger RM, Cascieri MR, Strader CD (1993) J Am Chem Soc 115:12 550
144. von Roedern E, Kessler H (1994) Angew Chem Int Ed 33:687
145. von Roedern E, Lohof E, Hessler G, Hoffmann M, Kessler H (1996) J Am Chem Soc 118:10 156
146. Gruner SAW, Locardi E, Lohof E, Kessler H (2002) Chem Rev 102:491
147. Scarborough RM, Rose JW, Hsu MA, Phillips DR, Fried VA, Campbell AM, Nannizzi L, Charo IF (1991) J Biol Chem 266:9359
148. Scarborough RM, Naughton MA, Teng W, Rose JW, Phillips DR, Nannizzi L, Arfsten A, Campbell AM, Charo IF (1993) J Biol Chem 268:1066
149. Fisher MJ, Gunn B, Harms CS, Kline AD, Mullaney JT, Nunes A, Scarborough RM, Arfsten AE, Skelton MA, Um SL, Utterback BG, Jakubowski JA (1997) J Med Chem 40:2085
150. Samanen J, Ali F, Romoff T, Calvo R, Sorenson E, Vasko J, Storer B, Berry D, Bennett D, Strohsacker M, Powers D, Stadel J, Nichols A (1991) J Med Chem 34:3114
151. Kopple KD, Baures PW, Bean JW, Dambrosio CA, Hughes JL, Peishoff CE, Eggleston DS (1992) J Am Chem Soc 114:9615
152. Ali FE, Bennett DB, Calvo RR, Elliott JD, Hwang SM, Ku TW, Lago MA, Nichols AJ, Romoff TT, Shah DH, Vasko JA, Wong AS, Yellin TO, Yuan CK, Samanen JM (1994) J Med Chem 37:769
153. Xue CB, DeGrado WF (1995) J Org Chem 60:946
154. Jackson S, DeGrado W, Dwivedi A, Parthasarathy A, Higley A, Krywko J, Rockwell A, Markwalder J, Wells G, Wexler R, Mousa S, Harlow R (1994) J Am Chem Soc 116:3220
155. Müller G (1996) Angew Chem Int Ed 35:2767
156. Greenspoon N, Hershkoviz R, Alon R, Varon D, Shenkman B, Marx G, Federman S, Kapustina G, Lider O (1993) Biochemistry 32:1001
157. Corbett JW, Graciani NR, Mousa SA, DeGrado WF (1997) Bioorg Med Chem Lett 7:1371
158. Xue CB, Wityak J, Sielecki TM, Pinto DJ, Batt DG, Cain GA, Sworin M, Rockwell AL, Roderick JJ, Wang SG, Orwat MJ, Frietze WE, Bostrom LL, Liu J, Higley CA, Rankin FW, Tobin AE, Emmett G, Lalka GK, Sze JY, DiMeo SV, Mousa SA, Thoolen MJ, Racanelli AL, Hausner EA, Reilly TM, DeGrado WF, Wexler RR, Olson RE (1997) J Med Chem 40:2064
159. Blackburn BK, Lee A, Baier M, Kohl B, Olivero AG, Matamoros R, Robarge KD, McDowell RS (1997) J Med Chem 40:717
160. Rockwell AL, Rafalski M, Pitts WJ, Batt DG, Petraitis JJ, DeGrado WF, Mousa S, Jadhav PK (1999) Bioorg Med Chem Lett 9:937

161. Olson RE, Sielecki TM, Wityak J, Pinto DJ, Batt DG, Frietze WE, Liu J, Tobin AE, Orwat MJ, Di Meo SV, Houghton GC, Lalka GK, Mousa SA, Racanelli AL, Hausner EA, Kapil RP, Rabel HR, Thoolen MJ, Reilly TM, Anderson PS, Wexler RR (1999) J Med Chem 42:1178
162. Batt DG, Petraitis JJ, Houghton GC, Modi DP, Cain GA, Corjay MH, Mousa SA, Bouchard PJ, Forsythe MS, Harlow PP, Barbera FA, Spitz SM, Wexler RR, Jadhav PK (2000) J Med Chem 43:41
163. Pitts WJ, Wityak J, Smallheer JM, Tobin AE, Jetter JW, Buynitsky JS, Harlow PP, Solomon KA, Corjay MH, Mousa SA, Wexler RR, Jadhav PK (2000) J Med Chem 43:27
164. Gibson C, Sulyok GAG, Hahn D, Goodman SL, Hölzemann G, Kessler H (2001) Angew Chem Int Ed 40:165
165. Duggan ME, Hutchinson JH (2000) Expert Opin Ther Pat 10:1367
166. Biltresse S, Attolini M, Dive G, Cordi A, Tucker GC, Marchand-Brynaert J (2004) Bioorg Med Chem 12:5379
167. Keenan RM, Miller WH, Lago MA, Ali FE, Bondinell WE, Callahan JF, Calvo RR, Cousins RD, Hwang SM, Jakas DR, Ku TW, Kwon C, Nguyen TT, Reader VA, Rieman DJ, Ross ST, Takata DT, Uzinskas IN, Yuan CCK, Smith BR (1998) Bioorg Med Chem Lett 8:3165
168. Keenan RM, Lago MA, Miller WH, Ali FE, Cousins RD, Hall LB, Hwang SM, Jakas DR, Kwon C, Louden C, Nguyen TT, Ohlstein EH, Rieman DJ, Ross ST, Samanen JM, Smith BR, Stadel J, Takata DT, Vickery L, Yuan CCK, Yue TL (1998) Bioorg Med Chem Lett 8:3171
169. Keenan RM, Miller WH, Barton LS, Bondinell WE, Cousins RD, Eppley DF, Hwang SM, Kwon C, Lago MA, Nguyen TT, Smith BR, Uzinskas IN, Yuan CCK (1999) Bioorg Med Chem Lett 9:1801
170. Miller WH, Bondinell WE, Cousins RD, Erhard KF, Jakas DR, Keenan RM, Ku TW, Newlander KA, Ross ST, Haltiwanger RC, Bradbeer J, Drake FH, Gowen M, Hoffman SJ, Hwang SM, James IE, Lark MW, Lechowska B, Rieman DJ, Stroup GB, Vasko-Moser JA, Zembryki DL, Azzarano LM, Adams PC, Salyers KL, Smith BR, Ward KW, Johanson KO, Huffman WF (1999) Bioorg Med Chem Lett 9:1807
171. Miller WH, Alberts DP, Bhatnagar PK, Bondinell WE, Callahan JF, Calvo RR, Cousins RD, Erhard KF, Heerding DA, Keenan RM, Kwon C, Manley PJ, Newlander KA, Ross ST, Samanen JM, Uzinskas IN, Venslavsky JW, Yuan CCK, Haltiwanger RC, Gowen M, Hwang SM, James IE, Lark MW, Rieman DJ, Stroup GB, Azzarano LM, Salyers KL, Smith BR, Ward KW, Johanson KO, Huffman WF (2000) J Med Chem 43:22
172. Sulyok GAG, Gibson C, Goodman SL, Hölzemann G, Wiesner M, Kessler H (2001) J Med Chem 44:1938
173. Miller WH, Manley PJ, Cousins RD, Erhard KF, Heerding DA, Kwon C, Ross ST, Samanen JM, Takata DT, Uzinskas IN, Yuan CCK, Haltiwanger RC, Gress CJ, Lark MW, Hwang SM, James IE, Rieman DJ, Willette RN, Yue TL, Azzarano LM, Salyers KL, Smith BR, Ward KW, Johanson KO, Huffman WF (2003) Bioorg Med Chem Lett 13:1483
174. Scarborough RM (1999) Curr Med Chem 6:971
175. Yasuda N, Hsiao Y, Jensen MS, Rivera NR, Yang CH, Wells KM, Yau J, Palucki M, Tan LS, Dormer PG, Volante RP, Hughes DL, Reider PJ (2004) J Org Chem 69:1959
176. Wityak J, Sielecki TM, Pinto DJ, Emmett G, Sze JY, Liu J, Tobin AE, Wang SG, Jiang B, Ma P, Mousa SA, Olson RE, Wexler RR (1997) J Med Chem 40:1292
177. Ku TW, Ali FE, Barton LS, Bean JW, Bondinell WE, Burgess JL, Callahan JF, Calvo RR, Chen LC, Eggleston DS, Gleason JG, Huffman WF, Hwang SM, Jakas DR, Karash CB,

Keenan RM, Kopple KD, Miller WH, Newlander KA, Nichols A, Parker MF, Peishoff CE, Samanen JM, Uzinskas I, Venslavsky JW (1993) J Am Chem Soc 115:8861
178. Ku TW, Miller WH, Bondinell WE, Erhard KF, Keenan RM, Nichols AJ, Peishoff CE, Samanen JM, Wong AS, Huffman WF (1995) J Med Chem 38:9
179. Samanen JM, Ali FE, Barton LS, Bondinell WE, Burgess JL, Callahan JF, Calvo RR, Chen WT, Chen LC, Erhard K, Feuerstein G, Heys R, Hwang SM, Jakas DR, Keenan RM, Ku TW, Kwon C, Lee CP, Miller WH, Newlander KA, Nichols A, Parker M, Peishoff CE, Rhodes G, Ross S, Shu A, Simpson R, Takata D, Yellin TO, Uzsinskas I, Venslavsky JW, Yuan CK, Huffman WF (1996) J Med Chem 39:4867
180. Davis HM, Hwang SM, Freed MI (1997) Blood 90:1322
181. Wong A, Hwang SM, Johanson K, Samanen J, Bennett D, Landvatter SW, Chen WT, Heys JR, Ali FE, Ku TW, Bondinell W, Nichols AJ, Powers DA, Stadel JM (1998) J Pharmacol Exp Ther 285:228
182. Okumura K, Shimazaki T, Aoki Y, Yamashita H, Tanaka E, Banba S, Yazawa K, Kibayashi K, Banno H (1998) J Med Chem 41:4036
183. Fisher MJ, Arfstan AE, Giese U, Gunn BP, Harms CS, Khau V, Kinnick MD, Lindstrom TD, Martinelli MJ, Mest HJ, Mohr M, Morin JM, Mullaney JT, Nunes A, Paal M, Rapp A, Ruhter G, Ruterbories KJ, Sall DJ, Scarborough RM, Schotten T, Sommer B, Stenzel W, Towner RD, Um SL, Utterback BG, Vasileff RT, Voelkers S, Wyss VL, Jakubowski JA (1999) J Med Chem 42:4875
184. Fisher MJ, Giese U, Harms CS, Kinnick MD, Lindstrom TD, McCowan JR, Mest HJ, Morin JM, Mullaney JT, Paal M, Rapp A, Ruhter G, Ruterbories KJ, Sall DJ, Scarborough RM, Schotten T, Stenzel W, Towner RD, Um SL, Utterback BG, Wyss VL, Jakubowski JA (2000) Bioorg Med Chem Lett 10:385
185. Miller WH, Keenan RM, Willette RN, Lark MW (2000) Drug Discov Today 5:397
186. Duggan ME, Duong LT, Fisher JE, Hamill TG, Hoffman WF, Huff JR, Ihle NC, Leu CT, Nagy RM, Perkins JJ, Rodan SB, Wesolowski G, Whitman DB, Zartman AE, Rodan GA, Hartman GD (2000) J Med Chem 43:3736
187. Kling A, Backfisch G, Delzer J, Geneste H, Graef C, Hornberger W, Lange UEW, Lauterbach A, Seitz W, Subkowski T (2003) Bioorg Med Chem 11:1319
188. Keenan RM, Miller WH, Kwon C, Ali FE, Callahan JF, Calvo RR, Hwang SM, Kopple KD, Peishoff CE, Samanen JM, Wong AS, Yuan CK, Huffman WF (1997) J Med Chem 40:2289
189. Marinelli L, Gottschalk K-E, Meyer A, Novellino E, Kessler H (2004) J Med Chem 47:4166
190. Marinelli L, Meyer A, Heckmann D, Lavecchia A, Novellino E, Kessler H (2005) J Med Chem 48:4204
191. Gottschalk K-E, Kessler H (2004) FEBS Lett 557:253
192. Li R, Mitra N, Gratkowski H, Vilaire G, Litvinov R, Nagasami C, Weisel JW, Lear JD, DeGrado WF, Bennett JS (2003) Science 300:795
193. Gottschalk KE, Adams PD, Brunger AT, Kessler H (2002) Protein Sci 11:1800
194. Schneider D, Engelman DM (2004) J Biol Chem 279:9840
195. Wester HJ, Kessler H (2005) J Nucl Med 46:1940
196. Tilley JW, Sidduri A (2001) Drug Future 26:985
197. Tilley JW (2002) Expert Opin Ther Pat 12:991
198. Hemler ME (1990) Annu Rev Immunol 8:365
199. Postigo AA, Sanchezmateos P, Lazarovits AI, Sanchezmadrid F, Delandazuri MO (1993) J Immunol 151:2471
200. Berlin C, Berg EL, Briskin MJ, Andrew DP, Kilshaw PJ, Holzmann B, Weissman IL, Hamann A, Butcher EC (1993) Cell 74:185

201. Kilger G, Holzmann B (1995) J Mol Med 73:347
202. Yusuf-Makagiansar H, Anderson ME, Yakovleva TV, Murray JS, Siahaan TJ (2002) Med Res Rev 22:146
203. von Andrian UH, Engelhardt B (2003) New Engl J Med 348:68
204. Holzmann B, McIntyre BW, Weissman IL (1989) Cell 56:37
205. Picarella D, Hurlbut P, Rottman J, Shi XJ, Butcher E, Ringler DJ (1997) J Immunol 158:2099
206. Elices MJ (2003) Curr Opin Invest Drugs 4:1354
207. Noseworthy JH, Kirkpatrick P (2005) Nat Rev Drug Discov 4:101
208. Steinman L (2005) Nat Rev Drug Discov 4:510
209. Renz ME, Chin HH, Jones S, Fox J, Kim KJ, Presta LG, Fong S (1994) J Cell Biol 125:1395
210. Jones EY, Harlos K, Bottomley MJ, Robinson RC, Driscoll PC, Edwards RM, Clements JM, Dudgeon TJ, Stuart DI (1995) Nature 373:539
211. Shroff HN, Schwender CF, Dottavio D, Yang LL, Briskin MJ (1996) Bioorg Med Chem Lett 6:2495
212. Newham P, Craig SE, Seddon GN, Schofield NR, Rees A, Edwards RM, Jones EY, Humphries MJ (1997) J Biol Chem 272:19 429
213. Dando J, Wilkinson KW, Ortlepp S, King DJ, Brady RL (2002) Acta Crystallogr D 58:233
214. Tan K, Casasnovas JM, Liu JH, Briskin MJ, Springer TA, Wang JH (1998) Structure 6:793
215. Singh J, Adams S, Carter MB, Cuervo H, Lee WC, Lobb RR, Pepinsky RB, Petter R, Scott D (2004) Curr Top Med Chem 4:1497
216. Singh J, van Vlijmen H, Liao YS, Lee WC, Cornebise M, Harris M, Shu IH, Gill A, Cuervo JH, Abraham WM, Adams SP (2002) J Med Chem 45:2988
217. Jackson DY (2002) Curr Pharm Des 8:1229
218. Yang GX, Hagmann WK (2003) Med Res Rev 23:369
219. Tilley JW, Chen L, Sidduri A, Fotouhi N (2004) Curr Top Med Chem 4:1509
220. Hagmann WK (2004) Curr Top Med Chem 4:1461
221. Huryn DM, Konradi AW, Ashwell S, Freedman SB, Lombardo LJ, Pleiss MA, Thorsett ED, Yednock T, Kennedy JD (2004) Curr Top Med Chem 4:1473
222. Gottschling D, Boer J, Marinelli L, Voll G, Haupt M, Schuster A, Holzmann B, Kessler H (2002) ChemBioChem 3:575
223. Shroff HN, Schwender CF, Baxter AD, Brookfield F, Payne LJ, Cochran NA, Gallant DL, Briskin MJ (1998) Bioorg Med Chem Lett 8:1601
224. Rockwell A, Melden M, Copeland RA, Hardman K, Decicco CP, DeGrado WF (1996) J Am Chem Soc 118:10 337
225. Gottschling D, Boer E, Schuster A, Holzmann B, Kessler H (2001) Bioorg Med Chem Lett 11:2997
226. Gottschling D, Boer J, Schuster A, Holzmann B, Kessler H (2002) Angew Chem Int Ed 41:3007
227. Belanger PC, Dufresne C (1986) Can J Chem 64:1514
228. Farmer PS (1980) Drug Design. Academic, New York
229. Papageorgiou C, Haltiner R, Bruns C, Petcher TJ (1992) Bioorg Med Chem Lett 2:135
230. Hirschmann R, Sprengeler PA, Kawasaki T, Leahy JW, Shakespeare WC, Smith AB (1993) Tetrahedron 49:3665
231. Olson GL, Bolin DR, Bonner MP, Bos M, Cook CM, Fry DC, Graves BJ, Hatada M, Hill DE, Kahn M, Madison VS, Rusiecki VK, Sarabu R, Sepinwall J, Vincent GP, Voss ME (1993) J Med Chem 36:3039

232. Hirschmann R, Hynes J, Cichy-Knight MA, van Rijn RD, Sprengeler PA, Spoors PG, Shakespeare WC, Pietranico-Cole S, Barbosa J, Liu J, Yao WQ, Rohrer S, Smith AB (1998) J Med Chem 41:1382
233. Le GT, Abbenante G, Becker B, Grathwohl M, Halliday J, Tometzki G, Zuegg J, Meutermans W (2003) Drug Discov Today 8:701
234. Boer J, Gottschling D, Schuster A, Holzmann B, Kessler H (2001) Angew Chem Int Ed 40:3870
235. Chen L, Tilley JW, Huang TN, Miklowski D, Trilles R, Guthrie RW, Luk K, Hanglow A, Rowan K, Schwinge V, Wolitzky B (2000) Bioorg Med Chem Lett 10:725
236. Chen L, Tilley JW, Guthrie RW, Mennona F, Huang TN, Kaplan G, Trilles R, Miklowski D, Huby N, Schwinge V, Wolitzky B, Rowan K (2000) Bioorg Med Chem Lett 10:729
237. Harriman GCB, Schwender CF, Gallant D, Cochran NA, Briskin MJ (2000) Bioorg Med Chem Lett 10:1497
238. Hagmann WK, Durette PL, Lanza T, Kevin NJ, de Laszlo SE, Kopka HE, Young D, Magriotis PA, Li B, Lin LS, Yang G, Kamenecka T, Chang LL, Wilson J, MacCoss M, Mills SG, Van Riper G, McCauley E, Egger LA, Kidambi U, Lyons K, Vincent S, Stearns R, Colletti A, Teffera J, Tong S, Fenyk-Melody J, Owens K, Levorse D, Kim P, Schmidt JA, Mumford RA (2001) Bioorg Med Chem Lett 11:2709
239. Tilley JW, Kaplan G, Rowan K, Schwinge V, Wolitzky B (2001) Bioorg Med Chem Lett 11:1
240. Locardi E, Boer J, Modlinger A, Schuster A, Holzmann B, Kessler H (2003) J Med Chem 46:5752
241. Yun Z, Menter DG, Nicolson GL (1996) Cancer Res 56:3103
242. Chorev M, Dresnerpollak R, Eshel Y, Rosenblatt M (1995) Biopolymers 37:367
243. Engleman VW, Nickols GA, Ross FP, Horton MA, Griggs DW, Settle SL, Ruminski PG, Teitelbaum SL (1997) J Clin Invest 99:2284
244. Friedlander M, Theesfeld CL, Sugita M, Fruttiger M, Thomas MA, Chang S, Cheresh DA (1996) Proc Natl Acad Sci USA 93:9764
245. Matsuno H, Stassen JM, Vermylen J, Deckmyn H (1994) Circulation 90:2203
246. Choi ET, Engel L, Callow AD, Sun S, Trachtenberg J, Santoro S, Ryan US (1994) J Vasc Surg 19:125
247. Noiri E, Gailit J, Sheth D, Magazine H, Gurrath M, Muller G, Kessler H, Goligorsky MS (1994) Kidney Int 46:1050
248. Romanov V, Noiri E, Czerwinski G, Finsinger D, Kessler H, Goligorsky MS (1997) Kidney Int 52:93
249. Goligorsky MS, Noiri E, Kessler H, Romanov V (1997) Kidney Int 51:1487
250. Brooks PC, Montgomery AMP, Rosenfeld M, Reisfeld RA, Hu TH, Klier G, Cheresh DA (1994) Cell 79:1157
251. Eskens F, Dumez H, Verweij J, Brindley C, Perschl A, Kovar A, Wynendaele W, van Oosterom A (2000) Clin Cancer Res 6:4524s
252. Auernheimer J, Zukowski D, Dahmen C, Kantlehner M, Enderle A, Goodman SL, Kessler H (2005) ChemBioChem 6:2034
253. Elmengaard B, Bechtold JE, Søballe K (2005) Biomaterials 26:3521
254. Schliephake H, Scharnweber D, Dard M, Rossler S, Sewing A, Meyer J, Hoogestraat D (2002) Clin Oral Implants Res 13:312
255. Hersel U, Dahmen C, Kessler H (2003) Biomaterials 24:4385
256. Kantlehner M, Schaffner P, Finsinger D, Meyer J, Jonczyk A, Diefenbach B, Nies B, Hölzemann G, Goodman SL, Kessler H (2000) ChemBioChem 1:107
257. Kantlehner M, Finsinger D, Meyer J, Schaffner P, Jonczyk A, Diefenbach B, Nies B, Kessler H (1999) Angew Chem Int Ed 38:560

258. Dahmen C, Auernheimer J, Meyer A, Enderle A, Goodman SL, Kessler H (2004) Angew Chem Int Ed 43:6649
259. Haubner R, Wester HJ, Reuning U, Senekowitsch-Schmidtke R, Diefenbach B, Kessler H, Stöcklin G, Schwaiger M (1999) J Nucl Med 40:1061
260. Haubner R, Wester HR (2004) Curr Pharm Des 10:1439
261. Gottschalk K-E, Kessler H (2004) Structure 12:1109
262. Li R, Gorelik R, Nanda V, Law PB, Lear JD, DeGrado WF, Bennett JS (2004) J Biol Chem 279:26 666
263. Haubner R, Wester HJ, Burkhart F, Senekowitsch-Schmidtke R, Weber W, Goodman SL, Kessler H, Schwaiger M (2001) J Nucl Med 42:326
264. Haubner R, Wester HJ, Weber WA, Mang C, Ziegler SI, Goodman SL, Senekowitsch-Schmidtke R, Kessler H, Schwaiger M (2001) Cancer Res 61:1781
265. Thumshirn G, Hersel U, Goodman SL, Kessler H (2003) Chem Eur J 9:2717
266. Poethko T, Schottelius M, Thumshim G, Hersel U, Herz M, Henriksen G, Kessler H, Schwaiger M, Wester HJ (2004) J Nucl Med 45:892
267. Chen X, Tohme M, Park R, Hou Y, Bading JR, Conti PS (2004) Mol Imaging 3:96
268. Chen X, Sievers E, Hou Y, Park R, Tohme M, Bart R, Bremner R, Bading JR, Conti PS (2005) Neoplasia 7:271
269. Line BR, Mitra A, Nan A, Ghandehari H (2005) J Nucl Med 46:1552
270. Nasongkla N, Shuai X, Ai H, Weinberg BD, Pink J, Boothman DA, Gao JM (2004) Angew Chem Int Ed 43:6323
271. Burkhart DJ, Kalet BT, Coleman MP, Post GC, Koch TH (2004) Mol Cancer Ther 3:1593
272. Chen X, Plasencia C, Hou Y, Neamati N (2005) J Med Chem 48:5874
273. Schiffelers RM, Ansari A, Xu J, Zhou Q, Tang Q, Storm G, Molema G, Lu PY, Scaria PV, Woodle MC (2004) Nucleic Acids Res 32:e149
274. Raddatz P (1995) Expert Opin Ther Pat 5:1163
275. Ojima I, Chakravarty S, Dong Q (1995) Bioorg Med Chem 3:337
276. Wityak J, Sielecki TM (1996) Expert Opin Ther Pat 6:1175
277. Coller BS (1997) J Clin Invest 99:1467

Top Curr Chem (2007) 272: 51–82
DOI 10.1007/128_2006_090

Published online: 20 October 2006

Dynamics and Thermodynamics of Ligand–Protein Interactions

S. W. Homans

Astbury Centre for Structural Molecular Biology, University of Leeds, Leeds LS2 9JT, UK
s.w.homans@leeds.ac.uk

Abstract Protein–ligand interactions are of fundamental importance in a great many biological processes. However, despite enormous advances in the speed and accuracy of the three-dimensional structure determination of proteins and their complexes, our ability to predict binding affinity from structure remains severely limited. One reason for this dilemma is that affinities are governed not only by energetic considerations arising from the precise spatial disposition of interacting groups (loosely, enthalpy), but also by the dynamics of these groups (loosely, entropy) including solvent effects. In this work I will review current methodology for unravelling this complex problem, including X-ray crystallography, NMR, isothermal titration calorimetry and theoretical free energy perturbation methods.

Keywords ITC · Ligand · NMR · Protein · Solvation · Thermodynamics

1 Introduction

1.1 Biomolecular Interactions

Biomolecular interactions are of fundamental importance in biological processes. However, despite enormous advances in the speed and accuracy of the three-dimensional structure determination of proteins and their complexes, our ability to predict binding affinity from structure (i.e. whether and how strongly two molecules will interact) remains severely limited. One reason for this dilemma is that affinities are governed not only by energetic considerations arising from the precise spatial disposition of interacting groups (loosely, enthalpy), but also by the dynamics of these groups (loosely, entropy). Thus, in order to predict accurately the affinity of a protein for a given ligand, it is essential to have knowledge of both of these factors. While arguably it is possible to obtain a good estimate of the enthalpy of binding on a per-residue basis from a high-resolution crystal structure of the complex together with molecular mechanical energy calculations, it is not possible to obtain the entropy from this static model. Moreover, a simple static picture provides no information on the role of solvent reorganization to the binding affinity, the importance of which is still a subject of much debate. The following pages describe recent experimental and theoretical developments aimed at understanding the various individual contributions to the overall free energy of binding.

2 Thermodynamic Principles

2.1 Free Energy, Enthalpy and Entropy

Enthalpy and entropy are encompassed in the fundamental thermodynamic equation that describes the Gibbs free energy of binding, ΔG_b:

$$\Delta G_\mathrm{b} = \Delta H_\mathrm{b} - T\Delta S_\mathrm{b}\,, \tag{1}$$

where ΔH_b and ΔS_b represent the enthalpy and entropy of binding, respectively, and T is the absolute temperature. In any spontaneous process the free energy is minimized, hence a negative ΔG implies that a reaction or process will proceed in the direction as written. Thus, a ligand L will only associate with a protein P if ΔG_b for the following process is negative:

$$P + L \rightarrow PL\,. \tag{2}$$

Thermodynamic arguments enable us to determine the strength of binding via the affinity (association constant) K_a by use of a second fundamental relationship:

$$\Delta G_b^o = -RT \ln K_a . \quad (3)$$

The term on the left-hand-side of Eq. 3 is the *standard* Gibbs free energy of binding, which is the free energy of binding measured under certain defined standard conditions. This term is often confused with the free energy of binding. This confusion is not based upon semantics, since in general these two terms will differ numerically. This can be understood with reference to Eq. 4, which directly relates the free energy of binding and the standard free energy of binding:

$$\Delta G - \Delta G_b^o = RT \ln Q . \quad (4)$$

Here, Q is the reaction quotient, which is expressed in terms of the concentrations of products and reactants. Thus, for the process shown in Eq. 2:

$$Q = [PL] / [P] [L] . \quad (5)$$

At equilibrium, Q is equivalent to the equilibrium constant for the process, which in turn is equivalent to the affinity K_a, and since ΔG is zero at equilibrium, Eq. 3 follows directly from Eq. 4. Thus, while ΔG is zero at equilibrium, ΔG_b^o in general is not, except in the case that $K_a = 1$.

It is important to appreciate the molecular interpretation of Eqs. 3 and 4, which lies in the concept of the entropy of mixing. Consider a hypothetical case in which species A that exists in a specified standard state is converted to B, also in the standard state, such that there is no mixing of A and B. The change in free energy with respect to the mole fraction of B is described by the dotted line in the upper panel of Fig. 1. However, in all practical cases there will be a mixing of reactants and products, resulting in an increase of the entropy of the system. The Gibbs free energy of mixing is:

$$\Delta G = RT \left[x_a \ln x_a + x_b \ln x_b \right] , \quad (6)$$

where x_a and x_b are the mole fractions of A and B. This function is plotted in the lower panel of Fig. 1 for a process at 300 K. The free energy of the system will consist of one part derived from the free energy of the pure components multiplied by their mole fractions, and a second part derived from the free energy of mixing. Thus,

$$\Delta G = x_b \Delta G^o + RT \left[\left(1 - x_b\right) \ln \left(1 - x_b\right) + x_b \ln x_b \right] . \quad (7)$$

This function is represented by the solid line in the upper panel of Fig. 1.

At equilibrium, the free energy is zero, i.e. when the slope of the solid line in the upper panel of Fig. 1 is zero:

$$\mathrm{d}\Delta G / \mathrm{d}x_b = \Delta G^o + RT \left[-\ln \left(1 - x_b\right) + \ln x_b \right] = 0 , \quad (8)$$

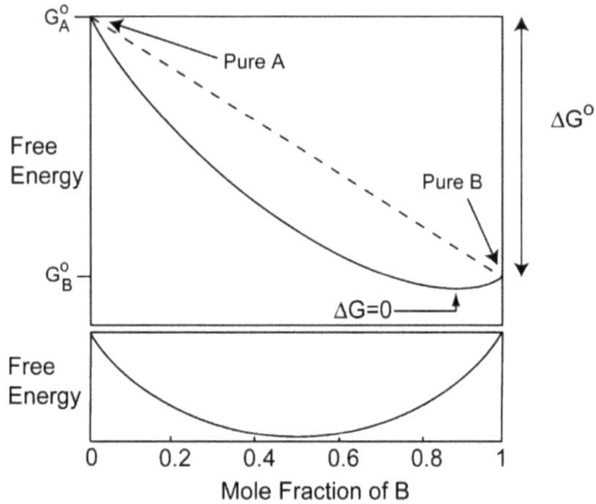

Fig. 1 Standard free energy changes in the conversion of species A to species B. In the hypothetical case that A is converted to B without mixing of the two components, the change in the standard free energy of binding will be linear with respect to the mole fraction of B, and is described by the *dashed line*. However, in any practical case there will be a mixing of A and B as A converts to B. A change in free energy is associated with this process resulting from the entropy of mixing (represented by the *solid line* in the *lower panel*), which is minimal at equal mole fractions of A and B, i.e. where the entropy of mixing is maximal. The observed free energy change is given by the *solid line* in the *upper panel* and equals the sum of the standard free energy change and the contribution from mixing. At equilibrium $\Delta G = 0$, i.e. where the slope of the free energy curve is at a minimum

which is equivalent to Eq. 3:

$$\Delta G^o = -RT \ln \left(x_b^{eq} / \left(1 - x_b^{eq}\right) \right) = -RT \ln K \,, \tag{9}$$

where x_b^{eq} is the mole fraction of B at equilibrium and K is the equilibrium constant for the process. Thus, the free energy change for a given process can be thought of as comprising the standard free energy change plus a term related to the entropy of mixing. The position of equilibrium for processes with a large negative standard free energy change will lie towards products, whereas processes with a large positive standard free energy will lie towards reactants.

2.2 The Hydrophobic Interaction and Heat Capacity

The hydrophobic interaction arises from the low solubility of nonpolar compounds in water. At physiological temperature the driving force derives from the unfavourable decrease in entropy of the hydrating waters. Water molecules are unable to hydrogen-bond to nonpolar solutes, resulting in a disruption of the favourable hydrogen-bonding network of bulk water. Those water molecules that are in contact with the solute compensate by bonding more strongly to their neighbours, resulting in an ordering of water molecules around the solute. This ordering has variously been described as "clathrate-type", "icelike", icebergs and flickering clusters [1–3]. The minimization of exposed surface area of solute results in a release into bulk solvent of a proportion of these ordered water molecules (Fig. 2). This is an entropically favourable process which therefore results in the burial of nonpolar surface area from water. A further consequence of this process is that the number of water–water contacts increases and more water–water hydrogen bonds can form. Thus, it might be anticipated that the hydrophobic interaction would be accompanied by a decrease in enthalpy. However, at physiological temperatures, the stronger bonding of water molecules in clathrate structures surrounding the nonpolar solute compensates for the smaller number of hydrogen bonds that can be formed, and the net enthalpy change is close to zero. Despite this interpretation, many "hydrophobic interactions" possess an enthalpy-driven thermodynamic signature, for subtle reasons that will be discussed in Sect. 3.2.1. Although we are only concerned below with thermodynamic processes occurring at physiological temperature, it is worth noting that at higher temperatures the hydrophobic effect becomes enthalpy driven, to the point where the entropy of nonpolar surface burial is approximately zero [4].

Large changes in heat capacity (C_p) are often taken as evidence for the existence of the hydrophobic interaction in biomolecular recognition. This interpretation derives from data for the transfer of hydrophobic solutes from nonaqueous to aqueous environments, which is generally accompanied by a positive ΔC_p. A simple molecular interpretation of this phenomenon is that the progressive "melting" of hydrogen-bonded water structure around

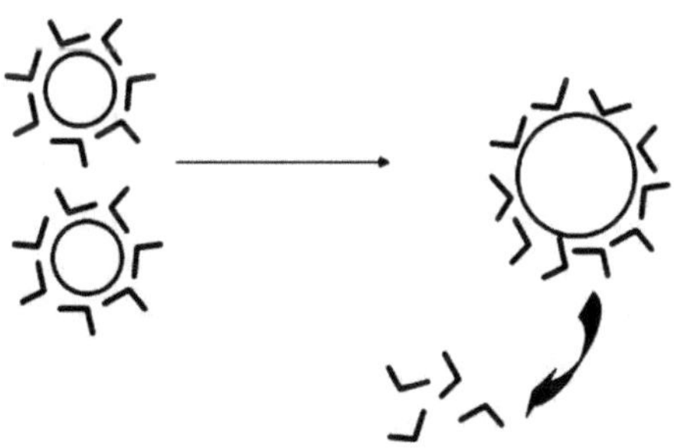

Fig. 2 "Clathrate" model of the hydrophobic effect. Water molecules that are in contact with the hydrophobic solute compensate by bonding more strongly to their neighbours, resulting in an ordering of water molecules around the solute. Minimization of exposed surface area of the solute is proposed to result in a release into bulk solvent of a proportion of these ordered water molecules. This is an entropically favourable process that therefore derives the association of hydrophobic species at physiological temperature

nonpolar solutes as the temperature is increased "absorbs" the increase in thermal energy, thus resulting in increased heat capacity. Arnett and co-workers were the first to demonstrate that the heat capacity increment for the dissolution of nonpolar solutes is a unique property of water, and is not observed in any other solvent [5]. In the case of protein-protein and ligand-protein interactions, the burial of hydrophobic surface area conversely gives rise to a significant negative ΔC_p, which is thus taken as a further thermodynamic signature of the hydrophobic effect.

It is important to note that the above view of the hydrophobic effect, while predominant, is not universally accepted. Indeed, despite much effort, there is little physical evidence for the existence of water clathrates surrounding nonpolar solutes. The "Small-Size Model" [6–9] has been proposed as an alternative to the "iceberg" model. This model is concerned with free energies of solvation (Sect. 3.2.2) rather than entropy and heat capacity. The high free energy cost of incorporation of a nonpolar solute into water is argued to arise from the absence of an appropriate cavity in water due to its small size. Large cavities are more likely in solvents comprised of large molecules, and since water molecules are amongst the smallest solvent molecules, the free energy cost of creating a cavity is greater in water than other solvents. In general the creation of a cavity by the coalescence of a number of smaller volumes throughout the solvent will lead to an entropic cost, together with an enthalpic cost of breaking solute–solute intermolecular interactions. In the "special" case of solvent water, the Small-Size Model suggests that the additional entropic cost of ordering waters in the solvation shell and the favourable enthalpic contribution due to stronger hydrogen bonding in this shell compensate almost perfectly (i.e. enthalpy-entropy compensation), and thus does not contribute to the free energy of cavity formation [7]. However, using the so-called "MB" model of water, Dill and co-workers [10, 11] showed that such compensation is limited to small nonpolar solutes, and suggested that "water's complexities appear to be important in most other circumstances".

More recent theoretical treatments further support the view that hydrophobicity manifests different characteristics depending on whether small molecular units or large clusters are involved, either alone or in combination [12]. In the small molecule case (such as methane), a cavity is created in the solvent that excludes water molecules from a spherical volume less than 0.5 nm in diameter. This volume is sufficiently small to permit hydrogen-bonding patterns to go around the solute, and thus the extent to which hydrogen bonds are broken at any given time is similar to that in pure water. However, in larger complexes, where the solute surface extends over areas larger than 1 nm^2, it is impossible for adjacent water molecules to maintain a complete hydrogen-bonding network. As a result, water tends to move away from the large solute and forms an interface that bears a similarity to that between a vapour and a liquid. This phenomenon provides a physical basis for the understanding of hydrophobic effects in that statistical thermodynamical calculations on the formation of small cavities in water accurately reproduce the entropy and heat capacity of solvation of, for example, methane. Moreover, the driving force for hydrophobic association is readily understood in terms of the dependence of hydrophobic solvation on solute size—when solute molecules cluster to form a hydrophobic complex, the overall solvation energy can be shown to change from a linear dependence on solvent volume to a linear dependence on solvated surface area. Thus, if the number of solute molecules in the cluster is sufficiently large, the volume to surface area ratio will result in a solvation free energy that is lower than the solvation free energy of the individual components. This results in a favourable driving force for association. A compelling advantage of this model is that it can be observed in simulations, unlike the "clathrate" model.

3 Dynamics and Thermodynamics of Biomolecular Associations

3.1 Overview of the Binding Process

Before embarking on a detailed analysis of the factors responsible for biomolecular interactions, it is worthwhile initially to examine the qualitative aspects of the process. The interaction between a protein and a ligand is often described schematically by Eq. 2. However, this is a gross over-simplification for two reasons.

First, Eq. 2 does not describe all of the partners in the interaction. Since we are concerned with biomolecular interactions, all such processes take place within an aqueous *milieu*. The individual species will thus interact with solvent water in some manner before the association. For example, as discussed in Sect. 2.2, hydrophobic ligands will be surrounded by icelike cages of water molecules or water-vapour interfaces (provided they are of sufficient size), and binding pockets within proteins will also typically contain solvent water molecules that may be ordered to some degree. When the two species associate, their interaction with solvent will certainly be entirely different. For example, some or all of the solvent water molecules within the binding pocket are likely to be expelled as the ligand binds. Moreover, the expelled solvent molecules will make new interactions with bulk solvent. Since the free energy of binding is defined as the difference between the free energy of the system (solutes plus solvent) in the complexed state versus the uncomplexed state, it follows that solvent reorganization can have a dramatic influence on binding.

Second, Eq. 2 assumes that the structures of the interacting species before and after association will be equivalent. However, it is very likely that structural changes will occur following complexation, and hence formally we are dealing with different species before and after complexation, in which case Eq. 2 should more formally be written [13]:

$$P + L \rightarrow P'L' . \tag{10}$$

It is important to be aware that the contribution of solvation or structural changes to binding affinity can be dramatic. A simple calculation using Eq. 3 shows that the difference between micromolar and nonamolar affinity results from a change in free energy at 300 K of $\sim$ 17 kJ/mol, which is on the order of the strength of a hydrogen bond. Thus, the loss of an ordered water molecule from a protein binding pocket or the reorientation of an amino-acid residue side-chain may be sufficient to alter binding affinity by orders of magnitude. It follows that a quantitative description of the thermodynamics of ligand-protein association requires full account to be taken of all interacting species at every stage of the process. A convenient formalism involves the use of Born–Haber cycles.

3.1.1
Born–Haber Cycles

The representation of ligand-protein association in the form of a Born–Haber cycle offers a rigorous conceptual framework which includes all the interacting species. A typical cycle is shown in Fig. 3. The "intrinsic" standard free energy of binding between protein P and ligand L is represented by ΔG_i^{o}, whereas the standard free energy of binding that would typically be determined experimentally is represented by $\Delta G_{\mathrm{obs}}^{\mathrm{o}}$. In addition, two further processes can be defined which are represented by $\Delta G_{\mathrm{su}}^{\mathrm{o}}$ and $\Delta G_{\mathrm{sb}}^{\mathrm{o}}$. These are the standard free energies of solvation of the uncomplexed species and the complex, respectively. Since free energy is a state function, it is independent of the path taken from one state of the system to another, and we can therefore write:

$$\Delta G_i^{\mathrm{o}} + \Delta G_{\mathrm{sb}}^{\mathrm{o}} = \Delta G_{\mathrm{su}}^{\mathrm{o}} + \Delta G_{\mathrm{obs}}^{\mathrm{o}} , \tag{11}$$

which can be rearranged:

$$\Delta G_{\mathrm{obs}}^{\mathrm{o}} = \Delta G_i^{\mathrm{o}} + \left\{\Delta G_{\mathrm{sb}}^{\mathrm{o}} - \Delta G_{\mathrm{su}}^{\mathrm{o}}\right\} . \tag{12}$$

Equation 12 shows how the observed standard free energy of binding can be decomposed into the intrinsic contribution plus a solvation term shown in curly braces. Equivalent expressions can be written for the standard enthalpy and standard entropy of binding since these parameters are also state functions.

The advent of sensitive isothermal titration calorimetry [14] has enabled the accurate determination of $\Delta G_{\mathrm{obs}}^{\mathrm{o}}$, $\Delta H_{\mathrm{obs}}^{\mathrm{o}}$ and $\Delta S_{\mathrm{obs}}^{\mathrm{o}}$ for a wide variety of biomolecular complexes in aqueous solution. However, this technique measures the global thermodynamics of binding including solvation effects, as

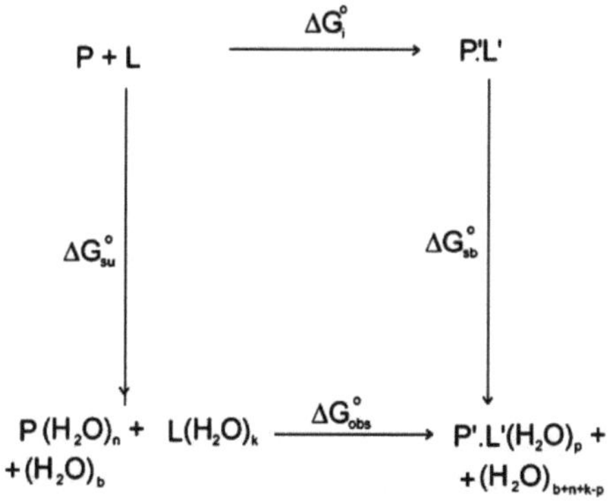

Fig. 3 A typical Born–Haber cycle relating the "intrinsic" (solute-solute) standard free energy of binding $\Delta G_{\mathrm{i}}^{\mathrm{o}}$ with the observed standard free energy of binding $\Delta G_{\mathrm{obs}}^{\mathrm{o}}$ for a protein *P* interacting with a ligand *L*. The vertical processes represent the solvation free energies of the uncomplexed species $\left(\Delta G_{\mathrm{su}}^{\mathrm{o}}\right)$ and of the complex $\left(\Delta G_{\mathrm{sb}}^{\mathrm{o}}\right)$

defined in Eq. 12. In many cases it is therefore practically impossible to delineate the factors responsible for the association process. The key to a complete quantitative understanding of the thermodynamics of ligand-protein association requires a deconvolution of the overall thermodynamics of binding into the terms on the right-hand side of Eq. 12. For this purpose it is convenient to consider the enthalpic and entropic contributions separately, as detailed below.

3.2 Enthalpic Contributions to Binding

3.2.1 Intrinsic Contributions to Binding Enthalpy

Intrinsic contributions to binding enthalpy result from differences in non-bonded interactions within each species prior to complexation versus those present in the complex. (Strictly, we should also include covalent interactions, since complexes exist involving the formation of solute–solute covalent bonds, but we will not consider such systems here). Thus, the intrinsic standard enthalpy of binding ΔH_i^{o} can be thought of as the total change in internal energy of the interacting species in vacuo.

If we first consider the uncomplexed ligand, this will exist in one or more conformations whose populations are governed by the Boltzmann distribution at a given temperature. The internal energies of these conformers will depend upon a complex interplay between dispersive interactions, Coulombic interactions and hydrogen bonding. On binding to the protein receptor, typically only one of these conformations will be selected, or indeed none if there is conformational strain on binding. In general, the result will be an unfavourable contribution to binding enthalpy. The size of this unfavourable contribution will obviously depend upon the nature of the ligand. As examples, methyl-methyl gauche interactions represent approximately 1 kJ/mol of unfavourable enthalpy, whereas the barrier to rotation about the carbon-carbon bond of ethane is approximately 12 kJ/mol. Thus, even small deviations from minimum energy conformations can result in significant unfavourable contributions to the binding free energy.

Similar considerations exist for the protein receptor. Penel and Doig have quantified strain energies for amino-acid residues in α-helices in the context of protein folding [15]. The mean change in rotameric strain energy (i.e. the energy resulting from a side-chain that is not in its lowest-energy rotamer) was 1.75 kJ/mol, whereas the mean dihedral strain energy (i.e. the energy resulting from a shift of a dihedral angle from the most stable conformation of a rotamer) was reported as 2.67 kJ/mol. It is anticipated that unfavourable enthalpic contributions of this magnitude will exist in the context of ligand-protein interactions also, and this contribution may be very significant due to the many additional degrees of freedom within a typical protein binding site.

Paradoxically, evidence exists that the many degrees of freedom and resulting interactions within the protein structure can in principle result in improved ligand binding as a result of "tightening" of the protein structure on association. Williams and co-workers have obtained experimental support for this hypothesis in, for example, the binding of biotin to the streptavidin tetramer [16]. The binding of these species is about 1000-fold stronger than anticipated on the basis of the sum of the individual interactions. Williams and co-workers used mass spectrometry to measure the extent of hydrogen/deuterium exchange of the amide protons in streptavidin in the absence and presence of biotin. Twenty-two amide protons per streptavidin monomer were found to be protected upon binding of biotin. The latter reduced the solvent accessibility in much of the structure, indicating a global tightening of the structure rather than a localized effect at the ligand-protein interface. This phenomenon is further emphasized in the thermal stability of streptavidin evaluated using differential scanning calorimetry [17]—biotin binding increases the thermal denaturation of streptavidin from $T_m = 75\,^\circ C$ to $112\,^\circ C$, indicating that the complex is more stable than the uncomplexed protein. These data support positively cooperative binding whereby the ligand reduces the dynamic behaviour of the receptor. Williams has proposed [16] that this does not necessarily involve the formation of new noncovalent interactions within the receptor, rather tightening of existing interactions. Major conformational changes need not thus be invoked. The generality of this phenomenon remains to be determined, but examples certainly exist where this phenomenon is not the driving force for association (see below).

In addition to the modification of nonbonded interactions that pre-exist in the protein and ligand prior to association, nonbonded interactions form at the solute–solute interface. Naively, it might be thought that each of these "new" interactions will contribute favourably to the enthalpy of binding. For example, ab initio quantum chemical calculations suggest stabilization of the core of the small protein rubredoxin resulting from dispersion interactions by approximately – 200 kJ/mol [18]. Correspondingly, large intrinsic stabilization enthalpies are expected in ligand-protein complexes. However, it must be remembered that prior to the association each binding partner will form dispersive interactions with solvent water. Indeed, it is commonplace to assume that the change in dispersion energy on association of two shape complementary molecules in solution is negligible, since new solute–solute dispersion interactions following association have "exchanged" for solute–solvent dispersion interactions that exist prior to the association (Fig. 4). However, recent evidence suggests that this assumption may not generally be true for some ligand-protein associations. Very recently, Barratt et al. examined the binding thermodynamics of the mouse major urinary protein (MUP), a promiscuous binder of small hydrophobic ligands (see Sect. 4). Despite the fact that the binding site is extremely hydrophobic, the association with a variety of different hydrophobic ligands is invariably strongly enthalpy driven. This is counter-intuitive, based upon the expected thermodynamic signature of the hydrophobic effect [19]. However, detailed scrutiny of the binding pocket of this protein by site-directed mutagenesis, X-ray crystallography and molecular dynamics simulations, revealed that the binding pocket is poorly solvated in the absence of ligand. Thus, in this particular system the gain in solute–solute dispersion energy will not be compensated by interactions between binding-site residues and solvent prior to the association. A significant fraction of the favourable intrinsic solute–solute enthalpy is thus expected to appear in the free energy of binding, which thus accounts for an enthalpy-driven thermodynamic signature. This conclusion is indeed supported by molecular mechanical energy calculations, and is reminiscent of the interaction of small organic molecules in nonaqueous solvents where the geometry of the binding pocket prevents solvation [20, 21].

It must be emphasized that these observations are not necessarily at variance with current models of the hydrophobic effect. Rather, they may offer an explanation for the fact that many "hydrophobic associations" in solution do not possess the anticipated entropy driven thermodynamic signature—many,

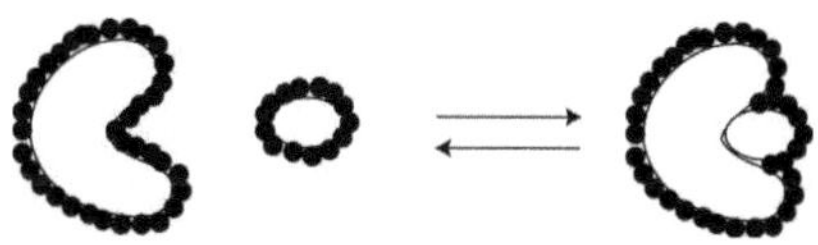

Fig. 4 Prior to association, both ligand and protein partake in dispersive interactions with solvent molecules. Following association, it is generally assumed that these solute–solute dispersive interactions exchange for solute–solute dispersive interaction, with a negligible net contribution to the binding enthalpy

but by no means all. For example, the binding of a series of hydrophobically modified benzamidinium chloride inhibitors to trypsin is strongly entropy driven at a number of temperatures [22]. Notably, the trypsin binding site is a cleft that is presumably heavily solvated. These data are thus consistent with a model whereby the degree of expression of the intrinsic solute–solute enthalpic contribution to the free energy of binding is dependent on the solvation of the protein binding pocket.

In addition to dispersive interactions, it might be anticipated that the formation of hydrogen bonds at the solute–solute interface will result in a favourable contribution to the enthalpy of binding. However, again it must be remembered that the relevant hydrogen bond donors and acceptors will interact with solvent water molecules prior to the association. Nonetheless, evidence exists that solute–solute hydrogen bonds are stronger than solute–solvent hydrogen bonds, giving rise to a nett favourable enthalpy of binding. Daranas et al. [23] determined the global thermodynamics of binding of galactose and various deoxy derivatives to the arabinose binding protein (Table 1). Binding of all ligands was found to be enthalpy driven, and with the exception of charged ligands such as heparin and heparin sulphate, both ΔH^{o} and particularly $T\Delta S^{o}$ are significantly larger than typical values reported for the vast majority of carbohydrate-protein interactions [24], including oligosaccharides. The reason for these anomalously large values could not be delineated with certainty from global thermodynamics measurements. However, the enthalpy of binding of galactose compared with 1-deoxy, 2-deoxy or 3-deoxy galactose was found to be favourable by $\sim$ 30 kJ/mol. As mentioned above, it might be considered unlikely that the more favourable enthalpy of binding of galactose compared with deoxy-derivatives arises from the additional hydrogen bond(s) that form due to the additional hydroxyl group in the complex, since prior to binding, the ligand is hydrogen bonded to solvent water. The enthalpic component of such solute–solvent hydrogen bonds is contained within the free energy of solvation of these ligands, which is not known with any degree of accuracy. However, intuitively, the solvation enthalpy of deoxy-analogues of galactose must be less favourable than those of

Table 1 Thermodynamics of binding of galactose and derivatives to the arabinose-binding protein derived from ITC measurements at 308 K

Ligand	K_d (μM)	Error μM	ΔG^{o} (kJ/mol)	Error	ΔH^{o} (kJ/mol)	Error	$T\Delta S^{o}$ (kJ/mol)	Error
galactose	2.2	0.02	− 33.36	0.3	− 95	0.6	− 61	0.6
1-deoxy	14600	730	− 10.8	0.5	− 63	3	− 52	3
2-deoxy	780	60	− 18.3	1.4	− 61	4.8	− 43	4.8
3-deoxy	29620	2620	− 9.0	0.8	− 57	5	− 48	5

galactose. If one assumes momentarily that the enthalpic contribution from ligand-protein hydrogen bonds is effectively zero, since hydrogen bonds to solvent exist prior to the association, on the basis of the above data one must conclude that the enthalpies of solvation of the 1-deoxy, 2-deoxy and 3-deoxy analogues of galactose are each more favourable than galactose by $\sim -30\,\mathrm{kJ/mol}$, which is counter-intuitive. Therefore, the conclusion that interactions between the various hydroxyl groups of galactose and the protein are enthalpically significantly more favourable than those with solvent would appear to be inescapable. Whether this favourable enthalpy can be attributed in the main to hydrogen bonding is difficult to ascertain, since the loss of an hydroxyl group removes van der Waals' interactions with that group in addition to the hydrogen bonding contribution.

3.2.2 Solvation Contribution to Binding Enthalpy

The solvation contribution to binding enthalpy, $\{\Delta H^{\mathrm{o}}_{\mathrm{sb}} - \Delta H^{\mathrm{o}}_{\mathrm{su}}\}$, is essentially comprised of the enthalpy of solvation of the complex minus the enthalpy of solvation of species prior to the association. The enthalpy of solvation of any species is defined as the heat gained or lost when that species is transferred from the gas phase into solution. When the solvent is water this parameter is sometimes called the enthalpy of hydration. This definition may appear to be somewhat abstract or esoteric in the context of ligand-protein association phenomena, but is nonetheless an integral part of the rigorous description of binding thermodynamics as is apparent from Fig. 3. In many respects, the binding process can be thought of as desolvation. Qualitatively, we can think of this as the stripping of some or all solvent water molecules from the surface of the ligand and from the protein binding pocket as the partners associate. Quantitatively, the desolvation process is represented by the negative of the solvation enthalpy term in the above expression.

The solvation enthalpies of a number of small organic molecules have been measured, and have been catalogued by Cabani et al. [25] and more recently by Plyasunov and Shock [26]. Solvation enthalpies of such molecules are invariably negative, i.e. desolvation upon association contributes unfavourably to the enthalpy of binding. That solvation is an exothermic process was originally regarded as counter-intuitive prior to current models of the hydrophobic effect—the large heat of vaporization and surface tension of water implied that the creation of a cavity to accommodate the solute would require a significant energy input. However, increased or strengthened hydrogen bond formation in the solvent surrounding nonpolar solutes would be anticipated in the models of the hydrophobic effect described in Sect. 2.2. The overall enthalpy of solvation will thus comprise a positive term associated with the formation of a cavity, which is more than compensated by a negative term associated with solvent ordering [27].

In theory, the standard free energy of solvation of any ligand is straightforward to measure, since it is necessary only to determine the concentrations of the ligand in aqueous solution and in the vapour phase in a closed system at equilibrium. The solvation free energies of volatile nonpolar compounds have been determined in this manner [28]. The standard enthalpy (and thus entropy) of solvation can then be estimated from the temperature dependence of the free energy. However, such measurements are only possible if the volatility is sufficient to offer a measurable concentration in the vapour phase. This excludes a number of biologically important ligands such as polar species or carbohydrates, for example. For these reasons there has recently been a surge of interest in the computation of solvation free energies of such molecules (reviewed by Orozco and Luque [29]).

The solvation process can conveniently be decomposed into three steps: (i) creation of a cavity in the solvent; (ii) van der Waals interactions and (iii) electrostatic contributions. The first step is clearly the creation of a cavity in the solvent that is large enough to accommodate the solute. Since this will involve breakage of the forces maintaining cohesion within the solvent, the free energy contribution to cavitation (ΔG_c) will be unfavourable. In contrast, the van der Waals contribution (ΔG_{vdW}) is favourable, since the solute cavity is created in regions of the solvent where the dispersion term is larger than the repulsion term. The third step (ΔG_{ele}) involves two components, namely the work necessary to create the gas-phase charge distribution of the solute in solution, and the work required to polarize this charge distribution by the solvent. Thus, the overall solvation free energy can be described by:

$$\Delta G^{o}_{solv} = \Delta G^{o}_{c} + \Delta G^{o}_{vdW} + \Delta G^{o}_{ele} \, . \tag{13}$$

The breakdown of the solvation process in this manner facilitates theoretical approaches to the computation of solvation free energies. Explicit solvent models provide the most complete description of solvation, but they are however extremely computationally demanding in view of the large number of atoms involved and the requirement to average over many solvent configurations. A particularly useful approach involves free energy perturbation (FEP) techniques [30], which have been shown to reproduce experimental solvation free energies of small organic molecules with impressive accuracy [31–33]. The conceptual basis for FEP calculations lies in the now familiar Born–Haber cycle (Sect. 3.1.1) for the conversion of a given ligand molecule A into a related molecule B (Fig. 5). The free energy of solvation can be defined as the difference between the free energies associated with the annihilation of a molecule in the gas phase and solution (Fig. 5a). Alternatively, the method can be used to determine the difference in solvation free energy between two related ligands (Fig. 5b). While it is difficult to compute directly the free energy difference between either ligand in the gas phase versus the aqueous phase (vertical processes in Fig. 5b), it is relatively straightforward to compute the free energy difference between molecules A and B in the gas

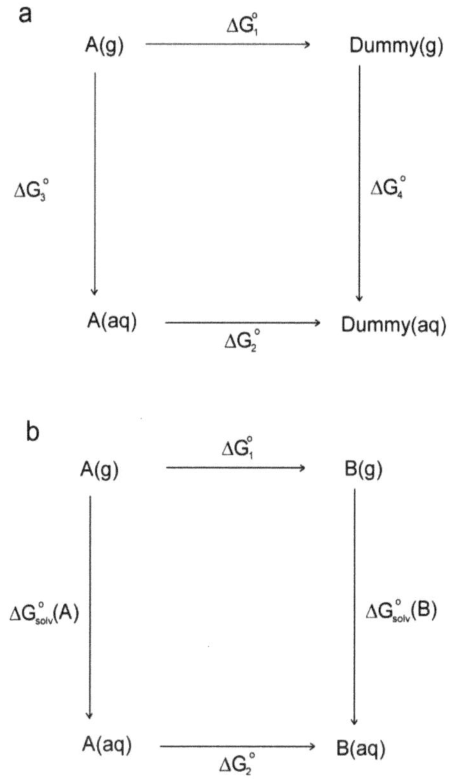

Fig. 5 The conceptual basis for FEP calculations can be described by Born–Haber cycles (Sect. 3.1.1) for the conversion of a given ligand molecule A into a related molecule B. The free energy of solvation can be defined as the difference between the free energies associated with the annihilation of a molecule in the gas phase and solution (**a**) or, alternatively, the method can be used to determine the difference in solvation free energy between two related ligands (**b**)

phase and in the solution phase (horizontal processes in Fig. 5b). Essentially, molecule A is either slowly annihilated (Fig. 5a) or "mutated" to molecule B in the gas and solution phases. In the latter case, since G is a state function, we can compute the difference in solvation free energy between two related ligands according to:

$$\Delta G^{\mathrm{o}}_{\mathrm{solv}}(A) - \Delta G^{\mathrm{o}}_{\mathrm{solv}}(B) = \Delta G^{\mathrm{o}}_1 - \Delta G^{\mathrm{o}}_2 \,. \tag{14}$$

The free energy differences ΔG^{o}_1 and ΔG^{o}_2 between related systems A and B represented by Hamiltonian H_{A} and H_{B} can be computed in a variety of ways. This free energy difference can be represented as:

$$\Delta G = -RT \ln \left\langle e^{-\Delta H/RT} \right\rangle_{\mathrm{A}} \,, \tag{15}$$

where $\Delta H = H_B - H_A$ and $\langle \rangle_A$ refers to an ensemble average over a system represented by Hamiltonian H_A. If the difference between systems A and B is other than trivial, then the resulting free energy difference in Eq. 15 will not be realistic. In the FEP approach, the calculation is therefore decomposed into a number of discrete windows, each involving a very small perturbation that allows the free energy to be determined accurately. However, this is only effective if molecules A and B are structurally similar, otherwise statistical noise becomes a major problem. Provided this limitation is met, it is possible to obtain reasonably accurate values for not only the solvation free energy, but also solvation enthalpies and entropies from the temperature dependence of the free energy using, for example, finite-difference methods [34, 35]. The full details of free energy perturbation methods are outside the scope of this work, and the reader is referred to several excellent reviews on the topic [29, 30].

In contrast to small organic molecules, quantitative data for the solvation free energies of proteins and protein–ligand complexes are in general lacking. Nonetheless, it is clear from the Born–Haber cycle of Fig. 3 that this information is required for a complete understanding of the binding process. Because of the huge computational cost, it is currently not realistic to compute solvation free energies for proteins and protein–ligand complexes using explicit solvation models such as those used in FEP calculations. Implicit solvent models offer an attractive alternative. In principle, to a first approximation the free energy of solvation can be derived from the intrinsic solvation properties of the constituent groups in the molecule. This "fractional approach" has however seldom been used for the practical computation of solvation free energies. Instead, methods based on a solvent-accessible surface have commonly been employed. In the early work of Chothia [36], a general empirical solvation parameter, σ, equal to 24 cal mol^{-1} A^{-2} was derived to compute protein hydrophobicity from the solvent accessible surface area of exposed residues. The solvation free energy of the protein is thus given by:

$$\Delta G_{\text{sol}} = \sum_{k=1}^{N} \sigma_k A_k , \tag{16}$$

where σ_k and A_k respectively represent the solvation parameter and solvent accessible surface area of residue k.

While useful, calculations based on solvent accessible surface are subject to a number of limitations. First, the suitability of data for small molecules extrapolated to proteins is questionable. Second, screening of intrasolute interactions by the solvent is ignored. Third, recent work suggests that a particular parametrization of the surface-area model is only applicable to a subset of the conformations of the molecule included in the parametrization [37]. Finally, and perhaps most importantly, electrostatic contributions to solvation are ignored. The intrinsic solvation properties of a given atom depend upon neighbours, whose effects can be included explicitly by considering their electrostatic contribution to solvation. Many empirical methods account for this contribution implicitly by defining different parameters according to the nature of neighboring groups. A promising, more complex strategy has been developed by Hawkins et al. [38], where atomic solvation parameters for each atom in a molecule are parameterized depending on their environment. The method reproduces the solvation free energies of small organic molecules to impressive accuracy. However, it remains to be seen whether the solvation free energies of macromolecules can be determined with similar accuracy.

Daranas et al. showed that the absence of reliable solvation thermodynamic data on proteins and protein–ligand complexes can be overcome in large part by considering the *relative* thermodynamics of association between closely related ligands binding to the same protein [23]. In this method knowledge of the solvation contribution from the free protein and from the protein–ligand complex is not required. The basis of the method involves a "three-dimensional" Born–Haber cycle as illustrated in Fig. 6, where P represents a given protein and $L1$ and $L2$ represent two closely related ligands. Since H is a state function, we can write:

$$\Delta H^{\mathrm{o}}_{\mathrm{obs2}} - \Delta H^{\mathrm{o}}_{\mathrm{obs1}} = \left[\Delta H^{\mathrm{o}}_{i2} - \Delta H^{\mathrm{o}}_{i1}\right] + \left\{\left[\Delta H^{\mathrm{o}}_{\mathrm{sb2}} - \Delta H^{\mathrm{o}}_{\mathrm{sb1}}\right] - \left[\Delta H^{\mathrm{o}}_{\mathrm{su2}} - \Delta H^{\mathrm{o}}_{\mathrm{su1}}\right]\right\}, \tag{17}$$

where the various terms are defined by analogy with Eqs. 11 and 12. The second of the solvation terms in curly braces represents the difference between the standard solvation enthalpies of the species prior to association, and since each ligand binds to the same protein, the solvation term for the free protein cancels. The first term in curly braces represents the difference between the solvation enthalpies of the complexes. If the same number and location of water molecules exists in the two complexes, then this term also cancels to first order. Thus, Eq. 17 simplifies to:

$$\Delta H^{\mathrm{o}}_{\mathrm{obs2}} - \Delta H^{\mathrm{o}}_{\mathrm{obs1}} = \left[\Delta H^{\mathrm{o}}_{i2} - \Delta H^{\mathrm{o}}_{i1}\right] - \left\{\left[\Delta H^{\mathrm{o}}_{\mathrm{sl2}} - \Delta H^{\mathrm{o}}_{\mathrm{sl1}}\right]\right\}, \tag{18}$$

where $\Delta H^{\mathrm{o}}_{\mathrm{sl1}}$ and $\Delta H^{\mathrm{o}}_{\mathrm{sl2}}$ are the standard solvation enthalpies of the ligands prior to association. Thus, provided that there are no structural changes in the protein in binding either ligand (which is most likely to be satisfied if the ligands are very similar), it is possible to obtain a value for difference between the intrinsic standard enthalpies of binding of each ligand from the observed standard enthalpies of binding and the standard solvation enthalpies of the relevant ligand. Clearly, intrinsic free energies and enthalpies of binding can also be obtained if the equivalent parameters are available.

An alternative approach for the experimental determination of the solvent contribution to the enthalpy of binding that does not require explicit knowledge of solvation terms, involves solvent isotope substitution methods introduced by Chervenak and Toone [39]. This method can be understood

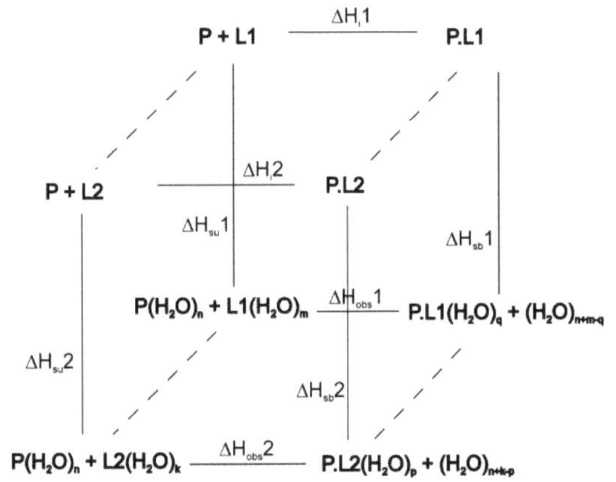

Fig. 6 "Three-dimensional" Born–Haber cycle for two ligands $L1$ and $L2$ binding to a given protein P, showing the relationship between the observed enthalpy of binding ΔH_{obs}, the "intrinsic" (solute–solute) enthalpy of binding ΔH_{i} and the solvation enthalpies of the unbound (ΔH_{su}) and bound (ΔH_{sb}) species

in terms of a formalism devised by Muller to account for the solvation properties of a solute in a bath of N solvent water molecules. In this formalism, the enthalpy of solvation of the solute under these circumstances is described by:

$$\Delta H_{\text{s}} = n^{\text{h}} \left[\left(1 - f_{\text{b}}\right) \Delta H_{\text{b}}^{\text{o}} - \left(1 - f_{\text{hs}}\right) \Delta H_{\text{hs}}^{\text{o}} \right] . \tag{19}$$

This equation essentially describes the solvation enthalpy in terms of an enthalpic contribution from both bulk water ($\Delta H_{\text{b}}^{\text{o}}$) and from water in a "hydration shell" around the solute ($\Delta H_{\text{hs}}^{\text{o}}$). The fraction of broken hydrogen bonds in bulk solution is denoted f_{b}, with f_{hs} representing the analogous quantity for the hydration shell. The number of hydrogen bonds in the hydration shell is denoted n^{k}, and Muller suggested that a plausible value for this parameter is $3N/2$, since for a given N this is the largest geometrically allowed number of bonds between neighbouring molecules within the same shell. Isotopic substitution of D_2O for H_2O affects f_{b} differently from f_{hs} and $\Delta H_{\text{b}}^{\text{o}}$ from $\Delta H_{\text{hs}}^{\text{o}}$, due to the lower zero-point energy of deuterium with respect to protium. In contrast, the enthalpy of binding derived solely from solute–solute interactions (ΔH_i^{o}) will be unaffected by isotopic substitution providing the same number of hydrogen bonds exist in both solvents, i.e. the structures of the individual species prior to the association and the complex are identical in H_2O and D_2O. Thus, the determination of thermodynamic binding parameters using, for example, isothermal titration calorimetry (ITC) enables the crucial separation of intrinsic versus solvation terms. In the original application of this work [39], the authors concluded "that 25–100% of the net measured enthalpy of binding is accounted for by solvent reorganization".

An interesting application of the solvent isotopic substitution methods has been described by Connelly et al. [40] on the binding of the macrocycles FK506 and rapamycin to the FK506 binding protein. A feature of this protein–ligand complex is a hydrogen bond between a binding-site tyrosine in FK506 (Tyr-82 Hζ) and the ligand. In a Y82F mutant of FK506, Connelly et al. recorded a significantly more favourable binding enthalpy to FK506 and rapamycin compared with the wild-type protein ($\Delta\Delta H^{\text{o}} = -17.6$ and -12.7 kJ/mol respectively). Moreover, significantly less favourable binding enthalpies were recorded on substitution of H_2O with D_2O for the Y82F mutant ($\Delta\Delta H^{\text{o}} = 18.0$ and 12.1 kJ/mol for FK506 and rapamycin, respectively). These observations were rationalized by noting that the crystal structure of the unliganded protein shows two solvent water molecules ordered around the Tyr-82 hydroxyl group. The more favourable binding enthalpy in the Y82F mutant was suggested to arise from the desolvation of the latter group which was considered to be a highly unfavourable enthalpic process.

3.3
Entropic Contributions to Binding

3.3.1
Intrinsic Contribution

Intrinsic contributions to the entropy of binding arise from differences in dynamics between the free and bound states of both binding partners.

Considering first the ligand, the translational and rotational degrees of freedom that exist prior to the association will be lost on binding, giving rise to an unfavourable entropy term. Of course, the protein also possesses translational and rotational entropy, but since the magnitude of this entropy varies with the logarithm of particle mass, the loss in entropy on binding is approximately equivalent to the translational and rotational entropy of the smaller particle. If all of the translational and rotational motion is removed on binding, then the entropic cost of association is approximately + 57 kJ/mol for a small ligand [41]. However, in typical ligand-protein complexes, the noncovalent interaction energies are comparable to the thermal energies at physiological temperature and hence a degree of motion of the ligand exists in the binding site. The degree of such motion is extremely difficult to determine experimentally, and accordingly estimates of the translational and rotational contribution to the overall entropy of binding varies enormously (see e.g. Table 6 in Burkhalter et al. [24]) and has been estimated to be as low as + 5.4 kJ/mol [42].

In addition to the loss of translational and rotational motion, internal degrees of freedom of the ligand will typically be restricted on binding. In particular, torsional degrees of freedom are typically substantially attenuated on binding. The entropy corresponding to internal rotation of a symmetric, free rotor (i.e. where any barrier to rotation is much less than kT) such as a methyl group can be calculated from the respective partition function which is given by:

$$Q_{\text{free}} = \left(\sigma_{\text{int}} h\right)^{-1} \left(8\pi^3 I_{\text{int}} kT\right)^{1/2} , \tag{20}$$

where σ_{int}, the internal symmetry number, is equal to the number of minima or maxima in the torsional potential, and I_{int} is the reduced moment of inertia for the internal rotation. The torsional entropy is given by:

$$S_{\text{free}} = R\left(0.5 + \ln Q\right) . \tag{21}$$

If the torsional barrier is comparable to kT (hindered rotor), then the torsional entropy is reduced by a factor that can be determined from tables compiled by Pitzer and Gwinn [43]. Typical values for the torsional entropy of hindered rotors such as a hydroxyl group or a hydroxymethyl group are 4.5 kJ/mol and 7.3 kJ/mol respectively. Once again, it is very difficult to determine experimentally the residual torsional motion that exists within a given ligand upon association, and consequently estimates of the contribution to the entropy of binding resulting from "freezing" of torsional degrees of freedom vary considerably. The entropy of fusion within homologous series of hydrocarbons provides an estimate of this entropic cost as – 1.6 to – 3.6 kJ/mol at 300 K within a hydrocarbon chain [44], suggesting that substantial torsional freedom is retained in the bound-state.

Turning now to the protein, restriction of degrees of freedom is also anticipated to occur upon ligand binding. A number of investigations have been reported within the last decade whereby protein dynamics has been correlated with binding thermodynamics using NMR relaxation techniques [45–54]. Since each resonance in an NMR spectrum corresponds to an individual nucleus or group of equivalent nuclei, relaxation measurements offer the potential to obtain thermodynamic parameters at discrete sites within a macromolecule. The details of such measurements are outside the scope of this work, and the reader is referred to several excellent reviews [55–64]. Essentially, the time decay of nuclear magnetization is determined as a function of time, from which characteristic relaxation rates can be determined. These can in turn be interpreted in terms of a formalism for the dynamic motions to which these relaxation rates are sensitive [65], giving rise to a generalized order parameter S that defines the extent of internal motions on the ps–ns timescale.

Early work by Akke et al. [45] described the derivation of free energies of binding from differences in the square of the NMR-derived generalized order parameter S^2 [65] determined from backbone ^{15}N relaxation data for calbindin in the "apo" and ligand (Ca^{2+}) bound states. In an important further advance, Li et al. [66] used a simple one-dimensional vibrator as a model for dynamic motion to illustrate the relationship between dynamics measured by NMR relaxation methods and the local residual entropy of proteins. They concluded that dynamics of methyl containing side-chains in proteins corresponds to a substantial entropic contribution to the free energy of ubiquitin of approximately 40 kcal/mol at 300 K. Subsequently, Yang and Kay [46, 67] examined the relation between the order parameter and conformational entropy from ns–ps bond vector dynamics considering a number of simple models describing bond vector motion. Although it was not possible to derive equations relating the order parameter to conformational entropy for the majority of models considered, an approximate relation was found to describe order parameters vs. entropy profiles extremely well:

$$S_p/k = A + \ln \pi \left[3 - (1 + 8S)^{1/2}\right] , \tag{22}$$

where A is a model-dependent constant.

The studies above suggest that the measurement of both the free energy and entropy of binding for a biomolecular association is possible on a per-residue basis. Moreover, the enthalpy of binding could thus be determined from Eq. 1. Unfortunately, however, as discussed by Yang and Kay [46], the free energy change between states derived from this approach, unlike the entropy change, is dependent upon differences in ground state energies. Since the latter are in general unavailable, NMR relaxation measurements are only able to offer reliable insight into the entropy of binding. There is a number of assumptions in the derivation of Eq. 22 [46, 56]. First, this equation contains the model dependent constant A, and in general the nature of the motional model is unknown. In the case of the entropy of binding this is not a severe limitation if the assumption is made that the motional models before and after association are similar, in which case the constant A cancels. Second, the order parameter measured from conventional heteronuclear relaxation measurements is sensitive only to motions on a time scale shorter than overall rotational diffusion (picoseconds to nanoseconds), and is sensitive only to reorientational motions of the relevant bond vector. Third, no account is taken of correlated motions between different bond vectors. However, despite these limitations, NMR relaxation measurements can provide reasonably accurate per-residue entropies for a variety of biomolecular associations (vide infra).

Backbone dynamics of proteins are typically probed by detecting the reorientation of the amide bond vector in ^{15}N-enriched proteins. Conformational entropies of NH groups for each amino-acid residue can be measured from ^{15}N relaxation data assuming a diffusion-in-a-cone model for NH vector motions [46]. In the case of side-chain dynamics measurements, the nucleus of choice is deuterium. The reasons for this choice are discussed at length elsewhere [68–70]. More recently, Millet et al. [71] have described an approach whereby five relaxation rates per deuteron can be obtained in ^{13}C-labelled and fractionally ^{2}H-enriched proteins, enabling self-consistency of the relaxation data to be established.

One of the first applications of the above approach considered the conformational entropy change associated with the folding-unfolding transition in the N-terminal SH3 domain of the Drosophila signal transduction protein drk [46]. The observed entropy change for the folding-unfolding transition averaged 12 J/mol K, compared with the average entropy change per residue estimated from alternative techniques of $\sim$ 14 J/mol K [72]. In a subsequent study, Wrabl et al. [73] used simulated order parameters for N – H bond vectors from nanosecond molecular dynamics simulations of staphylococcal nuclease, and compared per-residue entropies calculated using Eq. 4 with those estimated using quasiharmonic analysis [74]. A positive correlation between these parameters suggested that NMR-derived order parameters provide a reasonable estimate of the total conformational entropy change on protein folding.

A number of studies using NMR relaxation methods have shown that changes in the conformational entropy of the protein before and after ligand association can make significant contributions to the free energy of binding. For example, Bracken et al. [48] examined the dynamics of the leucine zipper domain of yeast transcription factor GCN4 on binding to DNA. In the absence of DNA, the N-terminal basic region adopts an ensemble of transient structures, but undergoes a transition to yield a stable α-helical structure on binding DNA. Thus, an unfavourable contribution to binding is anticipated from the change in conformational entropy of the protein backbone, which was estimated as – 0.6 kJ/mol/K, which agrees remarkably well with theoretical predictions based on calorimetric measurements for the same system (– 0.5 kJ/mol/K). At 300 K the contribution to the free energy of binding is thus between – 150 and – 180 kJ/mol. This contribution is likely offset by a number of other competing factors described in this section, but it illustrates that the unfavourable entropic contribution from freezing protein degrees of freedom on binding can be very significant.

Lee et al. examined the entropic contribution to binding from both backbone and side-chain degrees of freedom for calcium saturated calmodulin binding with a peptide model of the calmodulin-binding domain of myosin light chain kinase [52]. A remarkable result of these studies is that the protein effectively redistributes the side-chain entropy upon binding of the peptide. The side-chains of binding-site residues become more rigid upon association of the peptide as anticipated, whereas certain residues remote from the binding site become more flexible, thus offsetting in part the unfavourable entropic contribution from binding-site residues. Once again, the overall entropic contribution to binding free energy derived from NMR relaxation measurements is in qualitative agreement with calorimetric measurements.

More recently Bingham et al. [54] undertook a study of the binding of 2-methoxy-3-isobutylpyrazine (IBMP) and 2-methoxy-3-isopropylpyrazine (IPMP) to the major urinary protein. Backbone dynamics of certain regions of the protein exhibited increased flexibility on binding either ligand, whereas others displayed an overall reduction in flexibility (Fig. 7). The overall entropic contribution from backbone dynamics was unfavourable with $T\Delta S = -7.4 \pm 6.5$ kJ/mol. The overall contribution from side-chain methyl dynamics on binding IBMP was also unfavourable ($T\Delta S_b = -3.4 \pm 2.8$ kJ/mol), and in common with the calmodulin-peptide complex studied by Lee et al. [52], "entropy–entropy compensation" is observed, i.e. loss of dynamics for binding-site residues is offset by increased dynamics of side chains distal to the binding site.

The NMR measurements discussed thus far in this section probe dynamics on the ps–ns timescale. Dynamic changes on binding over timescales outside this regime, such as slower motions resulting from domain motions or substantial conformational rearrangement, will also clearly contribute to the entropy of binding. The internal motions that give rise to such phenomena typically have time constants in the microsecond to millisecond range, and the relaxation times $T_1\rho$ and T_2 are very sensitive to these motions since they contribute to resonance line-widths. Under the appropriate circumstances, relaxation dispersion experiments can be utilized to extract kinetic and thermodynamic parameters and chemical shift differences between the interconverting states [58, 75–85]. Experiments that have been most recently developed monitor transverse ^{15}N or ^{13}C relaxation during Carr-Purcell-Meiboom-Gill (CPMG) pulse trains [86], with effective relaxation rates measured as a function of the average CPMG radio frequency field strength. Loosely, the function of the CPMG pulse train can be thought of as suppressing chemical shift information, the extent of which depends on the applied CPMG field strength. At low field strengths, transverse relaxation rates are larger due to the presence of a contribution from conformational exchange. Conversely, at large field strengths, the exchange contribution is

Fig. 7 Structural details of residues that contribute to the entropy of binding of 2-methoxy-3-isobutylpyrazine to MUP-I. Backbone residues that exhibit an unfavourable entropic contribution to binding are *coloured yellow*, while those that exhibit a favourable contribution are *coloured red*. Similarly, residues whose methyl-containing sidechains exhibit an unfavourable contribution are *coloured light blue*, whereas those that exhibit a favourable contribution are *coloured magenta*. Reproduced with permission from J Am Chem Soc 2004, 126:1675–1681. Copyright 2004 Am Chem Soc

largely suppressed, since it depends, inter alia, upon the chemical shift difference between exchanging sites.

Yung et al. recently applied the relaxation dispersion technique to probe the influence of conformational exchange in the homopentameric B subunit (VTB) of the toxin from E. coli O157 on the thermodynamics of binding of a novel ligand known as "Pk dimer" [87]. VTB is known to possess three binding sites for the natural carbohydrate ligand globotriaosylceramide on each monomeric subunit [88]. Kitov et al. designed the potential chemotherapeutic agent Pk dimer to straddle two adjacent binding sites, in order to optimize binding affinity through multivalency [89]. Remarkably, the binding of this ligand to the B subunit gives rise to a narrowing of the line-width of a number of resonances in the $^{15}N - {}^{1}H$ heteronuclear single quantum correlation (HSQC) of the protein (Fig. 8). This is at first sight counter-intuitive since linewidths are generally expected to increase due to the larger rotational tumbling time of the complex. However, this phenomenon can be explained by the presence of conformational exchange in the homopentamer before complexation that is suppressed on ligand binding. Interestingly, an early crystal structure of VTB in the absence of ligand showed an asymmetric structure for the protein, where two adjacent monomers were displaced giving the appearance of a "lockwasher" [90]. In contrast, the NMR-derived average structure suggested a symmetric homopentamer [91]. Typical relaxation dispersion profiles for VTB are shown in Fig. 9. Notably, residues that displayed the most significant relaxation dispersion were located at the monomer-monomer interface. These data were qualitatively consistent with interconversion between the symmetric, lower energy state, and a higher energy state that might be related to that observed in the crystal structure. Quantitatively, relaxation dispersion profiles can be fit to suitable expressions for the exchange process. In the limit of fast exchange between two sites the relevant expression is [92]:

$$R_2\left(\nu_{\mathrm{CPMG}}\right) = R_2\left(\nu_{\mathrm{CPMG}} = \infty\right) + \left(p_{\mathrm{a}} p_{\mathrm{b}} \delta\omega^2 / k_{\mathrm{ex}}\right) \times \left(1 - \left(4\nu_{\mathrm{CPMG}}/k_{\mathrm{ex}}\right) \tanh\left(k_{\mathrm{ex}}/4\nu_{\mathrm{CPMG}}\right)\right) , \quad (23)$$

where ν_{CPMG} is the CPMG field strength, p_{a} and p_{b} are the populations of states a and b, and k_{ex} is the exchange rate. Fitting of relaxation dispersion profiles in Fig. 9 to Eq. 23 gave rise to a single effective value of k_{ex} over a number of sites, indicating that a concerted exchange process was taking place at the monomer-monomer interface. Relaxation dispersion was unobservable in the presence of the ligand, which straddles adjacent monomers and effectively quenches the conformational exchange. Correspondingly, binding curves for the titration of Pk dimer with VTB probed using either ^{15}N chemical shift perturbations or ITC experiments could not be fitted to a simple two-state binding model. However, these data could be well-fitted with a co-operative sequential binding model. Assuming that a sec-

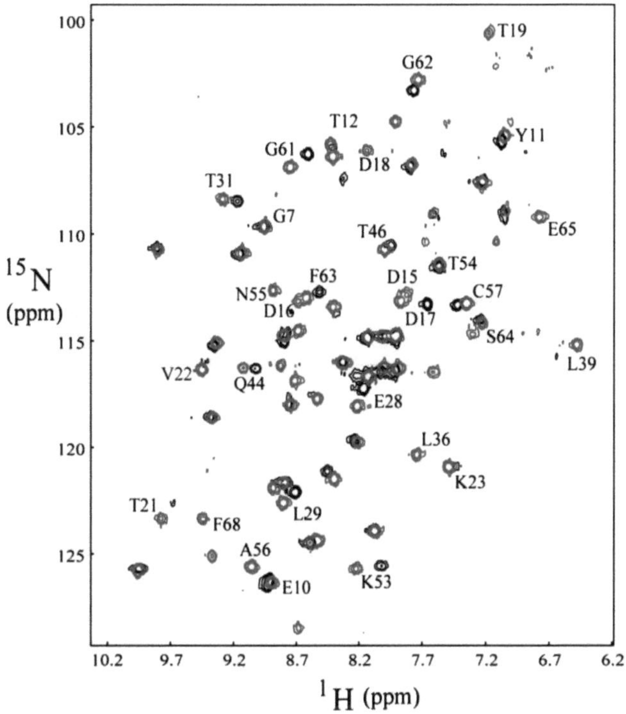

Fig. 8 $^{15}N - ^{1}H$ HSQC spectra of VTB in the absence (*black contours*) and presence (*red contours*) of bivalent inhibitor Pk-dimer. Resonance assignments of residues that experience a shift on inhibitor binding are labelled. Note that in a number of instances (e.g. L39 to the *right of the figure*) the resonance is broadened to the limit of, or below detection in the absence of inhibitor. Reproduced with permission from J Am Chem Soc 2003, 125:13058–13062. Copyright 2003 Am Chem Soc

ond Pk-dimer molecule bound to the homopentamer does not interact with the first, the entropic cost of suppressing conformational exchange by the first binding event was estimated to be – 68.5 kJ/mol at 45 °C. Thus, it is clear that conformational rearrangement can give rise to a very significant unfavourable entropic contribution to binding.

A conceptually different approach for characterizing protein dynamics involves the measurement of three-bond scalar couplings, which can report on rotamer distributions of amino-acid sidechains [93–95]. No information on the timescale of such motions is available from these measurements, but importantly scalar couplings are sensitive to motions over the entire range from picoseconds to milliseconds. Chou et al. [96] measured $^{3}J_{C'-C\gamma}$ and $^{3}J_{N-C\gamma}$ scalar couplings to determine the degree of side-chain order about the C^{α}–C^{β} bond ($\chi 1$ angle) for threonine, isoleucine and valine residue sidechains in ubiquitin. By use of the relevant Karplus parametrization, rotamer populations could be derived from a which a generalized order parameter, S_J^2, was

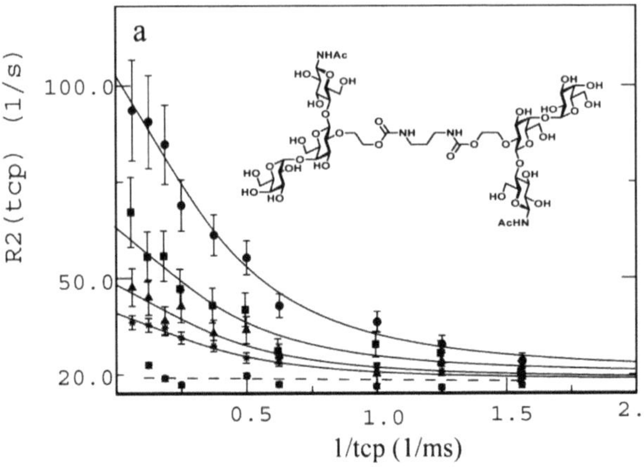

Fig. 9 Typical ^{15}N relaxation dispersion profiles for the amide nitrogens of Val 50 (•), Val 22 (■), and Lys 23 (▲) in VTB. The relaxation dispersion profile for Val 50 in the presence of 5-fold molar excess of inhibitor Pk-dimer (*inset*) is shown by the *broken line*. *Solid lines* represent the best-fit to the data using the equation appropriate for all exchange time-scales with $k_{ex} = 1000\ s^{-1}$. Reproduced with permission from J Am Chem Soc 2003, 125:13058–13062. Copyright 2003 Am Chem Soc

calculated according to the following expression:

$$S_J^2 = \sum_{i,j} P_i P_j \left(3\cos^2\theta_{ij} - 1\right)/2\,, \tag{24}$$

where the summation is over all pairwise combinations of the three rotamers, P_i is the population of rotamer i and θ_{ij} is the angle between the $C^\beta - C^\alpha$ bond vectors of rotamers i and j. Comparison of the order parameters thus obtained with those derived from methyl ^2H relaxation rates for the same residues gave a correlation coefficient of 0.81, which is remarkable given that the two approaches measure the order parameters over very different time-scales. To the knowledge of the author scalar coupling measurements have not been used to probe the contribution of side-chain motions to binding entropy, but clearly these are highly complementary to relaxation methods for this purpose.

3.3.2
Solvation Contribution

Entropies of solvation of small organic ligands are typically negative [25, 97]. This makes intuitive sense given the proposed solvent ordering around the solute in the current view of the hydrophobic effect (Sect. 2.2). To the extent that ligand binding is a desolvation process, it is therefore anticipated that solvent reorganization will offer a favourable entropic contribution to binding. The magnitude of this contribution has been a topic of considerable debate. As described by Dunitz, there are however limits for the entropic cost of bound water in biomolecules [98]. Comparison of the standard entropies at 298 K for anhydrous and hydrated inorganic salts shows that a bound water molecule in crystalline hydrates contributes approximately 42 J/mol/K to the standard entropy. Since the standard entropy at 298 K of liquid water is approximately 70 J/mol/K, the entropic cost of immobilizing a given water molecule in a crystalline lattice is thus approximately – 28 J/mol/K. Since a water molecule within the solvation cage of a nonpolar ligand or a protein binding site will not be bound more firmly than a crystalline hydrate, this value can be taken as the maximum entropic cost of ordering, which corresponds to a standard free energy of approximately 8.3 kJ/mol at 298 K. Thus, the release of a solvent water molecule from the solvation shell of a ligand or from a protein binding-site on binding is anticipated to correspond to a maximum – 8.3 kJ/mol favourable contribution to the free energy of binding.

Direct experimental measurement of the entropic contribution from bound water molecules in fraught with difficulty, and most conclusions have arisen from indirect observations. An interesting example is the study of Holdgate et al. [99] on the binding of the antibiotic novobiocin to a resistant mutant of DNA gyrase. Novobiocin binds to a 24 kDa fragment from the B subunit of DNA gyrase, and resistance to this antibiotic occurs from mutation of Arg-136 which hydrogen bonds to the coumarin ring of novobiocin. Holdgate et al. showed that an R136H mutant binds with a K_d that increases from 32 nM to 1200 nM compared with the wild-type protein at 300 K. This increased affinity was shown by isothermal titration calorimetry measurements to arise from a more favourable enthalpy of binding and a much less favourable entropy of binding. This is opposite to the expected thermodynamic signature given that the loss of the arginine residue is expected to reduce solute–solute hydrogen bonding. However, in the crystal structure of the mutant complex, an ordered water molecule is sequestered into the region vacated by the arginine guanidinium group. Holdgate et al. suggested that the resulting water-mediated protein-antibiotic hydrogen bonds give rise to a favourable enthalpic contribution, whereas the sequestration of a water molecule leads to an entropic cost and reduction in heat capacity of the system.

Further experimental evidence for an entropic contribution to binding derives from the work of Clarke et al. [100] on the binding of trimannoside oligosaccharides to the plant lectin concanavalin A (Con A). These authors characterized the thermodynamics of binding of Manα1-6(Manα1-3)Manα1-OMe and a derivative bearing a hydroxyethyl moiety at C-2 of the central mannose unit to Con A using ITC. Molecular dynamics simulations of the complexes of Con A with each of these ligands established that the hydroxyethyl moiety displaces a conserved water molecule present in the Con A binding site. Correspondingly, the binding of the hydroxyethyl derivative displayed a more favourable entropy and a relatively large unfavourable enthalpy term in comparison with the underivatized ligand. It was proposed that an indirect interaction through the water molecule in the complex with the latter provides a larger number of hydrogen bonds in the complex that have higher occupancies than in bulk solution, thus providing enthalpic stabilization. The more favourable entropy of binding of the hydroxyethyl derivative was suggested to derive from expulsion of the conserved water molecule that is present in the uncomplexed protein into bulk solvent.

The above studies probed the entropic contribution to binding from the solvent *indirectly* via changes in the thermodynamic signature of binding. It is very difficult experimentally to probe *directly* the entropic contribution to binding from the solvent. X-ray diffraction data often show bound water molecules in protein binding sites, but only if these are well-ordered and the occupancy is sufficiently high. Moreover, NMR methods typically report on the average properties of solvent water molecules, and since exchange with bulk solvent is typically fast on the NMR time-scale, it is difficult although not impossible [101] to observe ordered water molecules, but the extraction of accurate thermodynamic parameters is a different matter. Despite these difficulties, pioneering work by Halle and coworkers using ^{17}O, ^{2}H and ^{1}H NMR dispersion measurements has made significant progress in this direction [102, 103]. In principle, relaxation dispersion measurements on solvent water are similar to those described in Sect. 3.3.1. However, the exchange rates for solvent water molecules transiently bound to proteins are much greater than rates of conformational change within proteins, and it is necessary to vary the static magnetic field rather than a weak spin-locking field in order to record water relaxation dispersion profiles. Such measurements have however shown that the three buried water molecules in bovine pancreatic trypsin inhibitor exchange with bulk water on a time-scale of 15 ns to 1 μs and undergo librational motions of considerable amplitude in proteins [104]. Analysis of three independent order parameters provided by relaxation dispersion data for ^{17}O, ^{2}H and ^{1}H in terms of an anisotropic harmonic libration model, provided the amplitude and anisotropy of water rotation within the protein. Although each of the buried water molecules investigated engage in three or four hydrogen bonds, entropies were found to span the range from ice to bulk water, suggesting that the hydration of cavities in proteins with weaker hydrogen-bonding capacity may be entropically driven. Thus, these data contradict the conventional view (described in the previous two paragraphs) that ligand binding to proteins is entropically favored by release of ordered water.

Further insight into the entropic contribution of solvent water molecules has been obtained from atomistic molecular dynamics simulations. Li and Lazaridis used inhomogeneous fluid solvation theory to study the contribution of a bound water molecule in the binding site of HIV-1 protease to the energy, entropy and heat capacity of solvation [105]. The entropy loss in binding this water molecule was found to be 41 J mol^{-1} K^{-1} and the total contribution to the free energy of solvation was found to be $-63.5\,\mathrm{kJ\,mol^{-1}}$. Notably, the calculated entropy is significantly larger than that predicted by Dunitz [98]. Hamelberg and McCammon, using rigorous statistical mechanical molecular dynamics simulations, computed the standard free energy of releasing a bound water molecule from the binding pocket in a Trypsin/Benzylamine complex and an HIV-1/KNI-272 complex [106]. Values of $\sim -8\,\mathrm{kJ\,mol^{-1}}$ and $-13\,\mathrm{kJ\,mol^{-1}}$ respectively, were obtained, suggesting that localized water molecules stabilize ligand-protein interactions in both complexes.

4 Concluding Remarks

Complete dissection of the thermodynamics of binding of a ligand to a protein into enthalpic and entropic contributions from both partners and the solvent is a formidable task. However, it is a battle that must ultimately be won in order to make use of the rapidly increasing numbers of high-resolution protein structures in "rational" drug design. Although much remains to be done, significant progress has been made in the last decade. For example, high-resolution NMR methods have shed light on the contribution of protein degrees of freedom to the entropy of binding on a *per-residue* basis for a number of systems. While the approach is subject to a number of assumptions, a feature of the binding process in systems studied to date is the reduction in the unfavourable entropic contribution to binding resulting from freezing of binding site residues by melting of binding site residues at distal locations. This entropy–entropy compensation phenomenon may be a universal property of proteins. At present we know insufficient detail about protein dynamics to predict this phenomenon, and hence it will be difficult to exploit it for rational ligand design purposes. It is tempting to speculate that mutations distal to binding-site regions in various protein drug targets in resistant bacterial strains may derive from dynamic processes such as these, although this has not been examined experimentally to our knowledge. Turning to the enthalpy of binding, the notion of binding-site "shape" complementarity has long been a mainstay of computational approaches to lead compound design. The recent discovery that solute–solvent van der Waals interactions do not necessarily exchange for solute–solute van der Waals interactions in protein binding sites that are sub-optimally hydrated [107], offers the possibility to obtain dramatic increases in the free energy of binding by optimizing shape complementarity in such systems. Whether this approach is worthwhile will clearly be dependent on the degree of hydration of the binding pocket, which to our knowledge has not been systematically examined.

Despite these advances, there are still many gaps in our knowledge. Perhaps the most contentious area concerns the thermodynamic contribution from solvent water. For every "quantitative" measure of the enthalpic and entropic contribution from, for example, bound water molecules, it appears that there are conflicting data. The only common feature appears to be that the thermodynamic contribution from solvent water can be significant. There is general agreement that the ejection of bound water molecules from a binding pocket into bulk solvent on ligand binding is an entropically favourable process, but the enthalpic contribution from this process remains a matter for debate. Resolution of these issues is likely to be extremely difficult. NMR methods, except in certain special cases, only provide information on the average properties of solvent water molecules, which typically exchange rapidly between the protein-bound state and bulk water. X-ray diffraction methods can only detect bound water molecules if the occupancy or degree of order is sufficiently high. All atom molecular dynamics simulations perhaps provide the greatest hope of quantifying the solvation contribution, but the current level of accuracy of molecular mechanical forcefields leaves some concerns regarding the validity of the data thus obtained.

References

1. Frank HS, Evans MW (1945) J Chem Phys 13:507
2. Nemethy G, Scheraga HA (1962) J Chem Phys 36:3382
3. Tanford C (1980) The Hydrophobic Effect: Formation of Micelles and Biological Membranes. Wiley, New York
4. Privalov PL, Gill SJ (1988) Adv Protein Chem 39:191
5. Mirejovsky D, Arnett EM (1983) J Am Chem Soc 105:1112
6. Lee SH, Rossky PJ (1994) J Chem Phys 100:3334
7. Lee B (1991) Biopolymers 31:993
8. Lee B (1985) Biopolymers 24:813
9. Lucas M (1976) J Phys Chem 80:359
10. Southall NT, Dill KA, Haymet ADJ (2002) J Phys Chem B 106:521
11. Southall NT, Dill KA (2002) Biophys Chem 101:295
12. Chandler D (2005) Nature 437:640
13. Williams DH, Bardsley B (1999) Perspect Drug Discov Design 17:43
14. Wiseman T, Williston S, Brandts JF, Lin LN (1989) Anal Biochem 179:131
15. Penel S, Doig AJ (2001) J Mol Biol 305:961
16. Williams DH, Stephens E, Zhou M (2003) J Mol Biol 329:389
17. Hyre DE, Le Trong I, Freitag S, Stenkamp RE, Stayton PS (2000) Prot Sci 9:878
18. Vondrasek J, Bendova L, Klusak V, Hobza P (2005) J Am Chem Soc 127:2615
19. Ross PD, Subramanian S (1981) Biochemistry 20:3096
20. Chapman KT, Still WC (1989) J Am Chem Soc 111:3075
21. Hunter CA (2004) Angew Chem Int Ed 43:5310
22. Talhout R, Villa A, Mark AE, Engberts J (2003) J Am Chem Soc 125:10570
23. Daranas AH, Shimizu H, Homans SW (2004) J Am Chem Soc 126:11870
24. Burkhalter NF, Dimick SM, Toone EJ (2000) In: Ernst B, Hart GW, Sinay P (eds) Carbohydrates in Chemistry and Biology. Part I: Chemistry of Saccharides, Vol 2. Wiley, Weinheim, p 863
25. Cabani S, Gianni P, Mollica V, Lepori L (1981) J Sol Chem 10:563

26. Plyasunov AV, Shock EL (2000) Geochim Cosmochim Acta 64:439
27. Muller N (1990) Acc Chem Res 23:23
28. Buttery RG, Bomben JL, Guadagni DG, Ling LC (1971) J Agric Food Chem 19:1045
29. Orozco M, Luque FJ (2000) Chem Rev 100:4187
30. Kollman PA (1993) Chem Rev 93:2395
31. Carlson HA, Nguyen TB, Orozco M, Jorgensen WL (1993) J Comput Chem 14:1240
32. Orozco M, Jorgensen WL, Luque FJ (1993) J Comput Chem 14:1498
33. Jorgensen WL, Nguyen TB (1993) J Comput Chem 14:195
34. Wan S, Stote RH, Karplus M (2004) J Chem Phys 121:9539
35. Kubo MM, Gallicchio E, Levy RM (1997) J Phys Chem B 101:10527
36. Chothia C (1974) Nature 248:338
37. Levy RM, Zhang LY, Gallicchio E, Felts AK (2003) J Am Chem Soc 125:9523
38. Hawkins GD, Cramer CJ, Truhlar DG (1997) J Phys Chem B 101:7147
39. Chervenak MC, Toone EJ (1994) J Am Chem Soc 116:10533
40. Connelly P, Aldape RA, Bruzzese FJ, Chambers SP, Fitzgibbon MJ, Fleming MA, Itoh S, Livingston DJ, Navia MA, Thomson JA, Wilson KP (1994) Proc Natl Acad Sci 91:1964
41. Jencks WP (1981) Proc Natl Acad Sci 78:4046
42. Williams DH, Stephens E, O'Brien DP, Zhou M (2004) Angew Chem Int Ed 43:6596
43. Pitzer KS, Gwinn WD (1942) J Chem Phys 10:428
44. Searle MS, Williams DH (1992) J Am Chem Soc 114:10690
45. Akke M, Bruschweiler R, Palmer AG (1993) J Am Chem Soc 115:9832
46. Yang DW, Kay LE (1996) J Mol Biol 263:369
47. Spyracopoulos L, Gagne SM, Li MX, Sykes BD (1998) Biochemistry 37:18032
48. Bracken C, Carr PA, Cavanagh J, Palmer AG (1999) J Mol Biol 285:2133
49. Zídek L, Novotny MV, Stone MJ (1999) Nature Str Biol 6:1118
50. Sahu SC, Bhuyan AK, Udgaonkar JB, Hosur RV (2000) J Biomol NMR 18:107
51. Maler L, Blankenship J, Rance M, Chazin WJ (2000) Nature Str Biol 7:245
52. Lee AL, Kinnear SA, Wand AJ (2000) Nature Str Biol 7:72
53. Loh AP, Pawley N, Nicholson LK, Oswald RE (2001) Biochemistry 40:4590
54. Bingham R, Bodenhausen G, Findlay JHBC, Hsieh S-Y, Kalverda AP, Kjellberg A, Perazzolo C, Phillips SEV, Seshadri K, Turnbull WB, Homans SW (2004) J Am Chem Soc 126:1675
55. Palmer AG (1997) Curr Op Struct Biol 7:732
56. Cavanagh J, Akke M (2000) Nature Str Biol 7:11
57. Palmer AG (2001) Ann Rev Biophys 30:129
58. Palmer AG, Kroenke CD, Loria JP (2001) Meth Enzymol 339:204
59. Spyracopoulos L, Sykes BD (2001) Curr Op Struct Biol 11:555
60. Frueh D (2002) Progr NMR Spectrosc 41:305
61. Luginbuhl P, Wuthrich K (2002) Progr NMR Spectrosc 40:199
62. Case DA (2002) Acc Chem Res 35:325
63. Kempf JG, Loria JP (2003) Cell Biochem Biophys 37:187
64. Palmer AG (2004) Chem Rev 104:3623
65. Lipari G, Szabo A (1982) J Am Chem Soc 104:4546
66. Li ZG, Raychaudhuri S, Wand AJ (1996) Prot Sci 5:2647
67. Yang DW, Mok YK, FormanKay JD, Farrow NA, Kay LE (1997) J Mol Biol 272:790
68. Muhandiram DR, Yamazaki T, Sykes BD, Kay LE (1995) J Am Chem Soc 117:11536
69. Lee AL, Flynn PF, Wand AJ (1999) J Am Chem Soc 121:2891
70. Ishima R, Petkova AP, Louis JM, Torchia DA (2001) J Am Chem Soc 123:6164
71. Millet O, Muhandiram DR, Skrynnikov NR, Kay LE (2002) J Am Chem Soc 124:6439

72. Doig AJ, Sternberg MJE (1995) Prot Sci 4:2247
73. Wrabl JO, Shortle D, Woolf TB (2000) Proteins 38:123
74. Tidor B, Karplus M (1994) J Mol Biol 238:405
75. Millet O, Loria JP, Kroenke CD, Pons M, Palmer AG (2000) J Am Chem Soc 122:2867
76. Mulder FAA, van Tilborg PJA, Kaptein R, Boelens R (1999) J Biomol NMR 13:275
77. Akke M, Liu J, Cavanagh J, Erickson HP, Palmer AG (1998) Nature Str Biol 5:55
78. Davis DG, Perlman ME, London RE (1994) J Magn Reson B 104:266
79. Allerhand A, Gutowsky HS (1965) J Chem Phys 42:1587
80. Mulder FAA, Hon B, Muhandiram DR, Dahlquist FW, Kay LE (2000) Biochemistry 39:12614
81. Skrynnikov NR, Mulder FAA, Hon B, Dahlquist FW, Kay LE (2001) J Am Chem Soc 123:4556
82. Ishima R, Louis JM, Torchia DA (1999) J Am Chem Soc 121:11589
83. Ishima R, Torchia DA (1999) J Biomol NMR 14:369
84. Mulder FAA, Mittermaier A, Hon B, Dahlquist FW, Kay LE (2001) Nature Str Biol 8:932
85. Mulder FAA, Hon B, Mittermaier A, Dahlquist FW, Kay LE (2002) J Am Chem Soc 124:1443
86. Carr HY, Purcell EM (1954) Phys Rev 94:630
87. Yung A, Turnbull WB, Kalverda AP, Thompson GS, Homans SW, Kitov P, Bundle DR (2003) J Am Chem Soc 125:13058
88. Ling H, Boodhoo A, Hazes B, Cummings MD, Armstrong GD, Brunton JL, Read RJ (1998) Biochemistry 37:1777
89. Kitov PI, Shimizu H, Homans SW, Bundle DR (2003) J Am Chem Soc 125:3284
90. Stein PE, Boodhoo A, Tyrrell GJ, Brunton JL, Read RJ (1992) Nature 355:748
91. Richardson JM, Evans PD, Homans SW, DonohueRolfe A (1997) Nature Str Biol 4:190
92. Luz Z, Meiboom S (1963) J Chem Phys 39:366
93. Dzakula Z, Westler WM, Edison AS, Markley JL (1992) J Am Chem Soc 114:6195
94. Dzakula Z, Edison AS, Westler WM, Markley JL (1992) J Am Chem Soc 114:6200
95. Perez C, Lohr F, Ruterjans H, Schmidt JM (2001) J Am Chem Soc 123:7081
96. Chou JJ, Case DA, Bax A (2003) J Am Chem Soc 125:8959
97. Ben-Naim A, Marcus Y (1984) J Chem Phys 81:2016
98. Dunitz J (1994) Science 264:670
99. Holdgate GA, Tunnicliffe A, Ward WHJ, Weston SA, Rosenbrock G, Barth PT, Taylor IWF, Pauptit RA, Timms D (1997) Biochemistry 36:9663
100. Clarke C, Woods RJ, Gluska J, Cooper A, Nutley MA, Boons G-J (2001) J Am Chem Soc 123:12238
101. Otting G, Wuthrich K (1989) J Am Chem Soc 111:1871
102. Halle B, Andersson T, Forsen S, Lindman B (1981) J Am Chem Soc 103:500
103. Denisov VP, Halle B (1995) J Mol Biol 245:682
104. Denisov VP, Venu K, Peters J, Horlein HD, Halle B (1997) J Phys Chem B 101:9380
105. Li Z, Lazaridis T (2003) J Am Chem Soc 125:6636
106. Hamelberg D, McCammon JA (2004) J Am Chem Soc 126:7683
107. Barratt E, Bingham R, Warner DJ, Laughton CA, Phillips SEV, Homans SW (2005) J Am Chem Soc 127:11827

Top Curr Chem (2007) 272: 83–116
DOI 10.1007/128_068

Published online: 30 August 2006

The Fibroblast Growth Factor (FGF) – FGF Receptor Complex: Progress Towards the Physiological State

Nicholas J. Harmer

Department of Biochemistry, University of Cambridge, 80 Tennis Court Road, Cambridge CB2 1GA, UK
nic@cryst.bioc.cam.ac.uk

Abstract Signaling by the fibroblast growth factors (FGFs) and their receptors (FGFRs) has been implicated in a wide range of diseases including cancer and arthritis. The need to understand the mechanisms of these diseases, and the potential for the development of novel therapeutics, has driven the characterization of complexes of the FGFs, FGFRs, and the co-receptor heparin. These efforts have led to the proposal of two models, based on crystal structures, for the biological signaling complex: these models show considerable differences that are of great importance to the mechanism of signaling, but nevertheless share common themes. The merits of these models have been illuminated by a range of further crystal structures that have revealed the more relevant conformations of each model. The development of methods in ultracentrifugation and mass spectrometry has allowed the analysis of both complexes in solution, and has suggested that both architectures bind only one molecule of heparin. New methods for sequencing heparin and preparing heparin derivatives have allowed the affinity of FGFs for heparins to be determined. Finally, evidence has accumulated for complexes involving more than two FGFRs, and tantalizing hints have emerged of how both crystallographic models may contribute to a larger "signalosome".

Keywords Ligand–receptor complexes · Structure-function relationships · X-ray crystallography

Abbreviations

FGF Fibroblast growth factor
FGFR Fibroblast growth factor receptor
HS Heparan sulfate
Ig Immunoglobulin

1 Perspective on the FGF System

Fibroblast growth factors are found in all multicellular animals [1]. Signaling associated with FGFs has been associated with a wide range of cellular behaviors, including cell proliferation, differentiation, migration, and survival. This range of behaviors has led to the FGFs being co-opted into a wide range of pathways involved in the development and homeostasis of higher eukaryotes: indeed, there is no major organ or tissue in humans in which FGF signaling is not involved.

Following the commitment of a cell to express an FGF signal, there are many potential control steps before a response occurs in the target cell(s). For some FGFs (especially FGFs 1 and 2), some control occurs at the level of cellular release, as they are expressed without a conventional export sequence, and so require other mechanisms to effect their release [2]. The FGFs then bind heparan sulphate (HS) proteoglycans in the extracellular matrix, which both protect the FGF against degradation and sequester the FGF in the matrix [3]. Some FGFs have their activity potentiated by the FGF binding proteins, which release them from the matrix to allow them to interact with target cells [4, 5].

On reaching target cells, FGFs interact with HS as part of cellular proteoglycans (reviewed in [6]). Following this, the FGF-HS complex binds to the FGF receptor (FGFR), a transmembrane protein with an intracellular tyrosine kinase domain. Binding of the FGF to the FGFR leads to activation of the receptor kinase domain, and the activation of downstream effectors [7]. Following stimulation, clusters of activated FGFRs, together with the FGFs and HS, are endocytosed; FGFs are capable of escaping from the endosomal particles, and then translocate to the nucleus where they further activate cells [8,9]. Many of these steps are influenced by other proteins, demonstrating that the process of FGF signaling is a complex and highly regulated process.

This review will focus on the formation of the FGF-FGFR-HS complex that leads to the FGF signal being transmitted into the target cell, particularly on the synthesis of multiple crystal structures with biochemical data to generate a more complete picture of the signaling process.

2 Introduction

2.1 Fibroblast Growth Factors

The FGFs are a family of small (17–34 kDa) proteins containing a conserved core domain. Humans have 18 signaling FGFs; these can be divided by sequence similarity into six subfamilies (reviewed in [10]). The members of each subfamily exhibit similar, but subtly different, properties. The core domain contains the most significant sites for binding to both FGFRs and HS; N- and C-terminal extensions to this domain modulate the functions of the FGF, providing additional interactions with receptors and targeting signals. FGFs demonstrate a high affinity for both FGFRs and HS, with separate binding sites for the two molecules: this allows the formation of complexes that contain both HS and FGFR bound to the FGF.

2.2 FGF Receptors

FGFRs are the protein receptors for FGFs, and they bind to FGFs and HS to form a tight ternary complex [11,12]. Mammals have four signaling FGFRs (FGFR1–4), each of which consists of an ectodomain containing three immunoglobulin (Ig) domains, a single transmembrane helix, a short juxtamembrane region, and a split tyrosine kinase domain (Fig. 1 [13]). The FGF binding activity is located in the two membrane proximal Ig domains (Ig domains 2 and 3) [14–18], with both domains contributing significantly to the FGF binding interface. In spite of the high sequence identity of the four FGFRs [19], they show a wide variety in their affinities for the FGFs [20]. Furthermore, FGFRs1–3 undergo a significant alternative splicing event in Ig domain 3, where two exons are available for the C-terminal section of the domain, giving rise to the "IIIb" and "IIIc" forms of the receptors [1]. The sequence altered by this splicing event includes a part of the FGF binding

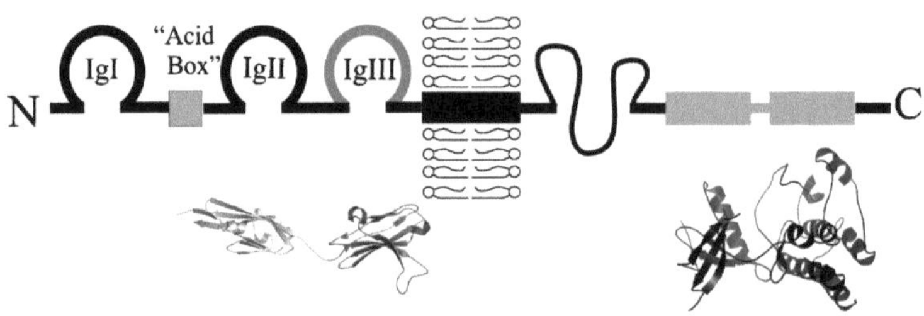

Fig. 1 Schematic of FGFR domain structure. Domains are shown for the most abundant forms of FGFRs1–4. The extracellular region (*left*) consists of three immunoglobulin-like domains, Ig1, Ig2, Ig3. Between Ig1 and Ig2, each FGFR has a sequence of four to eight acidic residues, termed the "acid box". Ig3 (*colored gray*) undergoes alternative splicing, leading to FGFRs with very different FGF binding properties. FGFRs have a single transmembrane helix (*black box*). The intracellular region (*right*) consists of a juxtamembrane region (*black, wavy line*) of undetermined structure, and a split tyrosine kinase domain (*gray boxes*). Cartoons are shown *below* for domains with experimental structures

site, and this leads to a dramatic change in the affinity of the receptor for FGFs [20]. There are therefore in effect seven mammalian FGFR proteins, and the range of affinities that these have for the FGFs constitutes the first level of specificity in the signaling system. FGFRs also have affinity for HS, which is located in an 18-residue loop rich in basic residues in Ig domain 2 [21]. Thus, there is a capacity for the FGFR to bind to both FGF and to the HS with which the FGF is complexed, forming a strong ternary complex.

2.3
Downstream Signaling from Tyrosine Kinase Receptors

Upon binding of the FGF ligand to FGFRs, kinase domains from activated FGFRs phosphorylate one another [22], and are then potentiated to initiate signaling cascades within the cell. FGFRs use a number of signaling pathways, most notably the Ras/MAP kinase, phospholipase C_γ, STAT1 and phosphatidylinositol 3-kinase pathways to transduce signals to the cell (reviewed in [7]). In addition to these, regulatory proteins such as phosphatases (Shp2) and ubiquitin ligases (Clb) are activated to downregulate the FGFR signal.

There is increasing evidence that, for many receptor types, including tyrosine kinases, the effect of the signal may also be dependent upon the formation of higher order complexes [23–26], and the duration and location [27] of the signal. Higher order complexes have been observed in a wide range of signaling systems, where stimulation of a cell with a large dose of a ligand to one of the cell's receptors leads to the receptor altering its localization to a limited number of foci. This has been best demonstrated for the epidermal growth factor receptor (EGFR) [24]. Following this, activated tyrosine kinase receptors are rapidly endocytosed using both clathrin and clathrin-independent mechanisms. During this process of endocytosis, and before the proteins are sent to proteolytic organelles, the receptors continue to signal (as has been convincingly shown for the EGF receptor); indeed, it may prove that certain receptors perform the majority of their signaling in the endosomal pathway.

It is clear that a response from an activated receptor can include a single local signal produced at the membrane location of that receptor, an intense membrane localized signal from a cluster of activated receptors, and finally an internalized "signalsome" of strongly activated receptors moving through the cell. The extent of the signaling is likely to depend upon the dose of the protein ligand that is administered to the cell, and so in a physiological context a range of cell responses can be achieved when a gradient of ligand is present. However, there are as yet few clues as to how a cluster of tyrosine kinase receptors might be built.

2.4 Heparin and Heparan Sulfate (HS)

HS is a polysaccharide modification that is attached to a number of cellular and extracellular matrix proteins (Fig. 2) [28,29]. It consists of a linear saccharide, 100 to 300 saccharide units in length, composed of repeating disaccharides of *N*-acetyl glucosamine and uronic acid [30,31]. The HS chains are modified by a range of enzymes that (1) replace *N*-acetyl with *N*-sulphate, (2) epimerise glucuronic acid to iduronic acid, (3) add sulphate groups to the 2-*O* position of iduronic acid and (4) add sulphate groups to the 6-*O*, and more rarely, the 3-*O* positions of glucosamine (Fig. 3). Each of these modifications is carried out incompletely; however, they are co-ordinated, with each step in the biosynthesis more likely where the previous steps have been performed, with the result that on each HS chain, several domains are formed that have extensive modification over a 12–14 saccharide unit (S-domains), and these are interspersed with stretches of 14–18 saccharide units that display intermediate or low levels of modification [32]. Most experiments that have been carried out on FGFs and other heparin binding growth factors have used fragments of heparin, a mast cell derived analogue of the S-domains of HS. Heparin is sufficient to replace HS for FGF signaling in chlorate treated (HS free) cells [33–35]; however, heparin is far more homogeneously sulphated than most S-domains, and lacks the variety of complex sulfation patterns that are observed with HS.

The structure of heparin has been extensively studied using nuclear magnetic resonance [36–41]: heparin has less flexibility around its glycosidic linkages than most saccharides, due to its highly charged density, and so forms a rather rigid structure with a repeating unit of four saccharides, displaying groups of sulphate modifications on alternating sides of the molecule [40].

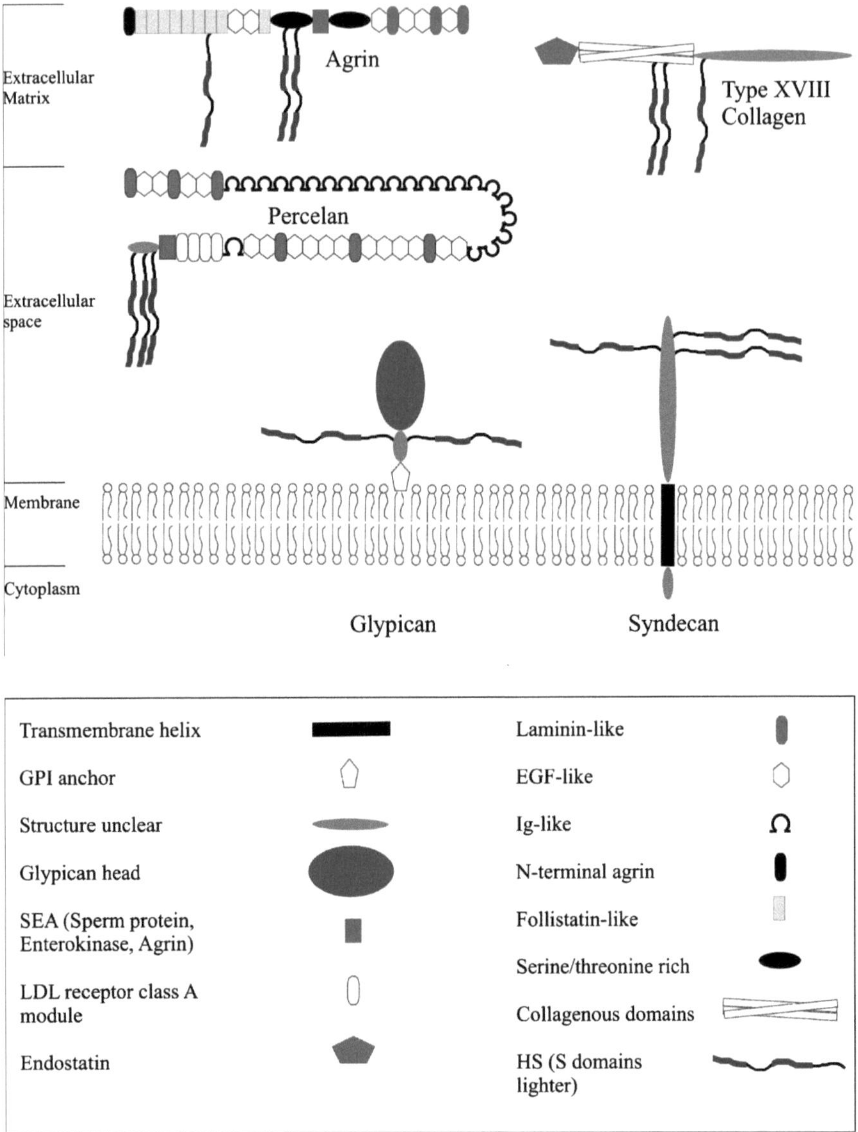

Fig. 2 HS carrying proteins. Domain structures are shown for the proteins that constitute the major HS attachment sites. After [28, 29]

HS, in contrast, cannot be studied using these techniques, owing to its heterogeneous nature. However, linkages similar to those that will be found outside S-domains have been found to have considerably more flexibility than heparin [42–44].

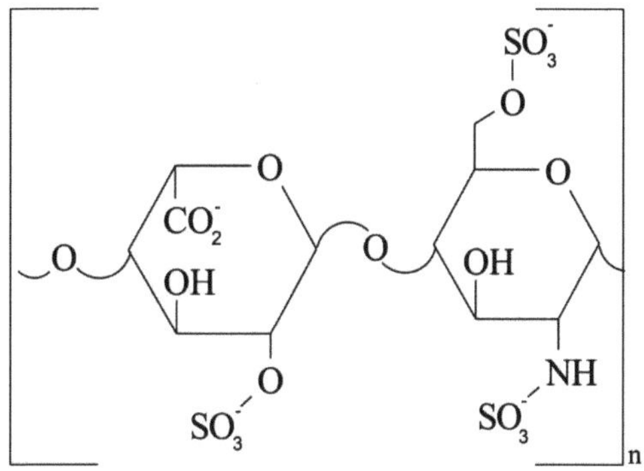

Fig. 3 Structure of heparin. The repeating disaccharide of fully modified heparin is shown. The structure is shown with the reducing end (attached to the protein) at the *right*, non-reducing end at the *left*

3 Structural Insights into the Complex of FGF-FGFR-HS

3.1 FGF-FGFR and FGF-Heparin Structures

A series of structures of complexes of FGFs, FGFRs and heparin have been solved crystallographically (Table 1). Firstly, the structures of the cores domains of FGFs 1 [45], 2 [46], 4 [47], 7 [48], 9 [49], and 19 [50] have shown that the structure of this domain is extremely well conserved across the family, indicating that many observations regarding FGF biology, especially in the context of complexes with receptors, will be generally applicable.

Several structures are available of complexes of one FGF bound to one FGFR [14, 16, 51–54]: these show the FGFs binding to the same sites on the FGFRs, comprising one face of both Ig domain 2 and 3, and the extended linker between these two domains (Fig. 4). Comparison of the structures reveals that the conformations of the FGF and the FGFR in the seven structures are strikingly similar (Fig. 5). The FGF components of these complexes bind to the receptors using very similar regions of the FGF, regardless of which FGFR, or splice form, is making the complex. This similarity in interacting regions is mirrored in the FGFRs, indicating that existing structural insights into the binding of FGFs to FGFRs are likely to be similar for FGF-FGFR pairs whose structure has not been determined.

Superposition of the structures shows that the location and orientation of the FGF, and of the FGFR Ig domain 3 with respect to the FGF is extremely well conserved between the seven structures: in contrast, while the orientation of Ig domain 2 shows small (10°) alterations in its conformation within six of the seven structures, the seventh shows a much larger (> 15°) rota-

Table 1 Currently available crystal structures of FGFs, FGFRs and heparin

FGF	FGFR	Heparin fragment	Resolution (Å)	Refs.
FGF1			1.1	[45, 105]
FGF2			1.6	[46, 106]
FGF4			1.8	[47]
FGF7			3.1	[48]
FGF9			2.2	[49]
FGF19			1.3	[50]
FGF1	FGFR1c		2.8	[51]
FGF1	FGFR2b		2.1	[54]
FGF1	FGFR2c		2.4	[52]
FGF1	FGFR3c		3.2	[14]
FGF2	FGFR1c		2.8	[16]
FGF2	FGFR2c		2.2	[51]
FGF10	FGFR2b		2.9	[53]
FGF1		10mer	2.9	[55]
FGF2		6mer	2.2	[56]
FGF1	FGFR2c	10mer	2.8	[15]
FGF2	FGFR1c	10mer	3.0	[57]

tion with respect to the other domains within the complex [53]. This complex involves an FGF from a separate subfamily to the other structures, and correlates with the loss of a key hydrogen bond between the FGF and FGFR Ig domain 2 in this FGF subfamily [38]. This suggests that the orientation of the FGFR domains with respect to the FGF is a variable that can be exploited by these proteins to ensure that the correct specificity between the ligand the receptor is achieved.

A further difference between these structures is observed in the conformation of the $\beta C'$–βE loop of the FGFR Ig domain 3. The conformation of this loop shows considerable flexibility, to the extent that it is even disordered in one structure. The intimate interactions that the loop makes with the FGF in some of the structures, together with this loop belonging in the region of Ig domain 3 that is alternatively spliced, have led to this loop being proposed as a "specificity loop", whose inherent flexibility and sequence diversity allows it to select a conformation for each FGF, and so disfavor the binding to FGFs that do not provide a sufficiently high enthalpic gain on binding to compensate for the loss of entropy in the loop [38].

These observations demonstrate how the availability of multiple crystal structures permits a comparison that highlights differences between the various structures, and so allows detailed hypotheses of key biological relevance

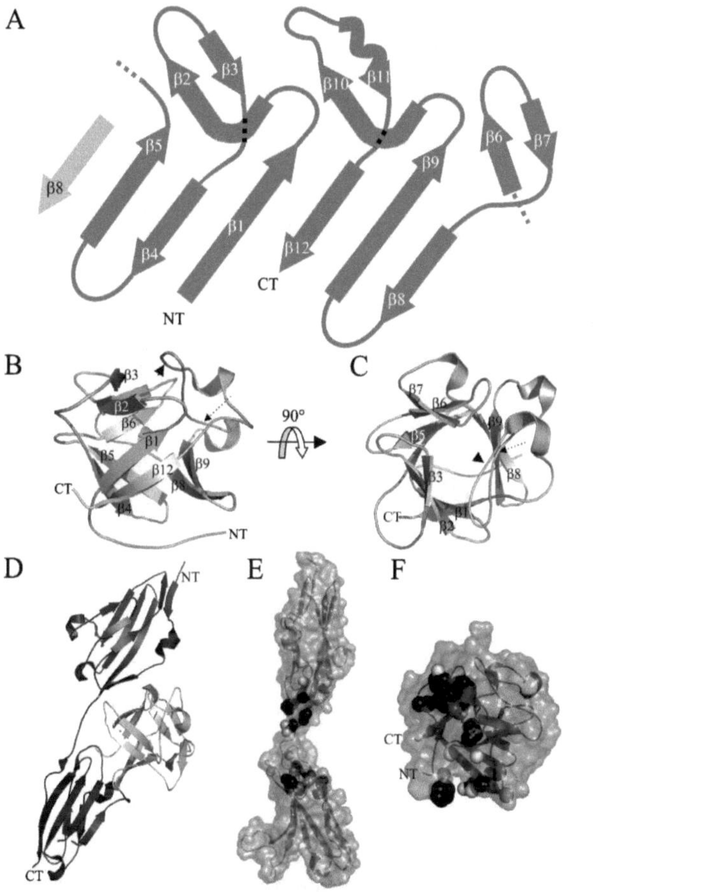

Fig. 4 FGF structure and FGFR binding site. **A** Cartoon of the structure of the FGF core. The twelve β-strands are shown. Strand $\beta 8$ is shown twice (second time in *light gray*) to illustrate the β-barrel at the base of the structure. "Strand $\beta 11$" is shown as a waved arrow to illustrate that it does not make conventional strand hydrogen bonds. **B,C** Orthogonal views of the FGF1 structure (PDB code: 1JQZ) shown in the cartoon representation. **B** Shows the six-stranded β-barrel at the base of the structure. **C** Shows the pseudo-threefold axis of the FGF fold. Selected β-strands are labeled. The regions corresponding to strands $\beta 10$ and $\beta 11$ are indicated by a *dashed black arrow* and a *black arrowhead*, respectively. **D** FGF2-FGFR1 complex, shown in cartoon format (structure 1CVS). FGF2 shown in *gray*, FGFR1 in *black*. FGFR1 is shown with Ig2 at the *top*, Ig3 at the *bottom*. **E,F** FGF2-FGFR1 binding interfaces. FGFR1 (**E**) and FGF2 (**F**) surfaces are shown *above* the cartoon. Atoms making hydrophobic interactions are shown in *black*, atoms making hydrogen bonds or charge–charge interactions are shown in *white*

to be formed. Crystal structures are particularly appropriate for this role, as the single conformations observed enhance the differences and similarities, particularly in cases such as loop conformations.

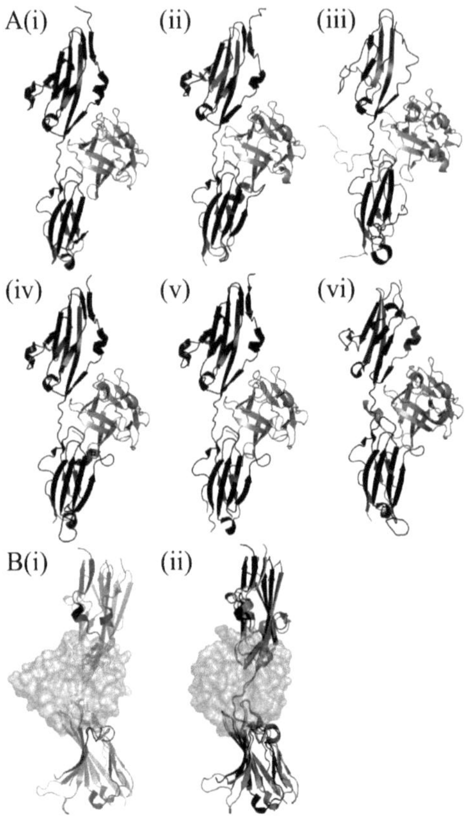

Fig. 5 Comparison of FGF-FGFR structures. **A** The structures of the six FGF-FGFR pairs deposited in public databases are shown as cartoons, with FGFR in *black*, FGF in *gray*. The structures were superimposed using the FGFs to guide the superimposition. All structures are shown in the same orientation, with FGFR Ig2 at the *top*, Ig3 at the *bottom*. The FGF-FGFR pairs and PDB codes are: (i) FGF1-FGFR1c (1EVT); (ii) FGF1-FGFR2c (1DJS); (iii) FGF1-FGFR3c (1RY7); (iv) FGF2-FGFR1c (1CVS); (v) FGF2-FGFR2c (1EV2); (vi) FGF10-FGFR2b (1NUN). **B** The FGF10-FGFR2b structure shows a change in the angle of Ig2. The FGF-FGFR complexes were superimposed as above. FGFRs shown as cartoons, FGFs as surface representation. (i) FGF10 shown, with FGFR2c (from FGF2-FGFR2c structure) in *black*, FGFR2b in *light gray*. (ii) FGF2 shown, with FGFR2c (from FGF2-FGFR2c structure) in *black*, FGFR2c (from FGF1-FGFR2c structure) in *light gray*. Note that Ig3 shows little change in angle between the structures, and that the Ig2 angle changes little on binding to FGF1 or FGF2, while there is a significant change on binding to FGF10

Two groups have presented structures of an FGF complexed with heparin (Fig. 6) [55, 56]. These structures demonstrate firstly that the conformation of the FGF does not noticeably alter upon binding to heparin: the changes observed are restricted to the heparin-binding site, and to necessary conformational alterations to accommodate the heparin. Both structures show

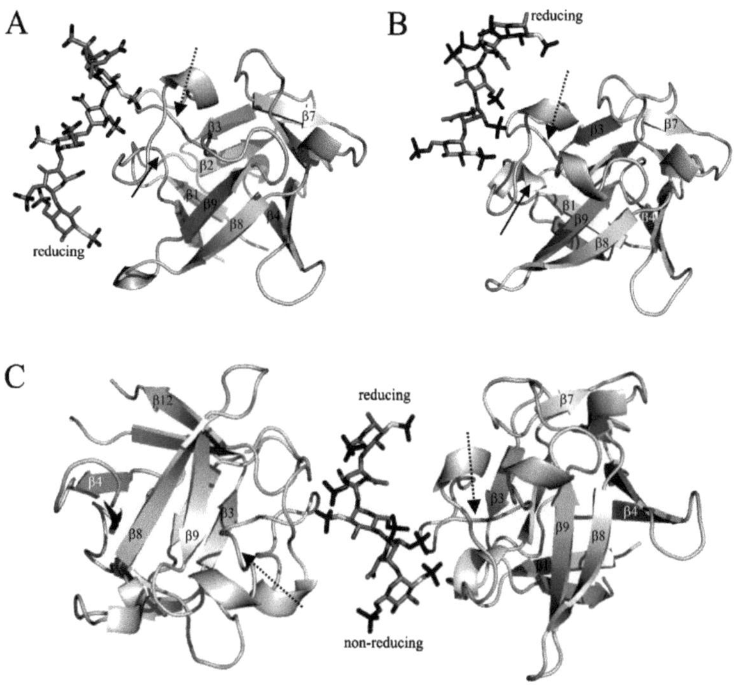

Fig. 6 FGF binding to heparin. FGFs shown as cartoon representation, with facing strands indicated. The $\beta 10$ and $\beta 11$ regions are indicated by a *dashed arrow* and a *solid arrow*, respectively. Heparin is shown as a sticks representation, with sulfate groups colored *black*. The reducing end is marked in each structure. **A** FGF2-heparin 6mer structure (1BFC) [56]. **B** Best-defined FGF molecule of the FGF1-heparin 10mer structure (1AXM) [55]. **C** Heparin 10mer linked FGF1 dimer [55]: the molecule represented in (**B**) is shown on the *left*

heparin binding to the same site on the FGF; however, one structure (Fig. 6B) shows two FGF molecules binding to the heparin. One of these molecules, (the better defined one) binds with the heparin molecule in the opposite orientation with respect to the FGF, compared to the second molecule in this structure, and the previous FGF-heparin structure [56] (Fig. 6). Further crystal structures [15, 57] have demonstrated that the orientation of heparin observed by DiGabriele et al. [55] in this better defined molecule is likely to be physiologically correct. These observations highlight the difficulties that crystallographers face when forming complexes of biological molecules: many of the most interesting complexes (especially those that mediate steps in signaling pathways) are rather weak and transitory. In cases such as heparin binding, where one molecule exhibits similar properties in multiple orientations, there is a risk that the best crystal lattice will be formed by a complex that is near native. Such complexes were observed by Faham et al. [56], and in the case of the second FGF molecule binding to heparin by DiGabriele et al. [55]. Nevertheless, these structures have been of great value in illuminating the heparin binding sites of the FGFs, and in highlighting the potential of HS to bind multiple molecules.

3.2
Ternary FGF-FGFR-Heparin Complex: Two Alternative Structures

One key goal of the work on the FGF signaling system has been to demonstrate how FGFs can lead to the juxtaposition of two (or more) FGFR molecules, which will lead to the activation of the tyrosine kinase domains and the transmission of the signal into the cytoplasm. Two structures have been proposed that show a complex that could achieve this (Fig. 7)—a 2 : 2 : 1 FGF1 : FGFR2 : heparin decamer structure [15] and a 2 : 2 : 2 FGF2 : FGFR1 : heparin decamer structure [57]. There are several significant differences between these complexes. They differ in the manner of the preparation of the crystallised complex: the Pellegrini complex (Fig. 7A,B) [15] was crystallised using a pre-formed

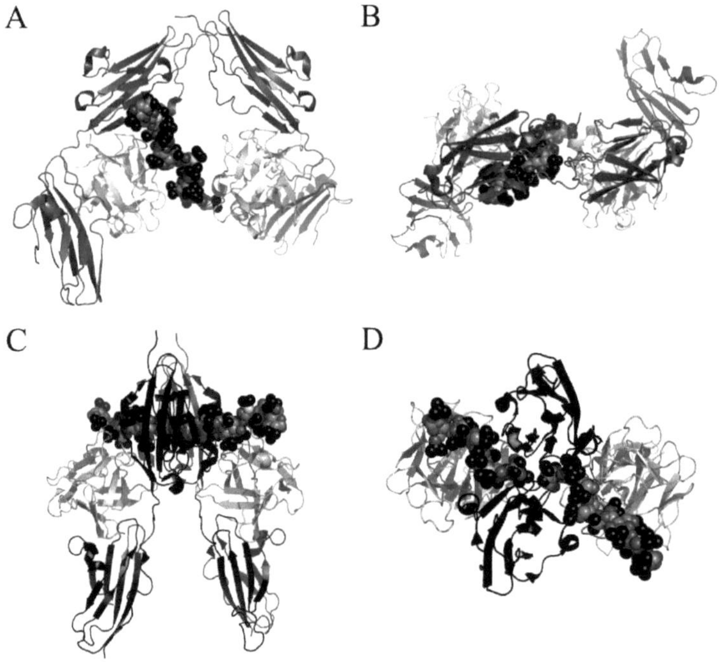

Fig. 7 FGF-FGFR-heparin ternary complexes. Structures are shown with FGFRs shown as a *black* cartoon, FGFs shown as *light gray* cartoon, heparin as all-atom structure, with sulphate and carboxyl groups in *black*, glucosamine amide nitrogens in *white*. FGFRs are shown with Ig2 at the *top*, Ig3 at the *bottom*. **A,B** Orthogonal views of the FGF1-FGFR2-heparin 10mer structure (1E0O) [15]. **C,D** Orthogonal views of the FGF2-FGFR1-heparin 10mer structure (1FQ9) [57]

complex of FGF-FGFR-heparin, while the Schlessinger–Mohammadi complex (Fig. 7C,D) [57] was prepared by first crystallising FGF-FGFR dimers [16] and then soaking heparin into the resulting crystal.

Each structure can be considered as two FGF-FGFR pairs, brought together by heparin to form a complex. The FGF-FGFR pairs are similar to those detailed above (Sect. 3.1), with the exception that the Ig domain 3 in the Pellegrini structure is in a different orientation to that observed in the other FGF-FGFR pairs. In the case of the Pellegrini structure, the two FGF molecules are dimerized upon a decamer heparin chain in a manner reminiscent to that observed by DiGabriele et al. [55]; however, in this case, the FGFs bind in the same orientation with respect to the heparin, confirming that this is likely to be the physiologically relevant orientation.

The complex observed by Schlessinger, Mohammadi and colleagues, in contrast, shows considerable protein–protein contacts between the two FGF-FGFR pairs, with each FGFR interacting with both of the two molecules from the other pair. In addition, this structure shows two heparin molecules (with eight saccharide units ordered for one molecule, and six saccharide units for the other), each making contacts primarily with one FGF-FGFR pair, but also with the other FGF unit. The orientation, conformation, and protein contacts of the heparin are consistent with those observed in both the better ordered molecule of the 2 : 1 FGF1-heparin dimer [55] and the Pellegrini complex. Thus, a unit comprising the FGF, Ig domain 2, and the heparin molecule are common to both complexes [58].

3.3 Critique of the Two Alternative Structures in the Light of Prior Evidence

The differences between these two crystal structures raise a number of important and potentially awkward questions. Foremost amongst these is how two structures with such seemingly irreconcilable differences can be determined from very similar components. Which structure represents the true, bioactive, conformation? Although the FGFs and FGFRs used for the two structures are different, they have considerable sequence similarity [10, 19], and it seems unlikely that this is the source of the discrepancy.

Analysis of previously existing data showed some support for each model. Many of the features observed in the conserved unit of the FGF, Ig domain 2 and the heparin were previously predicted: the heparin interacts primarily with a loop on the FGFR that was implicated in heparin binding and function [21], while a tripeptide whose deletion abrogated FGFR function [59] contacts both the heparin and the FGFR. The heparin binding sites are consistent with those observed in biophysical assays [60] and earlier crystal structures [55, 56]. Furthermore, the orientation of the FGF and the FGFR relative to one another are consistent with the previously observed FGF-FGFR structures [16, 51, 52]. The orientation of Ig domain 3 in the Schlessinger–

Mohammadi model showed a better fit to previous data, in that FGF residues identified as important in receptor binding by mutagenesis studies [61] interact with the receptor in this model, while they do not in the alternative orientation of this domain observed in the Pellegrini model.

Earlier biophysical analyses of the FGF-FGFR-heparin interaction had showed that FGFs and FGFRs tend to bind in a 1 : 1 stoichiometry [11, 62]; however, one of these studies had shown that a second FGFR molecule could bind to FGFs in the presence of HS, albeit at a dramatically lower affinity than the primary binding event (1.2 μM, compared to 5 nM for the primary site) [11]. This could be suggestive of the secondary FGFR binding observed in the Schlessinger–Mohammadi model, although the low affinity observed would suggest that this mechanism is unlikely to be physiologically relevant.

Finally, analysis of the heparin requirements for cells to respond to FGFs had shown that, in general, heparin fragments of eight to twelve saccharide units are required to generate similar cellular responses to HS [33]. However, a few reports had suggested some activity for very short fragments, including even disaccharides [63, 64]. These results suggest that any model must be compatible with some activity for short heparin fragments, with activity increasing with longer fragments. While the Schlessinger–Mohammadi model is consistent with activity for small fragments, this model cannot explain the greater activity of heparin dodecamers: in contrast, one can imagine a role for fragments as short even as a tetramer serving to dimerize two FGFs, as required by the Pellegrini model, while longer heparins will engage more of the FGF and FGFR, leading to greater activity for the longer fragments. Thus, the Pellegrini model appears to be more consistent with the heparin length requirements, although how disaccharides would stabilize this complex is unclear.

From these observations, it is clear that there is considerable support for many features of each model: however, for each model, there are key pieces of data that cannot be explained by that model. These structures have therefore served as a stimulus for further experiments aimed at reconciling each model with the data, refining the models, and determining the physiological relevance of each model.

4 Recent Insights into the FGF-FGFR-HS Complex

4.1 Recent FGF and FGF-FGFR Structures

Since the proposal of the two models for the FGF-FGFR-heparin ternary complex, the structures of several additional FGFs have been solved by X-ray crystallography [47–50]. These structures have confirmed that the core of the

FGFs is extremely well maintained across the entire family, and that there are no significant deviations in structure that should affect the FGF binding to the receptor. They have also given insights into novel features of the various FGF subfamilies that may influence the activity and biological roles of particular FGFs [49, 50], demonstrating how additions to a well-defined core structure can profoundly influence protein function.

Three additional FGF-FGFR dimeric complexes have been solved, which have provided significant insights into the nature of the FGF-FGFR interaction. The structure of FGF10-FGFR2b and FGF1-FGFR2b(P253R mutant) complexes [53, 54] demonstrated that the alternative splicing in the FGFR Ig domain 3 does not affect the conformation in which this domain binds FGFs: the conformation of this domain is highly similar to that found in the structure of the other FGF-FGFR pairs. The alternative splicing alters the side chains on a wide cleft in D3, leading to the formation of novel interactions with features of FGF10 that are not found in FGFs that do not bind to FGFR2b. In the FGF10-FGFR2b structure, the orientation of FGFR Ig domain 2 with respect to the rest of the structure showed a 15° rotation (Fig. 5B). This suggested that the orientation of this domain is induced by specific interactions with the different FGFs [53].

The FGF1-FGFR3c crystal structure [14], aside from confirming that FGFR3 adopts a similar conformation to the other FGFRs, powerfully demonstrated that the FGFR Ig domain 1 does not interact with the FGF. This study and others [17, 18] had shown that Ig domain 1 can be dispensed with for, and may even be inhibitory to, FGF binding. The structure was solved using a crystal in which the FGFR contained all three Ig domains: the data allowed FGFR Ig domains 2 and 3, and the FGF, to be positioned (in their expected conformations); Ig domain 1, however, could not be placed in the electron density. This strongly suggests that this domain has a highly flexible conformation, indicating that it is unlikely to be forming any meaningful interactions with the FGF. This structure highlights a strength and weakness of crystallography: regions of the protein that are not constrained, either by interactions with other molecules or crystal contacts, are often sampling a sufficiently large number of conformations that the averaged electron density is not interpretable. Although this can be highly frustrating, as no structural information can be determined for these regions even with excellent data, it nevertheless confirms that such regions are highly flexible, that they are not constrained in the biological unit, and that they are unlikely to contributing to the interactions in view.

4.2
Analysis of FGFR Mutants

A number of human developmental pathologies have been linked to the FGFR1-3 genes (reviewed in [65]). In the majority of these cases, the pathology is caused by point mutations in the FGFR that lead to (moderate) overactivation of the receptor due to either an increase in the affinity for its cognate FGF, the introduction of unpaired cysteines, or activation of the kinase domain. These mutations have provided the possibility to test aspects of the models for the ternary FGF-FGFR complex by crystallising complexes involving mutant FGFRs, and assessing the relevance of the interactions formed for the two models.

Amongst the most severe developmental effects associated with FGFRs are observed in Apert's syndrome [66]. This syndrome results from the overactivation of FGFR2, and is associated with two point mutations in the linker between FGFR2 Ig domains 2 and 3, S252W and P253R. Crystal structures of FGF2 in complex with the two Apert mutant FGFR2c molecules [67], and of FGF1 and FGF10 in complex with FGFR2b Apert mutant molecules [54], have provided a molecular basis for these mutations. These structures demonstrated that the mutant FGFRs provide additional contacts with FGFs (Fig. 8), explaining the additional affinity for FGFs. Biophysical analysis of the Apert mutant FGFR2 [54, 68] has confirmed that Apert FGFR2 shows an increased affinity for a range of FGFs, including in some reports an increase in affinity for FGFs that are not ligands for the wild-type protein [54]. Analysis of the equivalent mutations to P253R in FGFRs 1 and 3, which cause similar but less debilitating pathologies, confirmed that these form a similar interaction with the FGF [69]. These structures again provide considerable evidence that the conformation of the FGFR Ig domain 3 in the Schlessinger–Mohammadi model is correct, as the interactions observed in the structures can be explained by this conformation. The association of the structural, biophysical and clinical data provide strong evidence for this conformation being correct.

Further support for the physiological relevance of the Schlessinger–Mohammadi model of the ternary complex has been provided by analysis of an unusual FGFR2 mutant that causes Pfeiffer syndrome, a cranial developmental pathology [70]. One case of this disease has been mapped to an A172F mutant of FGFR2, a site in Ig domain 2 [71]. This site forms part of the FGFR-FGFR interface that is created between the two FGF-FGFR pairs in the Schlessinger–Mohammadi model. Crystallisation of a complex containing this mutation showed that the mutant FGFR is still capable of forming the 2 : 2 FGF : FGFR complex that forms the base of this model [72]. Indeed, the two introduced phenylalanine residues form a novel interaction in the complex, with the side chain rings stacking against one another. These data support the role of the secondary interactions formed between the FGFRs in bringing together FGF-FGFR pairs to form a ternary complex, as the additional interaction observed in the crystal structure correlates with a clinical phenotype.

The conclusions from these studies of mutations observed in human pathologies are that the orientation for Ig domain 3 proposed in the

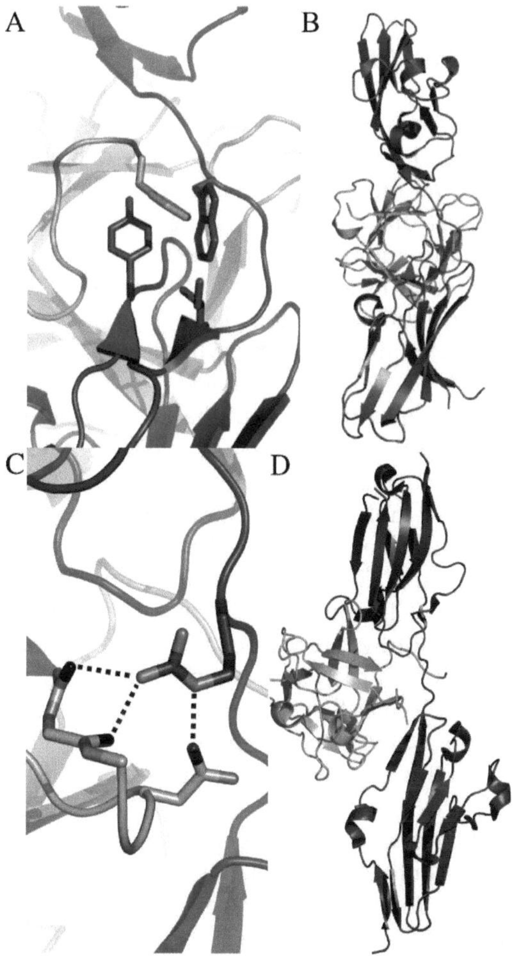

Fig. 8 Mutations that lead to Apert syndrome increase the affinity of FGFRs for FGFs. FGF2 shown as *light gray* cartoon, FGFR2 shown as *black* cartoon. Key side chains shown as sticks: FGFR2 nitrogen atoms are shown in *light gray*, FGF2 nitrogen atoms shown in *darker gray*, FGF2 oxygen atoms in *black*. Hydrogen bonds are shown as *dashed lines*. **A** FGFR2 S252W mutation creates a hydrophobic pocket (with FGFR2 I257 and Y281) that binds to FGF2 F21, providing an additional interaction. **B** View of the FGF-FGFR pair in the same orientation. **C** P253R mutation provides three additional hydrogen bonds with FGF2, with the carbonyl oxygens of L107 and E108, and the side chain of N111. **D** View of the FGF-FGFR pair in the same orientation

Schlessinger–Mohammadi complex is highly likely to be physiologically relevant, and that the secondary interactions formed between the FGFRs in this model seem to contribute to the formation of complexes that signal across the cell membrane.

4.3 Biophysical Analyses of the Alternative Models

The elucidation of the two models for the FGF-FGFR-heparin ternary complex has allowed researchers to apply a wider range of biophysical techniques to these molecules, as far more of the parameters are known and the discriminating factors between the complexes have been identified. Two key areas have been examined, in addition to those mentioned above: the unique interactions between the FGFs and FGFRs in the two models, and the role of heparin/HS in the formation of the complex.

Two sets of FGF-FGFR interactions have been probed in a series of elegant mutation studies [72]. These authors mutated two side chains in FGF10, one of which makes significant contacts with the FGFR Ig domain 3 only in the Pellegrini complex, and the other of which makes significant contacts with this domain only in the Schlessinger–Mohammadi complex. The former mutation resulted in FGF10 molecules that showed an apparently similar affinity for FGFR2 in vitro, while the other showed a significant reduction in affinity (~ 4 times increase in k_D). These data gave further weight to the supposition that the conformation of Ig domain 3 observed in the crystals by Pellegrini et al. [15] was driven not by interactions with the FGF, but by the requirements of crystal packing, in particular a very strong interaction between two FGFR2 Ig domain 3 molecules that was observed in this crystal. Furthermore, the orientation of the FGFR2 was stabilized by nickel ions observed in the crystal structure, and this may also have contributed to the favorability of this conformation [38].

In addition, Mohammadi and colleagues [72] examined mutations of two side chains that contribute to the formation of the secondary FGF-FGFR interaction site in the FGF10-FGFR2b structure. They demonstrated that the mutation of these residues does not affect the affinity of FGF10 for FGFR2b in the binary interaction, but that the mitogenic capacity of the FGFs was diminished nevertheless. This correlated with a reduction in the capacity of the FGF10 to form 2 : 2 complexes with FGFR2b in the presence of heparin in a semi-quantitative mass spectrometry experiment. The conclusion of these data is that the mechanism of multimerization of the FGFR proposed by Schlessinger, Mohammadi and colleagues [57] is likely to be relevant to signaling on the cell surface.

A key question has been whether the different crystal structures arise from a fundamental difference in the protein components. The discovery that complexes with in vitro properties consistent with the two proposed complexes could be prepared from identical protein components (as assessed by analytical ultracentrifugation and nanoflow mass spectrometry) [73] has confirmed that the difference is due to the manner of preparation of the complex, not the protein. Analytical ultracentrifugation of the two complexes showed that the Pellegrini complex dissociates most favorably with the loss of one FGFR subunit (as suggested by the presence of one FGFR that does not bind to heparin), while the Schlessinger–Mohammadi complex tends to dissociate into the two FGF-FGFR pairs (Fig. 9). These observations were confirmed by mass spectrometry of the two complexes, which again demonstrated significant differences in their dissociation. These observations crucially show that the both of the two complexes are stable in solution, and that neither are likely to be artifacts of the crystallisation process. Furthermore, this rules out the possibility that the architecture observed by Pellegrini and colleagues might be the result of FGFR misfolding (as had been previously suggested [67]), as this would considerably prejudice the formation of the Schlessinger–Mohammadi style complex. This also raises the intriguing problem of how the proteins might choose between the two possible complexes upon the cell surface.

In addition to confirming that both complexes could be formed from the same protein components, this study examined the stoichiometry with respect to heparin in the complexes in solution. This is a major difference between the two models, with the Pellegrini complex suggesting that one heparin molecule will bind to two molecules each of FGF and FGFR, while one molecule of heparin is required per FGF and FGFR in the Schlessinger–Mohammadi model. Mass spectrometry showed conclusively that the complexes prepared using either methodology contain one molecule only of heparin in the gas phase (Fig. 10). A small proportion of molecules from

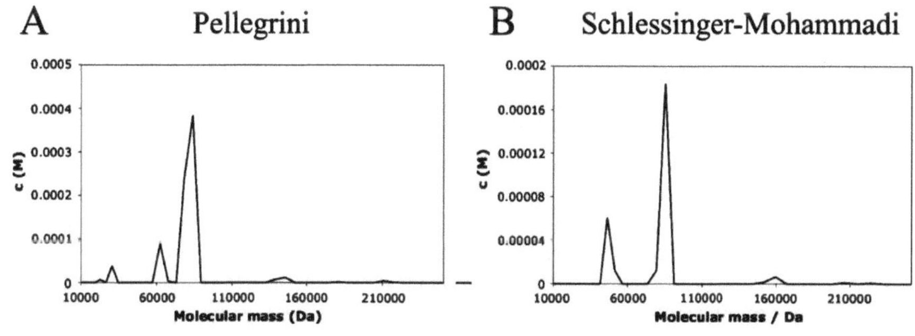

Fig. 9 Analytical ultracentrifugation of the Pellegrini and Schlessinger–Mohammadi complexes. Samples of FGF-FGFR-heparin complexes prepared according to Pellegrini and colleagues [15] (*left*) or Schlessinger and colleagues [57] (*right*) were used for analytical ultracentrifugation. Calculated distributions of solution species are shown. The Pellegrini complex shows a small level of dissociation into a 60 kDa (2 FGF1 : 1 FGFR2 : heparin) complex, and a 25 kDa species (FGFR2), as predicted from the uneven heparin binding of the FGFR2 species in this complex (Fig. 7). The Schlessinger complex shows dissociation into a 45 kDa complex (FGF-FGFR dimer, possibly also bound to heparin), consistent with a more symmetrical complex (Fig. 7). Reprinted from the Journal of Molecular Biology, Vol. 339, Harmer et al., Towards a Resolution of the Stoichiometry of the Fibroblast Growth Factor (FGF)–FGF Receptor–Heparin Complex, p 821–834, © 2004, with permission from Elsevier

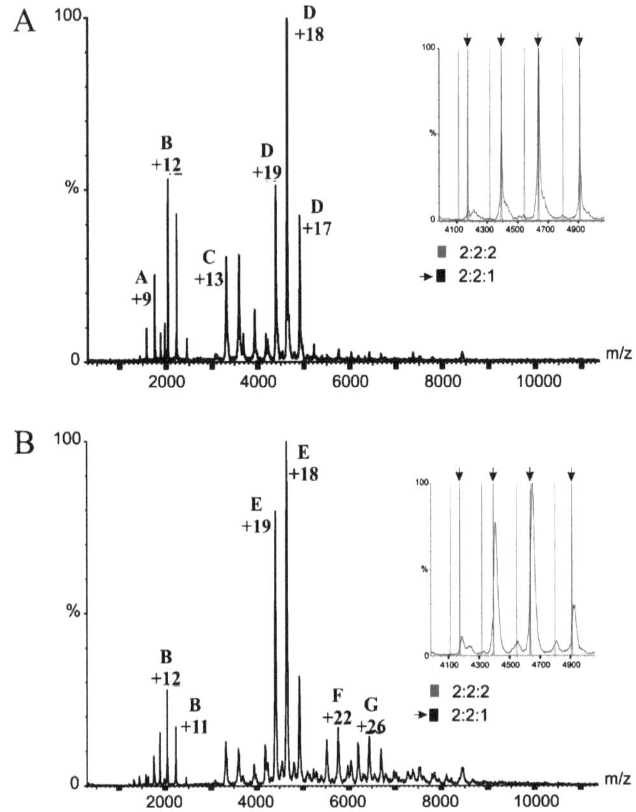

Fig. 10 Mass spectrometry of the Pellegrini and Schlessinger–Mohammadi complexes. **A** Mass spectrometry of complexes prepared according to Pellegrini and colleagues [15] shows FGF1 (**A**), FGFR2 (**B**), 1 : 1 : 1 FGF : FGFR : heparin complexes (**C**), and a 2 : 2 : 1 FGF1 : FGFR2 : heparin complex (**D**). *Inset*: comparison of experimental data with the calculated position of peaks for a 2 : 2 : 1 complex (*dark line*) or a 2 : 2 : 2 complex (*light line*) shows that the data is inconsistent with a 2 : 2 : 2 complex. **B** Mass spectrometry of complexes prepared according to Schlessinger and colleagues [57] shows FGFR2 (**B**), a 2 : 2 : 1 FGF1 : FGFR2 : heparin complex (**E**), and two larger species (**F**,**G**). *Inset*: comparison of experimental data with the calculated positions of peaks for a 2 : 2 : 1 complex (*arrowed dark line*) or a 2 : 2 : 2 complex (*light line*) shows that the data is most consistent with a 2 : 2 : 1 complex, although a small amount of 2 : 2 : 2 complex cannot be excluded. Reprinted from the Journal of Molecular Biology, Vol. 339, Harmer et al., Towards a Resolution of the Stoichiometry of the Fibroblast Growth Factor (FGF)–FGF Receptor–Heparin Complex, p 821–834, © 2004, with permission from Elsevier

the Schlessinger–Mohammadi sample showed two molecules of heparin per complex; however, this represented a small minority of the species. This result suggests that the stoichiometry suggested by crystallography in the Pellegrini complex, rather than the Schlessinger–Mohammadi complex, is correct.

Given that the biophysical data suggests that there is only one molecule of heparin in the solution complex consistent with the Schlessinger–Mohammadi model, it is necessary to consider why the crystal shows two molecules. The presence of two molecules in the structure presented by Schlessinger and colleagues might have been the result of a single heparin molecule soaking into each complex, creating disorder about the pseudo-symmetric axis of the complex that would be difficult to distinguish from two molecules binding per complex; or from two molecules of heparin binding per complex due to the extremely high protein concentration in the crystal, and the intrinsic low affinity for a second heparin suggested by the mass spectrometry. Both of these possibilities demonstrate difficulties that can be encountered in determining bioactive conformations by X-ray crystallography. Soaking experiments, which are an extremely powerful method for rapidly determining the structures of complexes, bring the risk of obtaining less favorable interactions due to the conformation of the protein that the crystal has fixed, or the concentration of the protein in the crystal. Nevertheless, the Schlessinger–Mohammadi method demonstrates also the power of the method, as no structure containing heparin could be obtained without soaking, and the structure, even with these caveats, is of immense value.

The stoichiometry that was obtained by mass spectrometry was more surprising, as a previous study of these complexes, using a series of careful native gel experiments, had suggested a 2 : 2 : 2 FGF : FGFR : heparin stoichiometry [74]. However, it is not clear that the complexes observed were active 2 : 2 : 2 complexes. FGF : FGFR : heparin complexes were observed with a number of heparin fragments that had previously been shown to be ineffective at supporting FGF signaling on cells. Furthermore, these experiments suggested that heparins sulphated at only the *N*- and 3-*O*-positions were capable of binding to FGFs and FGFRs, while the 2-*O*- and 6-*O*-positions have been shown to be important for FGF1 binding by a number of studies [75–77]. It therefore seems that this study might have been detecting 1 : 1 : 1 FGF : FGFR : heparin complexes, which have been observed by others with heparins that are not capable of activating cells [78].

The conclusion of these biophysical studies appears to be that the conformation of Ig domain 3 observed by Schlessinger, Mohammadi and colleagues is most likely to be the physiological conformation; that the Schlessinger–Mohammadi complex appears to have relevance for physiological signaling; that both complexes can be prepared from the same protein and saccharide samples; and that the stoichiometry with respect to heparin is most likely to be 2 : 2 : 1 FGF : FGFR : heparin. It is thus clear that while both models appear to have relevance to the conditions at the cell surface, there are also elements of each model that are inconsistent with the most reliable data.

5 Recent Advances in Other Areas of the FGF-FGFR-HS Model

5.1 Development of Methods for Assessing the Specificity of Heparin and HS

A major limitation in our understanding of the biology of the FGF-FGFR system (and indeed of many heparin-binding growth factors) has been the difficulty of working with HS. The overwhelming majority of studies have used heparin or depolymerised heparin fragments. These allow the use of well-defined fractions whose effects can be well understood, and a better comparison between different batches and studies. However, the polysaccharides that FGFs and FGFRs will encounter in the extracellular matrix and on the cell surface are considerably different to these species. Methodological and technological advances in the past few years have allowed a deeper study of the heparin/HS binding of FGFs and FGFRs, and have helped to advance our understanding of how FGFs and FGFRs might bind to HS at the cell surface.

One major advance has been the capacity to perform sequencing of HS derived oligosaccharides, using the small quantities of purified HS fragment that can typically be prepared experimentally [75, 79, 80]. Although these methods cannot always unambiguously determine the exact location of every sulfate group in every sample, the structure of individual HS or heparin fragments can be resolved to a very high level. Using these methods, the sequence specificity of a number of FGFs have been determined, by immobilizing FGFs, adding a range of HS fragments to the immobilized FGFs, and eluting fragments at a variety of sodium chloride concentrations. These experiments have suggested that FGFs bind to broadly similar sulfation patterns, but that there are differences between the exact sequences that are preferred by each FGF, and the overall affinity of each FGF for heparin [75, 81].

Similar methods have been used to elucidate the patterning of HS ends [82]. These studies have shown that the ends in FGF responsive cells are relatively homogeneous, that they are relatively highly sulfated (except at the terminal position), and that HS ends are not significantly degraded on the cell surface in vivo. This confirmed that the ends of HS chains might be competent to form signaling complexes, although the homogeneity of the structure, and the absence of modification at the terminal position, might prejudice against the very end of the chain being used for signaling.

A further advance has been the improvement in the technology for synthesizing heparins with specific sulfation patterns. Although it is not yet possible to produce large quantities of heparin with complex patterns, the removal of one type of sulfate group can be routinely achieved, using chemical desulfation [76], purified sulfotransferase enzymes [77], and a recently characterized 6-*O*-endosulfatase enzyme. The heparins thus prepared can then be assayed for FGF and FGFR binding activity in vitro or in vivo. The results of such experiments have again suggested that the FGFs have differences in their HS preferences, but these suggest a broader difference than has been suggested by the sequencing of FGF binding fragments of HS [76, 77, 83]. The causes of this difference are not at present clear. Furthermore, experiments using heparins with defined sulfation patterns in vivo have again shown that FGF-FGFR pairs have distinct HS preferences; and that the combination of FGF and FGFR is sufficiently important that the FGFR can overcome a reluctance of the FGF to bind to a heparin fragment in vitro [76].

The conclusion from these experiments is that, as had been predicted by the results of HS extracts from a variety of tissues, the sulfate sequence of the HS or heparin is important for FGF binding and the formation of active signaling complexes. It is therefore likely that different HS sequences will affect the tendency of FGFs and FGFRs to form complexes that can transmit signals, and so these must be borne in mind when considering the nature of the true signaling complex.

5.2 Possible Mechanisms for Multimerizing Complexes of FGFRs

Mass spectrometry of the FGF-FGFR-heparin complexes showed small amounts of complexes involving more than two FGFR units [73]. Although it cannot be ruled out that this is an artifact of the gas phase, it hints tantalizingly at a higher level of complexing of FGFs, FGFRs and HS. Such higher order complexes are expected if the FGFRs, like many other receptors including EGFR, are to form large clusters on the cell surface that will lead to "signalsomes" upon internalization. Mechanisms by which the FGFRs may multimerize have been suggested by recent observations.

Firstly, multiple FGF-FGFR complexes can form upon a single HS or heparin chain [84]. Increases in the length of heparin chains from 12 saccharide units to 24 units leads to four molecules of each of FGF1 and FGFR2 assembling on the chain, in a manner suggestive of two complexes after the Pellegrini model forming upon one saccharide chain. Preliminary experiments with full-length HS chains suggest that several complexes could form per chain. This suggests that one mechanism for multimerization would be the formation of multiple complexes upon a single HS chain, which would poise the intracellular domains for forming further interactions. This also strongly suggests that the non-reducing ends of heparin are not a requirement for the formation of FGF-FGFR-heparin complexes (as has been championed by Mohammadi and colleagues [38]): the complexes here form with two copies of the 2 : 2 FGF-FGFR complex, while only one non-reducing end is available. This implies that any model for FGFR dimerization should allow for the formation of multiple complexes upon one saccharide.

It has also become apparent that the two proposed models for FGFR dimerization are not mutually exclusive, and that they can be accommodated together. Indeed, the recent results that have tested aspects of the two models have tended to cast more doubt on the elements of the models that

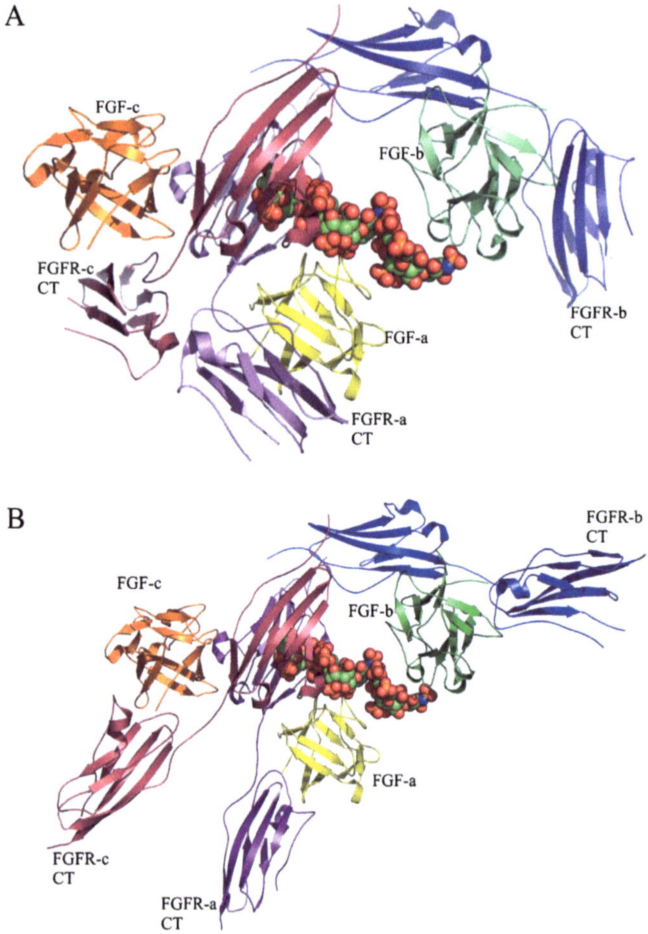

Fig. 11 **A** Schlessinger–Mohammadi like complex is observed in the crystal packing of the Pellegrini complex. FGF1 and FGFR2 are shown as cartoons, heparin as an all-atom model (carbon: *green*, oxygen: *red*, nitrogen: *blue*, sulfur: *orange*). The reported Pellegrini complex is formed by molecules FGF-a (*yellow*), FGF-b (*green*), FGFR-a (*violet*), FGFR-b (*blue*), while the Schlessinger–Mohammadi like complex in the crystal packing is formed by FGF-a, FGFR-a, FGF-c (*orange*) and FGFR-c (*raspberry*). Note that the crystal packing does not result in a second heparin molecule in the Schlessinger–Mohammadi like structure: there is no hint of a "two-ends" structure. **B** Placing the FGFR2 Ig3 in the orientation found in other FGF-FGFR structures highlights the Schlessinger–Mohammadi like complex

are less consistent with the co-existence of the complexes than those that are more consistent with co-existence. In the crystal structure solved by Pellegrini and colleagues [15], crystal packing contacts showed two copies of the Pellegrini complex packed in a manner that strongly resembles the Schlessinger–Mohammadi model (Fig. 11) [58]. In the crystal structure, the Schlessinger–Mohammadi like complex appears a rather unlikely structure, as the conformation of Ig domain 3 leaves the two C-termini oriented in a manner that is not consistent with a close contact with the membrane. However, upon superposition of the FGF-FGFR structure determined by Mohammadi and colleagues onto this structure, the conformation of these domains appears quite consistent with both complexes existing upon the cell membrane (Fig. 11B). It is also likely that the complexes, whose conformations have been selected for optimal crystal packing, would be remolded somewhat to provide a more energetically favorable complex in the context of membrane bound proteins. This is particularly likely in the case of the HS molecule, which is likely to have the capacity to sample more conformations than a heparin molecule, and which will have more flexibility to alter its conformation than the binary FGF-FGFR interactions. The true cellular conformations will have to be determined in the future by examining the complexes that form upon live cells.

6 Towards a Model for an FGFR "Signalosome" Initiating Complex

6.1 FGF-FGFR-HS Interactions in the Physiological State

The formation of complexes between FGFs, FGFRs and HS on the cell surface is likely to show a number of differences from the experimental conditions that have been generally used for in vitro experiments on these molecules. These differences will be briefly explored before an attempt is made to describe how an FGFR "signalsome" might be generated in vivo.

A first difference, as mentioned above, is that the HS complement of cells is quite different from heparin. In addition to the lower overall level of sulfation and uronic acid epimerization, the organization of HS into short highly sulfated S-domains interspersed with stretches of saccharide with a low level of sulfation [85] gives HS a far greater degree of conformational flexibility, as the linkages of poorly sulphated regions have far greater rotational freedom [39]. This will have considerable implications for the manner in which HS interacts with FGFs and FGFRs. Furthermore, the levels of the various biosynthetic enzymes are clearly regulated by cells to generate HS with distinct properties [86–89], possibly to take advantage of the subtle differences in affinity that the FGFs have for different HS structural patterns.

A further difference is that the HS on the cell surface is linked to the membrane by the HS carrying proteins, the syndecans and glypicans [28] (Fig. 2). These carry the cellular HS, and so will have an impact in determining the location of active HS chains. In particular, the glypican HS attachment sites are proximal to the membrane, while the syndecan attachment sites are distal to the membrane: this might lead to a series of FGF/FGFR binding events with HS, as the molecules bind to and are released by HS, searching for one another to form the start of a complex.

On the cell surface, HS will be in vast excess, due to the number of HS molecules that are found on an average cell in comparison to the number of FGFRs on the cell, and the number of FGF molecules required for activation [90]. This is consistent with the large number of signaling molecules that have affinity for and whose function requires HS [91]. One effect of this is that FGFs will be largely sequestered upon meeting a cell, as the concentration of HS will be sufficient to ensure that the FGF is likely to continue to bind further HS whenever it releases from one molecule. This will tend to increase the effective concentration of FGF on the cell surface as the FGFs are restricted to two dimensions. Furthermore, the FGFRs will tend to be associated with HS constitutively, due to the concentrations of the two molecules. One key consequence of the high concentration of HS will be that the favorability of HS-linked dimers of FGFs will be reduced by the vast excess of FGF binding sites. Although the 2 : 1 FGF : heparin and 2 : 2 : 1 FGF : FGFR : heparin complexes are stable in the presence of considerable excess of heparin [78, 84], it is not yet clear that these complexes are sufficiently stable for this excess of HS.

A key aspect of the membrane environment that has become clearer in the past few years is the effect of membrane microdomains [92, 93]. Some of these microdomains, such as caveolae, are highly enriched in signaling molecules, and receptors translocate to these locations upon activation [94]. This serves to increase the local concentrations of all the molecules involved in the interaction, increasing the likelihood of lower affinity, transient interactions becoming favorable. FGFRs [95], FRS-2 [96], glypicans and syndecans [97] have all been associated with localization to membrane microdomains in the basal states, and so it is likely that FGF signaling (which has been suggested to localize to rafts [98]) will be focused upon a number of local foci on the cell surface.

A final variable that must be borne in mind is the presence of other molecules that interact with FGFs and FGFRs. A number of other cell surface receptors and signaling molecules have been shown to interact with FGFRs and to have effects upon signaling via these receptors. Examples include N-CAMs, which several studies have implicated as FGFR co-receptors in certain cell types [99, 100], cadherins [101, 102], the tyrosine kinase FLRT3 [103], and integrins [104]. Although some of these interactions may be of low affinity in vitro, the effects of reduced dimensionality on the membrane, co-localization to microdomains, and bridging interactions with intracellular proteins may be sufficient to drive these molecules to participate in the clustering of FGFRs and FGFs.

6.2 How Might Signaling Clusters of FGFRs Form?

From all of these observations, one can speculate as to the steps that might lead to the formation of an FGFR signalsome (Fig. 12). The state of the cell in the absence of FGF signaling (although mammalian cells are unlikely ever to be totally devoid of some FGF signal) will have a pre-determined level of one or more FGFR isoforms, which will be distributed in the membrane between the various lipid microdomains and the bulk membrane. All of these are likely to be bound to HS S-domains, periodically releasing one HS chain and binding another. A significant proportion, if not all, of the FGFRs are also likely to be constitutively dimerized in an inactive, basal form.

On the production of FGF by a cell that will send a signal (perhaps in response to a change in differentiation state, or injury), the FGFs will form a gradient from the signaling cell. On encountering the target cell, the FGFs will rapidly form complexes with HS due to the high concentration of HS and the rapid speed of this interaction. This will have the effect of concentrating the FGF on the cell surface. Following these events, the FGFs will migrate to other HS molecules, binding to the most favorable sites for the FGF in question. At this stage, HS-linked dimers of the FGF are likely to form, driving the FGFs towards S-domains. The FGF-HS complexes will then form high affinity sites for FGFRs to bind, forming a complex after the Pellegrini model. The FGFRs may come from separate preformed dimers, or may be derived from a single preformed dimer dissociating to bind the two FGFs. In either event, the binding of an FGFR to the FGFs will create a potential binding site for another FGF-FGFR-HS complex to join, forming a Schlessinger–Mohammadi interaction, to nucleate a cluster of FGFRs. The local activation of the FGFRs appears to encourage the receptors to translocate to membrane microdomains (if they are not there already), as stimulation with FGF alters the localization of many types of receptors to this location. In this environment, with a higher concentration of FGF and FGFR, multiple complexes may form upon single HS chains, helping to drive the formation of a larger cluster. Activation of the tyrosine kinase domain will lead to the recruitment of signaling molecules to the associated FRS-2 adaptor, creating a multiprotein complex around the activated FGFR on the inner face of the membrane. Further signaling proteins that bind FGFRs may be recruited to join the complex and participate in signaling. Ultimately, the FGF-FGFR-HS complex will be internalized, with the FGFRs continuing to signal until they reach a late stage of the endosomal pathway and are degraded. Many of these effects are likely to be conditional upon the state of the cell when the FGF reaches it (as this

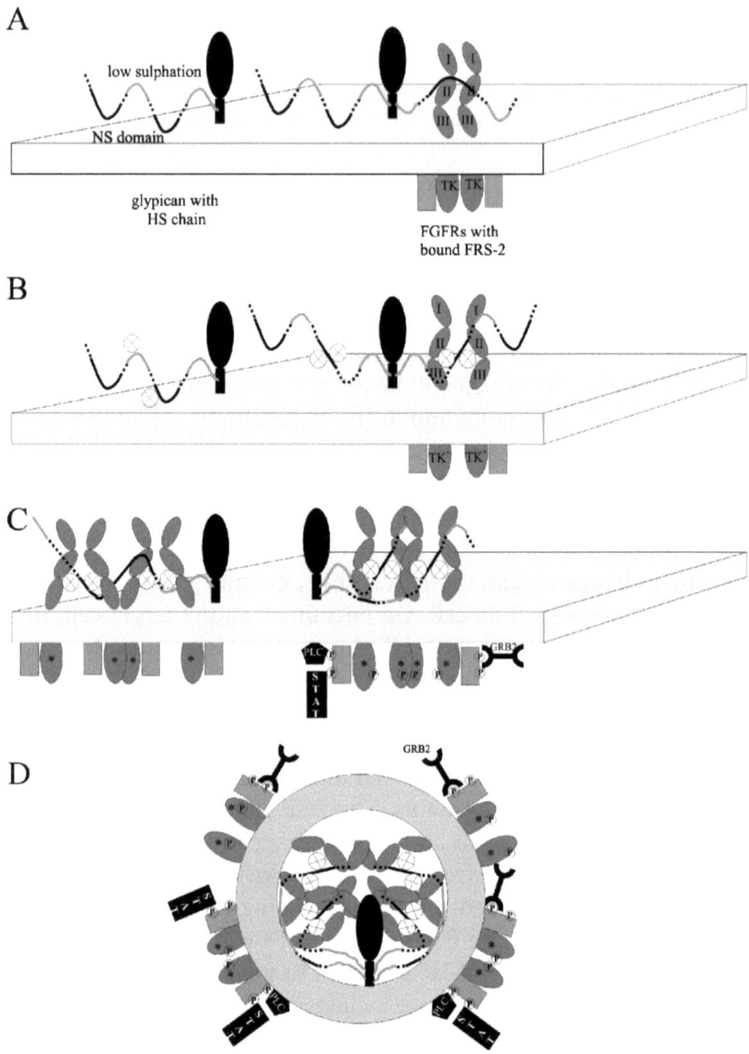

Fig. 12 Model for the formation of a signaling cluster. **A** Cell basal state. FGFRs are likely to be bound to HS, and may be dimerized in an inactive form in their intracellular domains. **B** FGFs exposed to the cell will rapidly bind to HS (*left*). Rapid binding and release of FGFs will allow them to find optimal sites, and form heparin-linked dimers (*center*). These will form optimal binding sites for FGFRs, and will drive the activation of the FGFRs (*right*: here, FGFRs are shown as dissociating from intracellular dimers). **C** The initial complexes will act as nucleating points for the formation of larger complexes, as more protein is driven into membrane microdomains. Routes to the formation of larger complexes that have been suggested experimentally include the formation of multiple complexes on single HS chains (*left*), or the formation of Schlessinger–Mohammadi like and Pellegrini like complexes linked together (*right*). **D** With high stimulation, the FGF-FGFR-HS complexes will be internalized into endosomes, from where they will continue signaling until the late endosomal stages

will affect the density of FGFR and other signaling molecules, and the level of crosstalk between signaling systems that affects the FGFR downstream pathway), and upon the dose of FGF. When smaller doses of FGF are present, the more transient signals such as those observed in the absence of HS, or with sub-optimal heparin replacement, may occur: at some stage in this process, the activation of the FGFR will fail to be sufficiently strong to drive the formation of a self-sustaining signaling cluster, and the FGFR will be degraded or dephosphorylated before a cluster can form. Conversely, in cells that are chronically stimulated by FGFs, new clusters will form as the old ones are endocytosed, ensuring that the signal continues and that the target cell is maintained in its correct state of differentiation, proliferation, or whatever other cellular behavior is appropriate.

7
Conclusions and Future Perspective

Since 2000, when the first experimental models for the active complex of FGFs, FGFRs and heparin were produced, a large number of insights have been made into the nature of the FGFR complexes that form in solution and in vivo. These have revealed the features of each model that are less tenable, and which may be consequences of the techniques and conditions that were used to prepare the crystals from which the structures were determined; and the features of each model that are consistent with the solution properties and in vivo activities of FGFs, and FGF-FGFR complexes. Revised views of each model, with the less reliable features altered to resemble the equivalent features of the other model (Figs. 11B, 13) show that the two models are indeed compatible with co-existence. These studies have shown both the power and the potential pitfalls of X-ray crystallography. Using this technique, the structures of large complexes of biological molecules can be accurately determined. In addition, once crystals have been obtained, ligands, cofactors, or potential inhibitors can be soaked into the crystals and often the resulting structural solution is facile. However, the requirements of crystal packing and the extremely high concentration of a single protein in the crystal can give rise to conformations that, in solution, are not greatly sampled, but which are selected by the experimental conditions. Furthermore, the high concentrations may give rise to interactions that are less favorable—for good or ill.

Although a great deal of progress has been made on the biophysical characterization of the FGF-FGFR-heparin complexes, there is still much to be done before the true bioactive conformation can be confidently modeled. Key areas include a movement away from heparin towards the natural co-receptor HS: this will require sensitive experimentation to determine the interactions of the more heterogeneous saccharides of HS. Further, experiments that de-

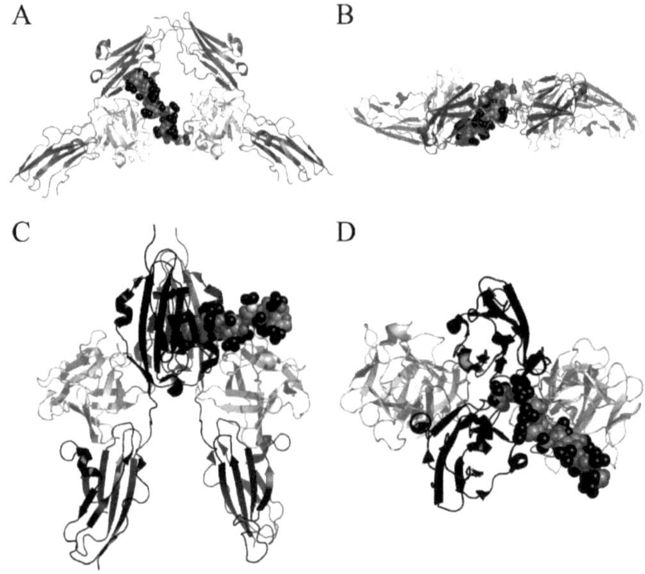

Fig. 13 Re-interpretation of the FGF-FGFR-heparin ternary complexes in the light of recent evidence. Structures are shown with FGFRs shown as a *black* cartoon, FGFs shown as *light gray* cartoon, heparin as all-atom structure, with sulphate and carboxyl groups in *black*, glucosamine amide nitrogens in *white*. FGFRs are shown with Ig2 at the *top*, Ig3 at the *bottom*. **A**,**B** Orthogonal views of the FGF1-FGFR2-heparin 10mer structure (1E0O) [15], with the orientation of Ig3 altered to conform to the other FGF-FGFR structures. Note how the new structure shows a greater splay of the FGFR Ig3 domains (**A**), and that it is flatter than the originally presented structure (**B**; compare Fig. 7). **C**,**D** Orthogonal views of the FGF2-FGFR1-heparin 10mer structure (1FQ9) [57], with the less well-defined heparin molecule removed

termine the effects on the interactions of tethering FGFRs to a membrane in which they are free to migrate will undoubtedly help to interpret much of the data. Finally, indications of how FGFRs multimerize to form the large clusters that generate the highest levels of signaling will add further dimensions to our understanding of the active conformations of the initial ternary FGF-FGFR-HS complexes.

Acknowledgements The molecular images in this article were prepared using the program PyMOL [107].

References

1. Powers C, McLeskey S, Wellstein A (2000) Endocrine Rel Cancer 7:165
2. Prudovsky I, Mandinova A, Soldi R, Bagala C, Graziana I, Landriscana M, Tarantini F, Duarte M, Bellum S, Doherty H, Maciag T (2003) J Cell Sci 116:4871

3. Flaumenhaft R, Rifkin D (1992) Mol Biol Cell 3:1057
4. Tassi E, Al-Attar A, Aigner A, Swift M, McDonnell K, Karavanov A, Wellstein A (2001) J Biol Chem 276:40247
5. Beer H-D, Bittner M, Niklaus G, Munding C, Max N, Goppelt A, Werner S (2005) Oncogene
6. Gallagher J (2001) J Clinical Invest 108:357
7. Eswanrakumar V, Lax I, Schlessinger J (2005) Cytokine Growth Factor Rev 16:139
8. Malecki J, Wiedlocha A, Wesche J, Olsnes S (2002) EMBO J 21:4480
9. Soulet F, Bailly K, Roga S, Lavigne A-C, Amalric F, Bouche G (2005) J Biol Chem 280:25604
10. Itoh N, Ornitz D (2004) Trends Genet 20:563
11. Pantoliano M, Horlick R, Springer B, Dyk DV, Tobery T, Wetmore D, Lear J, Nahapetian A, Bradley J, Sisk W (1994) Biochemistry 33:10229
12. Ibrahimi O, Zhang F, Hrstka S, Mohammadi M, Linhardt R (2004) Biochemistry 43:4724
13. Burke D, Wilkes D, Blundell T, Malcolm S (1998) Trends Biochem Sci 23:59
14. Olsen S, Ibrahimi O, Raucci A, Zhang F, Eliseenikova A, Yayon A, Basilico C, Linhardt R, Schlessinger J, Mohammadi M (2004) Proc Natl Acad Sci USA 101:935
15. Pellegrini L, Burke D, Delft Fv, Mulloy B, Blundell T (2000) Nature 407:1029
16. Plotnikov A, Schlessinger J, Hubbard S, Mohammadi M (1999) Cell 98:641
17. Wang F, Kan M, Xu J, Yan G, McKeehan W (1995) J Biol Chem 270:10222
18. Xu J, Nakahara M, Crabb J, Shi E, Matuo Y, Fraser M, Kan M, Hou J, McKeehan W (1992) J Biol Chem 267:17792
19. Coulier F, Pontarotti P, Roubin R, Hartung H, Goldfarb M, Birnbaum D (1997) J Mol Evol 44:43
20. Ornitz D, Xu J, Colvin J, McEwan D, MacArthur C, Coulier F, Gao G, Goldfarb M (1996) J Biol Chem 271:15292
21. Kan M, Wang F, Xu J, Crabb J, Hou J, McKeehan W (1993) Science 259:1918
22. Schlessinger J (2000) Cell 103:211
23. Franco R, Canals M, Marcellino D, Ferre S, Agnati L, Mallol J, Casado V, Ciruela F, Fuxe K, Lluis C, Canela E (2003) Trends Biochem Sci 28:238
24. Gillham H, Golding M, Pepperkok R, Gullick W (1999) J Cell Biol 146:869
25. Gulbins E, Grassme H (2002) Biochimica et Biophysica Acta 1585:139
26. Sacchettini J, Baum L, Brewer C (2001) Biochemistry 40:3009
27. Runyan C, Schnaper H, Poncelet A-C (2005) J Biol Chem 280:8300
28. Forsberg E, Kjellen L (2001) J Clin Invest 108:175
29. Iozzo R (2001) J Clin Invest 108:165
30. Sugahara K, Kitagawa H (2002) IUBMB Life 54:163
31. Varki A, Cummings R, Esko J, Freeze H, Hart G, Marth J (1999) Essentials in Glycobiology. Cold Spring Harbor Laboratory Press, New York
32. Murphy K, Merry C, Lyon M, Roberts I, Thompson J, Gallagher J (2004) J Biol Chem 279:27239
33. Ornitz D, Yayon A, Flanagan J, Svahn C, Levi E, Leder P (1992) Mol Cell Biol 12:240
34. Rapraeger A, Krufka A, Olwin B (1991) Science 252:1705
35. Walker A, Turnbull J, Gallagher J (1994) J Biol Chem 269:931
36. Mikhailov D, Linhardt R, Mayo K (1997) Biochem J 328:51
37. Mikhailov D, Mayo K, Vlahov I, Toida T, Pervin A, Linhardt R (1996) Biochem J 318:93
38. Mohammadi M, Olsen S, Ibrahimi O (2005) Cytokine Growth Factor Rev 16:107

39. Mulloy B, Forster M (2000) Glycobiology 10:1147
40. Mulloy B, Forster M, Jones C, Davies D (1993) Biochem J 293:849
41. Mulloy B, Forster M, Jones C, Drake A, Johnson E, Davies D (1994) Carbohydrate Res 255:1
42. Dowd M, French A, Reilly P (1992) Carbohydrate Res 233:15
43. Dowd M, Zeng J, French A, Reilly P (1992) Carbohydrate Res 230:223
44. Stevens E, Sathyanarayama B (1989) J Am Chem Soc 111:4149
45. Zhu X, Komiya H, Chirino A, Faham S, Fox G, Arakawa T, Hsu B, Rees D (1991) Science 251:90
46. Eriksson A, Cousens L, Weaver L, Matthews B (1991) Proc Natl Acad Sci USA 88:3441
47. Bellosta P, Iwahori A, Plotnikov A, Eliseenkova A, Basilico C, Mohammadi M (2001) Mol Cell Biol 21:5946
48. Ye S, Luo Y, Lu W, Jones R, Linhardt R, Capila I, Toida T, Kan M, Pelletier H, McKeehan W (2001) Biochemistry 40:14429
49. Plotnikov A, Eliseenkova A, Ibrahimi O, Shriver Z, Sasisekharan R, Lemmon M, Mohammadi M (2001) J Biol Chem 276:4322
50. Harmer N, Chirgadze D, Pellegrini L, Fernandez-Recio J, Blundell T (2004) Biochemistry 43:629
51. Plotnikov A, Hubbard S, Schlessinger J, Mohammadi M (2000) Cell 101:413
52. Stauber D, DiGabriele A, Hendrickson W (2000) Proc Natl Acad Sci USA 97:49
53. Yeh B, Igarashi M, Eliseenkova A, Plotnikov A, Sher I, Ron D, Aaronson S, Mohammadi M (2003) Proc Natl Acad Sci USA 100:2266
54. Ibrahimi O, Zhang F, Eliseenikova A, Itoh N, Linhardt R, Mohammadi M (2004) Human Mol Genet 13:2313
55. DiGabriele A, Lax I, Chen D, Svahn C, Jaye M, Schlessinger J, Hendrickson W (1998) Nature 393:812
56. Faham S, Hileman R, Fromm J, Linhardt R, Rees D (1996) Science 271:1116
57. Schlessinger J, Plotnikov A, Ibrahimi O, Eliseenkova A, Yeh B, Yayon A, Linhardt R, Mohammadi M (2000) Mol Cell 6:743
58. Pellegrini L (2001) Curr Opin Struct Biol 2001:629
59. Byers S, Amaya E, Munro S, Blaschuk O (1992) Devel Biol 152:411
60. Thompson L, Pantoliano M, Springer B (1994) Biochemistry 33:3831
61. Zhu H, Ramnarayan K, Menzel P, Miao Y, Zheng J, Mong S (1998) Protein Eng 11:937
62. Spivak-Kroizman T, Lemmon M, Dikic I, Ladbury J, Pinchasi D, Huang J, Jaye M, Crumley G, Schlessinger J, Lax I (1994) Cell 79:1015
63. Maccarana M, Casu B, Lindahl U (1993) J Biol Chem 268:23898
64. Ornitz D, Herr A, Nilsson M, Westman J, Svahn C, Waksman G (1995) Science 268:432
65. Wilkie A (2005) Cytokine Growth Factor Rev 16:187
66. Wilkie A, Slaney S, Oldridge M, Poole M, Ashworth G, Hockley A, Hayward R, David D, Pulleyn L, Rutland P, Malcolm S, Winter R, Reardon W (1995) Nature Genet 9:165
67. Ibrahimi O, Eliseenikova A, Plotnikov A, Yu K, Ornitz D, Mohammadi M (2001) Proc Natl Acad Sci USA 98:7182
68. Anderson J, Burns H, Enriquez-Harris P, Wilkie A, Heath J (1998) Human Mol Genet 7:1475
69. Ibrahimi O, Zhang F, Eliseenikova A, Linhardt R, Mohammadi M (2004) Human Mol Genet 13:69

70. Passos-Bueno M, WIlcox W, Jabs E, Sertie A, Alonso L, Kitoh H (1999) Human Mutation 14:115
71. Kan S-h, Elanko N, Johnson D, Cornejo-Roldan L, Cook J, Reich E, Tomkins S, Verloes A, Twigg S, Rannan-Eliya S, McDonald-McGinn D, Zackai E, Wall S, Muenke M, Wilkie A (2002) Amer J Human Genet 70:472
72. Ibrahimi O, Yeh B, Eliseenikova A, Zhang F, Olsen S, Igarashi M, Aaronson S, Linhardt R, Mohammadi M (2005) Mol Cell Biol 25:671
73. Harmer N, Lung L, Pellegrini L, Mulloy B, Robinson C, Blundell T (2004) J Mol Biol 339:821
74. Wu Z, Zhang L, Yabe T, Kuberan B, Beeler D, Love A, Rosenberg R (2003) J Biol Chem 278:17121
75. Kreuger J, Salmivirta M, Sturiale L, Giménez-Gallego G, Lindahl U (2001) J Biol Chem 276:30744
76. Ostrovsky O, Berman B, Gallagher J, Mulloy B, Fernig D, Delehedde M, Ron D (2002) J Biol Chem 277:2444
77. Ashikari-Hada S, Habuchi H, Kariya Y, Itoh N, Reddi A, Kimata K (2004) J Biol Chem 279:12346
78. Robinson C, Harmer N, Goodger S, Blundell T, Gallagher J (2005) J Biol Chem 280:42274
79. Vives R, Pye D, Salmivirta M, Hopwood J, Lindahl U, Gallagher J (1999) Biochem J 339:767
80. Loo B-M, Kreuger J, Jalkanen M, Lindahl U, Salmivirta M (2001) J Biol Chem 276:16868
81. Kreuger J, Jemth P, Sanders-Lindberg E, Elaihu L, Ron D, Basilico C, Salmivirta M, Lindahl U (2005) Biochem J 389:145
82. Wu Z, Lech M (2005) J Biol Chem 389:383
83. Loo B-M, Salmivirta M (2002) J Biol Chem 277:32616
84. Harmer N, Robinson C, Adam L, Ilag L, Robinson C, Gallagher J, Blundell T (2006) Biochem J 393:741
85. Merry C, Lyon M, Deakin J, Hopwood J, Gallagher J (1999) J Biol Chem 274:18455
86. Ford-Perriss M, Guimond S, Greferath U, Kita M, Grobe K, Habuchi H, Kimata K, Esko J, Murphy M, Turnbull J (2002) Glycobiology 12:721
87. Kato M, Wang H, Bernfield M (1994) J Biol Chem 269:18881
88. Maccarana M, Sakura Y, Tawada A, Yoshida K, Lindahl U (1996) J Biol Chem 271:17804
89. Turnbull J, Drummond K, Huang Z, Kinnunen T, Ford-Perriss M, Murphy M, Guimond S (2003) Biochem Soc Trans 31:343
90. Powell A, Fernig D, Turnbull J (2002) J Biol Chem 277:28554
91. Lin X (2004) Development 131:6009
92. Harris J, Werling D, Hope J, Taylor G, Howard C (2002) Trends Immunology 23:158
93. Mukherjee S, Maxfield F (2004) Ann Rev Cell Develop Biol 20:839
94. Zajchowski L, Robbins S (2002) Eur J Biochem 269:737
95. Toledo M, Suzuki E, Handa K, Hakomori S (2004) J Biol Chem 279:34655
96. Ridyard M, Robbins S (2003) J Biol Chem 278:13803
97. Chu C, Buczek-Thomas J, Nugent M (2004) Biochem J 379(Pt 2):331
98. Davy A, Feuerstein C, Robbins S (2000) J Neurochem 74:676
99. Anderson A, Kendal C, Garcia-Maya M, Kenny A, Morris-Triggs S, Wu T, Reynolds R, Hohenester E, Saffell J (2005) J Neurochem 95:570
100. Cavallaro U, Niedermeyer J, Fuxa M, Christofori G (2001) Nature Cell Biol 3:650

101. Suyama K, Shapiro I, Guttman M, Hazan R (2002) Cancer Cell 2:301
102. Utton M, Eickholt B, Howell F, Wallis J, Doherty P (2001) J Neurochem 76:1421
103. Bottcher R, Pollet N, Delius H, Niehrs C (2004) Nature Cell Biol 6:38
104. Toledo M, Suzuki E, Handa K, Hakomori S (2005) J Biol Chem 280:16227
105. Bernett M, Somasundaram T, Blaber M (2004) Proteins 57:626
106. Eriksson A, Cousens L, Matthews B (1993) Protein Sci 2:1274
107. DeLano W (2002) The PyMOL Molecular Graphics System, DeLano Scientific, San Carlos, CA, USA

Top Curr Chem (2007) 272: 117–167
DOI 10.1007/128_066

Published online: 20 October 2006

Characterization of Interactions Between Misfolding Proteins and Molecular Chaperones by NMR Spectroscopy

Bernd Reif[1] (✉) · Saravanakumar Narayanan[2]

[1]Forschungsinstitut für Molekulare Pharmakologie (FMP), Robert-Rössle-Str. 10, 13125 Berlin, Germany
reif@fmp-berlin.de

[2]Charité Universitätsmedizin, 10098 Berlin, Germany

Abstract Molecular chaperones are a group of proteins that bind transiently to nascent polypeptide chains and are thought to function by preventing aggregation by maintaining polypeptides in conformations competent for folding and subunit assembly. On the other hand, specific chaperones are considered to be involved in the mechanism of prion propagation and others are found to colocalize in plaques of patients suffering from neurodegenerative diseases. Solution-state NMR spectroscopy can provide insight into the interactions between the misfolding protein and the chaperone at atomic resolution. In particular, experimental results for Sup35, a yeast prion protein, and β-amyloid, which is responsible for Alzheimer's disease, and their interactions with Hsp104 and αB-crystallin, respectively, are discussed.

Keywords Alzheimer's disease · Chaperone · Heat shock proteins · Protein misfolding and aggregation · Saturation transfer difference (STD) · Solution-state nuclear magnetic resonance (NMR) · Yeast prions

1 Introduction

More than 16 proteins are known to be associated with a disease involving protein deposition [1]. Amyloid fibrils are formed by polymerization of abnormal states of normally soluble proteins and peptides [2]. The most prominent examples of peptides and proteins involved in protein aggregation are the β-amyloid peptide (Aβ) found in plaques in the brains of Alzheimer's disease patients [3], and the prion proteins assigned to transmissible spongiform encephalopathies (TSE) [4]. Dobson and coworkers found recently that not only disease-related proteins can form amyloid fibrils, but also fibril formation can be induced for almost any soluble protein under specific solubilization conditions [5–7].

In the last few years, rapid progress was made in the structural characterization of the fibril structure of the β-amyloid peptide using magic-angle spinning (MAS) solid-state NMR studies [8–10]. In a similar way, the fibril structure of a fragment of transthyretin [11, 12] as well as of the prion protein HET-s could be characterized [13, 14].

However, little is known about the structural mechanism of protein misfolding at atomic resolution. In some cases only specific residues seemed to be responsible for promoting fibril formation [15]. In other systems, stretches of amino acids were identified to contribute to the core of the fibril structure [16]. In a cell, a number of strategies have been evolved which prevent protein aggregation. In case a protein cannot be refolded properly, the misfolded protein will finally be targeted and degraded in the ubiquitin–proteasome pathway [17]. However, a large number of molecular chaperones are present in all types of cells and cellular compartments to prevent protein misfolding. Some chaperones interact with nascent chains as they emerge from the ribosome, whereas others are involved in guiding later stages of the folding process [18, 19]. The best-characterized chaperone so far is the bacterial complex of GroEL involving the co-chaperonin GroES [20, 21]. The X-ray structure of GroEL [22, 23] displays a cavity in which incompletely folded polypeptide chains can enter and undergo the final steps in the formation of their native structure. Solution-state NMR studies were carried out in order to study the interactions between GroEL and peptide substrates [24–27]. These studies focussed on the substrate. In addition, NMR experiments were performed to investigate the full-length protein GroEL [28].

The 70-kDa heat shock protein Hsp70 is another chaperone which plays a crucial role in protein (re-)folding in the mammalian cytosol. Solution-state NMR studies were carried out to better understand the structural mechanism due to which Hsp70 binds and disaggregates the target proteins [29–31]. Furthermore, it was demonstrated that Hsp70 inhibits, e.g., α-synuclein fibril formation via preferential binding to prefibrillar species [32], thereby altering the characteristics of toxic α-synuclein aggregates.

In the following, two examples will be given in which interactions between a misfolding peptide and a molecular chaperone are studied by solution-state NMR. In the first example (Sect. 2), interactions between the yeast prion protein Sup35 and Hsp104 are characterized. In the second example (Sects. 3 and 4), interactions between the Alzheimer's disease β-amyloid peptide and the small heat shock protein αB-crystallin are investigated.

2
Sup35 and Hsp104

The [PSI^+] factor of *Saccharomyces cerevisiae* is a protein-based genetic element that consists of the cytosolic translation termination factor Sup35. In [PSI^-] cells, Sup35 is a soluble protein, but in [PSI^+] cells, most of the Sup35 protein is found in a prion-like, aggregated conformation that can be propagated onto the daughter cells. The occurrence of this phenotype is modulated by the chaperone Hsp104. Hsp104 was first identified as a protein important for the thermotolerance of yeast [33, 34]. Like its homologue, ClpB from *Escherichia coli*, Hsp104 forms ring-shaped oligomers, and is thought to be involved in the dissociation of protein complexes and aggregates [35, 36]. Both deletion of the gene encoding Hsp104 and overexpression of Hsp104 can cure yeast cells from [PSI^+] and return Sup35 into its soluble state [37, 38]. The N-terminal domain of Sup35, which is rich in glutamine and asparagine, was attributed a critical role in the process of prion formation and propagation [39], hence the name prion domain (PrD). Especially, mutations in the N-terminal part of PrD not only alter the fibril formation properties, but can also cause curing of wild-type aggregates in vivo. In non-prion neurodegenerative disorders, like Huntington's disease, similar expansions of polyglutamine repeats appear to be responsible for amyloid formation and neurotoxicity [40, 41]. Fusion proteins consisting of huntingtin-type polyglutamine repeats and green fluorescent protein were found to form aggregates when expressed in yeast cells. Deletion of the Hsp104 gene, however, resulted in elimination of aggregation [42].

Two models have been proposed for [PSI^+] inheritance. Lindquist and coworkers suggest a model in which Hsp104 is required to put the polyglutamine repeats into an "aggregation competent" state, thereby inducing prion formation. Glutamine residues are a determining factor in interactions with Hsp104 [37, 38]. Hsp104 is thought to be necessary for the formation or maintenance (or both) of a partially unfolded form of Sup35. Therefore, an insufficient amount of the chaperone would prevent prion formation. If the amount of chaperone was too high, however, the prion "template" would dissociate from the unfolded intermediate, leading to the loss of the protein-modifying activity. Similarly, Cohen and Prusiner [43] have suggested that the conformational conversion from the wild-type to the prion form may require a partially unfolded intermediate. A prion template is supposed to bind to this intermediate, converting it into the prion conformation. Alternatively, Ter-Avanesyan and coworkers propose a model in which Hsp104 cleaves Sup35 [PSI^+] aggregates into smaller pieces, which is necessary for their stable segregation during cell division [44, 45].

In the past, the only biochemical evidence that soluble Sup35 directly interacts with Hsp104 came from circular dichroism (CD) experiments [46]. Recently, we used NMR spectroscopy to obtain a more quantitative understanding of the interaction between Sup35 and Hsp104 [47]. In our experiments, we show that the peptide Sup35$^{5-26}$, which corresponds to residues 5–26 of Sup35, instead of the full-length protein

$$\mathrm{N-N^{5}QGNNQ^{10}QNYQQ^{15}YSQNG^{20}NQQQG^{25}N-C}$$

interacts with Hsp104. This peptide was chosen since mutation studies reveal differences of prion formation propensity upon mutations in the respective section of the PrD [39]. More importantly, deletion of this fragment cures yeast cells from [PSI^+]. It was therefore hypothesized that binding of Sup35 to Hsp104 involves glutamine and asparagine side chains of Sup35$^{5-26}$. Although it was shown that other parts of Sup35 have an influence on prion formation as well [48, 49], this peptide constitutes a good model system to investigate the interaction of Sup35 with Hsp104 by NMR. The aim of our studies is to identify the chemical groups which are involved in interactions with the chaperone. We find that the peptide exists as a complex mixture of different oligomeric states ranging from monomers to octamers. Upon addition of Hsp104, the formation of a monomeric species can be observed that shows only weak interactions with Hsp104.

2.1 Aggregation of Sup35$^{5-26}$ is Suppressed by Hsp104

Upon dissolution of freeze-dried Sup35$^{5-26}$ in buffer, the peptide aggregates quickly, as shown by the rapid decay of observable ^{1}H NMR resonances (Fig. 1, triangle up). Only signals originating from soluble peptide

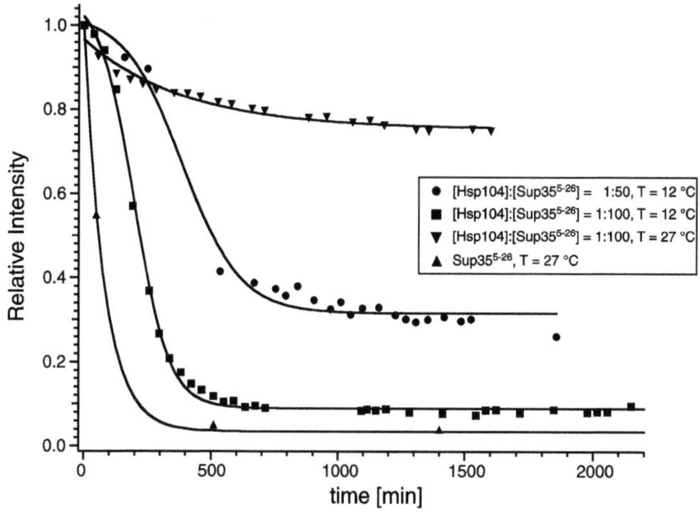

Fig. 1 Time dependence of the intensity of the ^{1}H NMR resonance line of tyrosine Hδ in $\mathrm{Sup35}^{5-26}$ for different molar ratios Hsp104 : $\mathrm{Sup35}^{5-26}$ [*triangles up*: without Hsp104 ($T = 27$ °C); *squares*: 1 : 100 ($T = 12$ °C); *circles*: 1 : 50 ($T = 12$ °C); *triangles down*: 1 : 100 ($T = 27$ °C)]. The initial concentration of $\mathrm{Sup35}^{5-26}$ was 1.5 mM in all experiments. The peptide was dissolved in 50 mM Na phosphate buffer, pH 7.7. $\mathrm{Sup35}^{5-26}$ and Hsp104 were mixed 3 min before recording the first data point

molecules are visible in the spectrum. At a peptide concentration of 1.5 mM and $T = 27$ °C, the characteristic time τ_{agg}, after which 50% of the peptide became insoluble, was found to be approximately 80 min. At 12 °C, aggregation was too fast to be monitored with this experimental setup. A possible explanation for this counterintuitive, temperature-dependent behavior is that the peptide consists of an ensemble of species with differing aggregation probabilities. At higher temperature, this distribution is shifted toward forms that have a smaller tendency to aggregate. Also, the aggregates may be less stable at elevated temperatures. This idea is supported by the observation that the aggregated peptide could be resolubilized by heating the sample to 60 °C (*vide infra*). Importantly, the aggregated peptide strongly increased the fluorescence of Thioflavin T, a dye that is commonly used to detect amyloids (*vide infra*).

Addition of Hsp104 substantially reduces the rate of aggregation (Fig. 1). Furthermore, aggregation is no longer quantitative in the presence of Hsp104. At a molar ratio of Hsp104 to $\mathrm{Sup35}^{5-26}$ of 1 : 50 (monomer : monomer, $T = 12$ °C), ~ 35% of the peptide remains soluble after incubating the sample for 800 min (Fig. 1, filled circles). τ_{agg} was determined to be in the range of 400 min. NMR resonances of protons stemming from Hsp104 are not observable in these experiments due to the low concentration and the large molecular weight of the hexameric Hsp104.

2.2 Reversibility of Aggregation

In order to understand the effect of temperature on Sup35$^{5-26}$, 1D ^{1}H NMR spectra were recorded at various temperatures. To start out with, a sample containing aggregated Sup35$^{5-26}$ was used. No Hsp104 was present in this case. The intensities of the NMR signals in ^{1}H NMR spectra are directly related to the fraction of soluble Sup35$^{5-26}$ in the NMR sample tube. Molecules involved in fibrillar structures cannot be detected by solution-state NMR. Figure 2 shows the reversibility of the aggregation process upon increase and decrease of temperature. Aggregated Sup35$^{5-26}$ could be partially resolubilized upon stepwise heating to 50 °C. Lowering the temperature again induced loss of signal intensity, and thus a change of equilibrium toward the aggregated state. After each temperature step, the system was given 20 min for equilibration. Notably, no additional resonance signals could be observed in the proton spectrum upon heating (data not shown), implying that no significant conformational changes are occurring prior to protein aggregation.

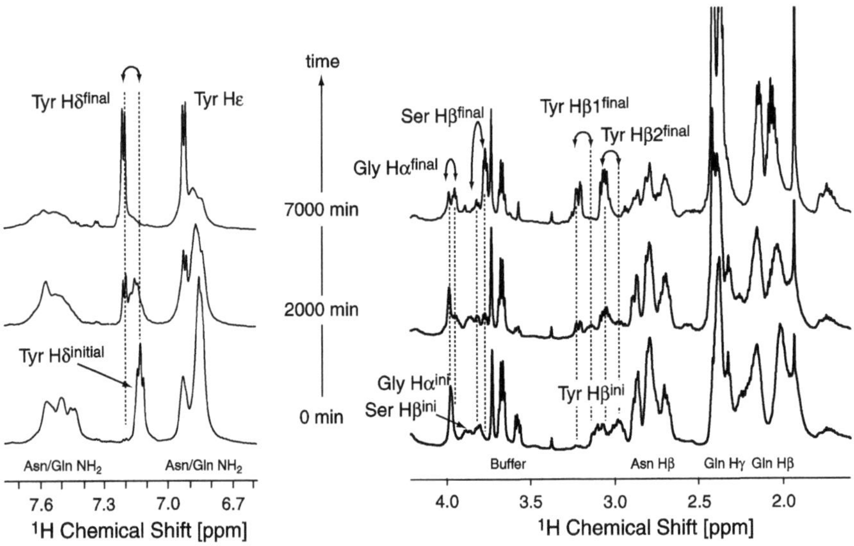

Fig. 2 ^{1}H NMR spectra of Sup35$^{5-26}$ incubated with Hsp104 at a molar ratio of Hsp104 : Sup35$^{5-26}$ = 1 : 25 at 27 °C as a function of time

2.3 Hsp104 Induces Conformational Changes in Sup35$^{5-26}$

To obtain a more detailed view of the interaction between Hsp104 and Sup35$^{5-26}$, we recorded 1D ^{1}H spectra at 27 °C and a molar ratio of Hsp104 to

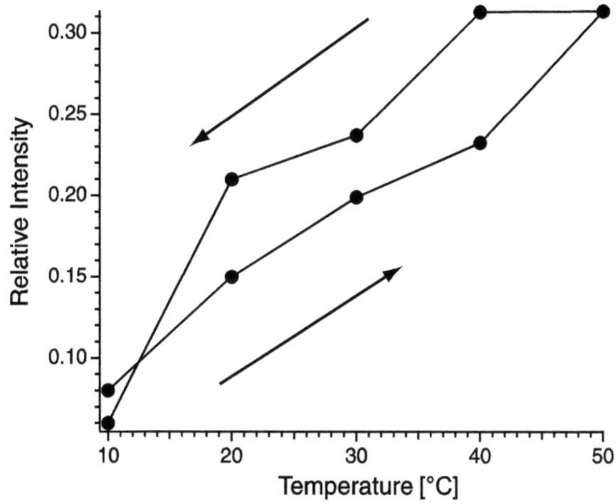

Fig. 3 Relative intensity of Sup35$^{5-26}$ as a function of temperature. The intensity directly reflects the fraction of Sup35$^{5-26}$ in solution (in contrast to the fraction of Sup35$^{5-26}$ which is in an aggregated/fibrillar state)

Sup35$^{5-26}$ of 1 : 25 (Fig. 3). Under these conditions, the peptide apparently can no longer aggregate. The intensities of nonexchangeable protons are constant throughout the experiment. After 300 min, however, new signals appear in the spectrum that cannot be found in the reference spectrum recorded in the absence of Hsp104 or in the experiments recorded at lower temperature (12 °C) in the presence of Hsp104. Matrix-assisted laser desorption/ionization mass spectrometry (MALDI-MS) was carried out to verify that the peptide had not become degraded or chemically modified during the experiment (data not shown). The new signals must therefore reflect a change in the conformation of Sup35$^{5-26}$. A detailed view of the Tyr aromatic region is displayed in Fig. 4. The decay of the initial resonance line (Tyr Hδ^{initial}) is accompanied by the rise of two intermediate species (Tyr H$\delta^{\text{intermediate(1,2)}}$). At later times, a fourth conformer (Tyr Hδ^{final}) accumulates (Fig. 5).

Sup35$^{5-26}$ contains two tyrosines. The scalar coupling between Hδ and Hε causes a splitting of the signal into a doublet. For clarity, the intensities of the doublets for intermediate 2 and the final conformer are divided by two, to take into account the overlap of the two resonance lines. The decay of the initial species occurs with a time constant of 1600 min. Concomitantly, intermediate 1 becomes populated and reaches its maximum concentration after ~ 2000 min. After 5000 min, the prevailing species of Sup35$^{5-26}$ is intermediate 2. At this point, the final conformer is only populated with ca. 25%; 100% population of this final Sup35$^{5-26}$ conformer was found only after 7 days.

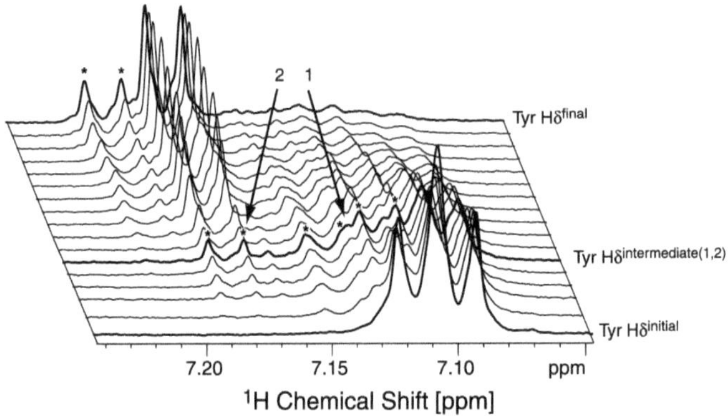

Fig. 4 1D ^{1}H NMR spectra displaying only the tyrosine Hδ spectral region. Spectra recorded between 0 and 5500 min are represented in equidistant steps. Transiently populated species (1,2) are indicated by *bold lines* and *asterisks*

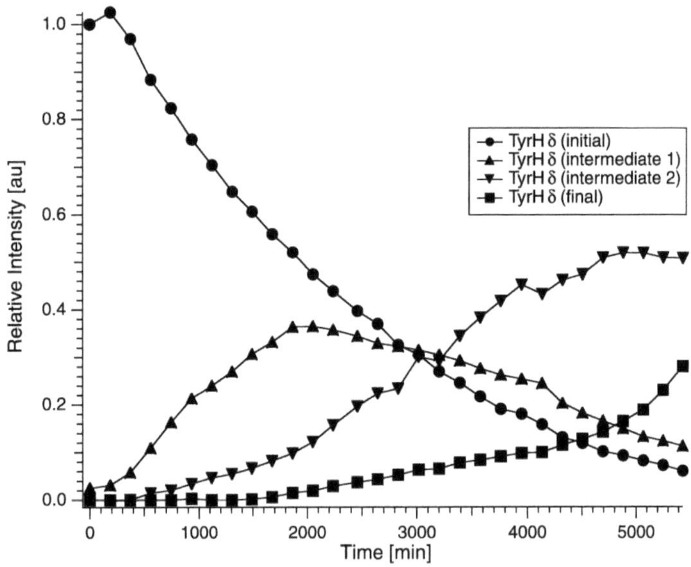

Fig. 5 Time dependence of the ^{1}H signal of tyrosine Hδ in Sup35$^{5-26}$ for the molar ratio Hsp104 : Sup35$^{5-26}$ = 1 : 25 at 27 °C for the initial (*circles*), intermediate (*triangles*), and final set (*squares*) of tyrosine resonances

2.4 Sup35$^{5-26}$ is a Mixture of Various Oligomeric Species

The differences in the aggregation behavior at 12 and 27 °C (Fig. 1), and the conformational changes observed in the presence of Hsp104 (Fig. 3), indicate that Sup35$^{5-26}$ consists of a variety of species with differing aggregation properties and affinities for Hsp104. To further characterize this ensemble, we determined the molecular mass of Sup35$^{5-26}$ with diffusion-ordered spectroscopy (DOSY) NMR experiments [50]. In these experiments, the size of a molecule can be estimated by encoding the diffusion of a molecule in a gradient echo. The measured diffusion constant $D = (k_B T)/(6\pi\eta F r_S)$ is related via the Stokes–Einstein relation to the hydrodynamic radius r_S of the molecule, and thus to the molecular weight at a given viscosity η of the solution. k_B denotes Boltzmann's constant, T the absolute temperature, and F the dimensionless Perrin factor. Only molecules that are not diffusing along a given axis are detected. The decay of magnetization can be analyzed in an analytical way, employing $I/I_0 = \exp\{-D(\Delta - \delta/3)q^2\}$, where $q = \gamma\delta g$. Δ refers to the separation of the gradient echo, δ to the duration of the gradient, γ to the gyromagnetic ratio of the nucleus, and g to the strength of the gradient.

Figure 6 shows the time dependence of the calculated molecular weight of Sup35$^{5-26}$ in the presence of Hsp104 at 27 °C. It is apparent that the average molecular weight of Sup35$^{5-26}$ changes due to a shift in equilibrium between different oligomeric states. At 12 °C, the molecular weight decreases

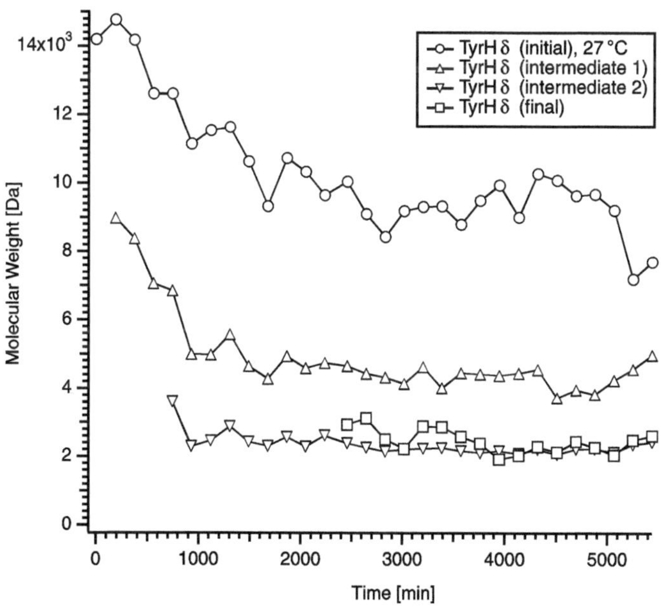

Fig. 6 NMR diffusion data (DOSY) for Sup35$^{5-26}$ in the presence of Hsp104 at a molar ratio of Hsp104 : Sup35$^{5-26}$ = 1 : 25 and $T = 27$ °C. The figure displays the calculated molecular weight of Sup35$^{5-26}$ as a function of time. Each *curve* thus reflects the diffusion coefficient of the tyrosine aromatic resonance as indicated in Fig. 2. The calculated molecular weight shifts from $\sim$ 15 kDa ($t = 0$ min; $I = I_0$) to 5.0 kDa ($t = 4000$ min; $I = 0.2\,I_0$)

from approximately 22 kDa ($t = 0$ min; $I = I_0$) to 10 kDa ($t = 280$ min; $I = 0.8$–$0.4\ I_0$) (data not shown). Under these conditions $\sim 65\%$ of the peptide molecules precipitate. Apparently, the initial distribution is centered around an octameric species. Upon longer incubation, the majority of the molecules form insoluble aggregates, whereas a fraction dissociates into smaller species which remain soluble. At a molar ratio of 1 : 25 for Hsp104 : Sup35^{5-26} at $T = 27\,°\text{C}$, we observe a different behavior. Initially, the average molecular weight of Sup35^{5-26} is ~ 14 kDa, corresponding to a population of a predominantly hexameric species. With longer incubation, this species dissociates into smaller oligomeric states, and the apparent molecular weight decreases (~ 8 kDa). Importantly, the chemical shift of the Tyr Hδ proton remains the same, indicating that the electronic and chemical environment of this proton does not change during this process. From the spectra displayed in Fig. 4, we deduce that Sup35^{5-26} is slowly converting into new species, indicated by changes in the chemical shifts. Using signals specific for the new species, their molecular weight could be determined in the same DOSY experiment. Intermediate 1 is mainly tetrameric at the beginning of the experiment. Upon incubation with Hsp104, the equilibrium is shifted to smaller oligomeric states, and Sup35^{5-26} is slowly converted into a dimeric species. Intermediate 2 as well as the final product are monomeric throughout the experiment.

These results suggest that at 27 °C, the Sup35^{5-26} peptide adopts a range of oligomeric states that is centered around hexameric species. Upon interaction with Hsp104, these hexamers are slowly converted to structurally closely related tetramers. Subsequently, these tetramers dissociate into monomers in a two-step process, which is accompanied by characteristic changes in the 1D ^{1}H NMR spectrum.

2.5
Binding of Sup35^{5-26} to Hsp104

In order to identify the chemical groups of Sup35 that interact with Hsp104, we carried out saturation transfer difference (STD) experiments. STD was successfully used in the past to screen compound mixtures for binding to a receptor protein [51]. Furthermore, it was used to characterize interactions between membrane channel proteins and neurotoxic peptides [52]. This way it is possible to identify the chemical groups involved in ligand binding. The experiment relies on the selective saturation of the ^{1}H resonances of the target protein. If a ligand binds weakly, saturation is transferred via cross relaxation to the bound ligand. An attenuated signal is then observed for the free ligand after dissociation. The method works best for small ligands (with fast correlation times) and large complexes, where the reorientation process is slow enough to give rise to large cross relaxation rates. The STD amplification factor α is defined as $\alpha = (I_0 - I_{sat}/I_0)c$, where I_0 and I_{sat} correspond to the intensity of the signal during off-resonance and on-resonance irradiation, respectively, and c refers to the relative concentration of the ligand with respect to the enzyme. The time dependence of the STD amplification factor for Sup35$^{5-26}$ in the presence of Hsp104 at a molar ratio of 50 : 1 at 12 °C is shown in Fig. 7.

At 12 °C, no STD signal is observed in the beginning, indicating that the initial ensemble of Sup35$^{5-26}$ species shows no interaction with Hsp104. Taking into account the results of the DOSY experiment, this observation suggests that octameric Sup35$^{5-26}$ does not bind to Hsp104. With further incubation, however, STD signals are increasing monotonously, suggesting that species are formed that bind to Hsp104. Side-chain resonances of Asn and Gln, as well as Tyr, show the fastest STD buildup. This may be expected, since mutation studies show that these residues are essential for prion propa-

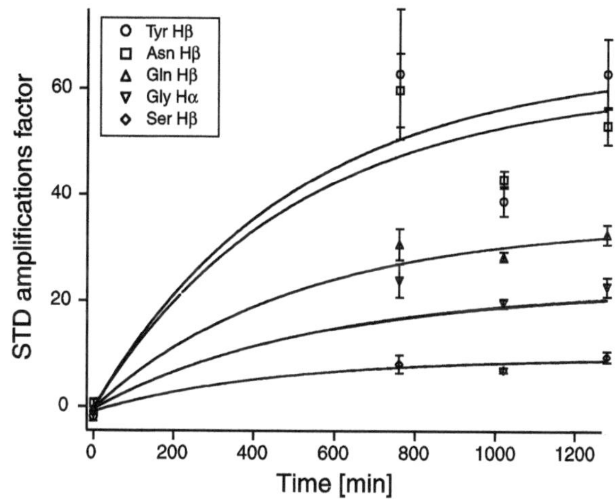

Fig. 7 Time dependence of the STD signal for Sup35$^{5-26}$ in the presence of Hsp104 for a molar ratio of Hsp104 : Sup35$^{5-26}$ = 1 : 50 at $T = 12$ °C. The assignment of the STD amplification factor for the various protons is given in the figure. Side-chain protons of Asn, Gln, and Tyr show the fastest buildup, corresponding to the strongest interaction with Hsp104. At this temperature, no interaction between Hsp104 and Sup35$^{5-26}$ is observed at $t = 0$. As a consequence, the affinity between high oligomeric states of Sup35$^{5-26}$ and Hsp104 must be small. Experimentally, a train of 300 square pulses of 12 ms duration (with 1.0 ms gap between the pulses) is applied in order to saturate the ^{1}H(methyl) resonances of Hsp104. Even (on-resonance irradiation at – 0.02 ppm) and odd scans (off-resonance irradiation at + 30.0 ppm) are recorded subsequently and subtracted from each other by incrementation of the receiver phase. 512 scans are accumulated at each time point, resulting in a total experimental time of 45 min per time point. STD experiments were always adjusted in the first place in a reference experiment on a sample containing no Hsp104 in order to minimize artifacts. The amount of the relative concentration of Sup35$^{5-26}$ was always taken into account in the evaluation by recording 1D reference spectra in parallel with the STD experiments

gation and interaction with Hsp104 [39]. On average, the time constant of the process monitored in the STD experiment is similar to the time constant observed for the decay of magnetization shown in Fig. 1 (ca. 400 min). Since no new ^{1}H resonances can be observed under these conditions, the decrease in signal intensity is therefore purely due to aggregation. Apparently, the molecular chaperone Hsp104 modulates an equilibrium between different conformers of Sup35$^{5-26}$. Initially, the Sup35 species that interact with Hsp104 are not populated and therefore no STD signal can be observed. Binding of Hsp104 to these molecules, however, could shift the distribution toward these conformers by the law of mass action. Their concentrations increase and hence we observe an STD signal.

At 27 °C, Sup35$^{5-26}$ behaves strikingly differently. Figure 8 displays the STD amplification factors of the tyrosine Hδ resonance lines for the initial, the intermediate, and the final species of Sup35$^{5-26}$. Other resonance lines show qualitatively the same behavior. The tyrosine signal has been selected

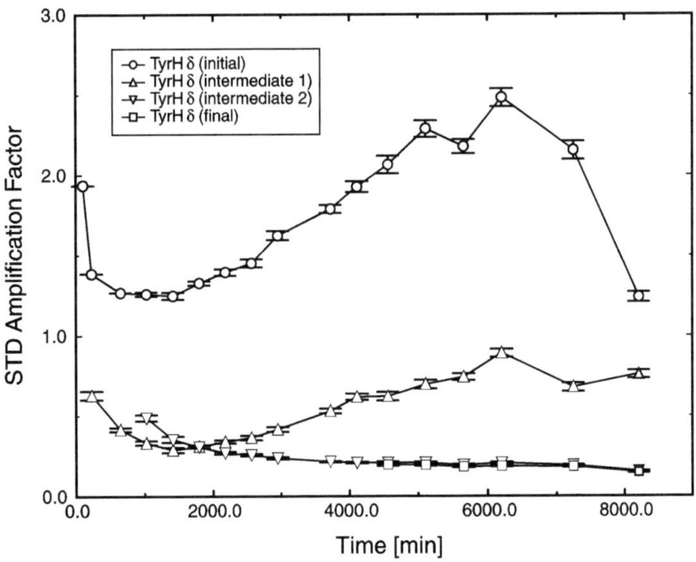

Fig. 8 Time dependence of the STD signal for Sup35$^{5-26}$ in the presence of Hsp104 for a molar ratio of Hsp104 : Sup35$^{5-26}$ = 1 : 25 and $T = 27$ °C. *Open circles*, *squares*, and *triangles* indicate the STD amplification factor of the Hδ tyrosine signal of the initial, intermediate, and final conformer, respectively. In contrast to $T = 12$ °C, first a decay and after $t = 2000$ min an increase of the STD amplification factor is observed. During the experiment, the equilibrium between high oligomeric and low oligomeric states of Sup35$^{5-26}$ is shifted toward the low oligomeric state of Sup35$^{5-26}$ (*circles*), which can both interact with hexameric Hsp104. The initial species shows strong interactions with Hsp104, whereas the "release" form shows a decreased interaction with Hsp104. The time constant of the rate of change is in agreement with the decay of magnetization due to the formation of a new species

in the analysis since it is not overlapping with other ^{1}H resonances. The initial species (circles) shows the strongest interaction with Hsp104, whereas the interaction of the intermediates 1 and 2 (triangles) is reduced. The final species (squares) only shows very weak binding to the chaperone. Error bars are based on the signal-to-noise ratio of the respective data sets. Generally, the error bars for the initial species are increasing in the course of the experiment, whereas the error bars for the final species are decreasing, due to the changing populations of the two states.

The STD amplification factor for the initial conformer shows a characteristic minimum after ca. 1500 min. Thereafter, the signal is slowly increasing, and finally dropping again at the end of the experiment. Combining the intensity data, to which all conformational states at a specific chemical shift contribute, and the DOSY results allows us to estimate the population of the individual Sup35$^{5-26}$ oligomer at a given time (assuming that there is conformational averaging between two predominantly populated species). The population of the initial hexameric conformer is decaying with a time constant of ca. 1500 min, whereas the population for intermediate 1 is first increasing and finally decaying at the end of the experiment. Since the relative STD amplification factor is related to the population of an individual species in a nonlinear manner, superposition of one exponentially decaying and one increasing STD curve leads to the observed amplification factor with an absolute minimum after ca. 1500 min. We conclude, therefore, that a hexamer as well as a tetramer of Sup35$^{5-26}$ can interact with Hsp104 at this temperature (circles).

A similar dip-type curve would also be expected for the STD amplification factor of intermediate 1 (triangles up), which consists of a tetrameric and a dimeric species (see above). The total population of this intermediate increases in the beginning, and then decreases after 2000 min. At the same time, the distribution is shifted from the tetrameric to the dimeric species, as monitored by DOSY experiments. Qualitatively, the STD signal of intermediate 1 is comparable to the curve shape of the initial species. The absolute intensity, however, is much less pronounced, indicating a weaker binding affinity of intermediate 1 compared to the initial conformer of Sup35$^{5-26}$ with respect to Hsp104. A decreased STD amplification factor might also be indicative of stronger binding. This possibility can be ruled out, as stronger binding would induce an increased aggregation rate of the remaining Sup35$^{5-26}$ peptides in solution. This is however, not observed. In principle, an increase of the STD amplification factor for $t < 800$ min would be expected. However, these time points are not accessible, since this species is not populated at the beginning of the experiment. For the second intermediate of Sup35$^{5-26}$, only a decrease in the STD amplification factor can be detected. Simultaneously, small changes in the apparent molecular weight for this conformer are observed which correspond to a shift of equilibrium from a dimeric to monomeric species of Sup35$^{5-26}$. Again, the initial points in the DOSY experiments for intermediate 2 are difficult to access for a reliable interpretation due to the low signal-to-noise ratio for intermediate 2 for $t < 1500$ min. No STD signal is observed for the resonance lines of the final state of Sup35$^{5-26}$ (squares), indicating that there is negligible interaction between the release form of Sup35$^{5-26}$ and Hsp104.

In these studies, care was taken to prepare Hsp104 in defined oligomeric states. Gradient SDS-PAGE performed at different concentrations of Hsp104, as well as with and without addition of peptide, shows that Hsp104 can be quantitatively cross-linked to a complex with a molecular mass of ~ 600 kDa corresponding to a hexamer. This is in agreement with other studies [53]. Additionally, dynamic light scattering (DLS) experiments were carried out to obtain information about the degree of oligomerization of Hsp104. Hsp104 was found to be hexameric under the conditions used in the NMR experiments (data not shown). Assuming a globular structure for Hsp104, a hydrodynamic radius of ca. 3.42 nm for the monomer is expected[1] [54]. The observed hydrodynamic radius of ca. 7.0 nm corresponds relatively well to a hexameric complex of Hsp104 at the two temperatures investigated (12 and 27 °C).

2.6 Analysis of the Sup35$^{5-26}$ Aggregation Process

The NMR experiments presented above are ambiguous with respect to the absolute oligomeric state of Sup35$^{5-26}$. The diffusion constants measured by DOSY experiments yield only the average molecular weight and do not give information on the population of different oligomeric states. In addition, the interactions of Hsp104 with fibrillar Sup35$^{5-26}$ could not be addressed due to molecular weight restrictions of solution-state NMR spectroscopic experiments. So far, it is controversial whether Hsp104 is able to resolubilize aggregated Sup35 [55–59]. In order to resolve these ambiguities, we carried out dialysis experiments using dialysis bags with varying molecular weight cutoff (MWCO) membranes and a molar ratio of 1 : 25 for Hsp104 : Sup35$^{5-26}$ (Fig. 9). We monitored the population of aggregated Sup35$^{5-26}$ within the dialysis bag as a function of time using Thioflavin T (ThT) fluorescence, and determined the amount of peptide which diffused out of the dialysis bag using tyrosine fluorescence [60].

Using a MWCO of 3.5 kDa, which only allows the release of monomeric Sup35$^{5-26}$, we observed a strong monotonous increase in ThT fluorescence in the absence of Hsp104. This increase is much reduced in the presence of Hsp104 (Fig. 10b). Tyrosine fluorescence analysis confirms that an increased amount of monomer is released from samples incubated with Hsp104

[1] The empirical relation between the number of residues N and the hydrodynamic radius of a native-state protein is given as $r_H = 4.75\,N^{0.29}$ Å.

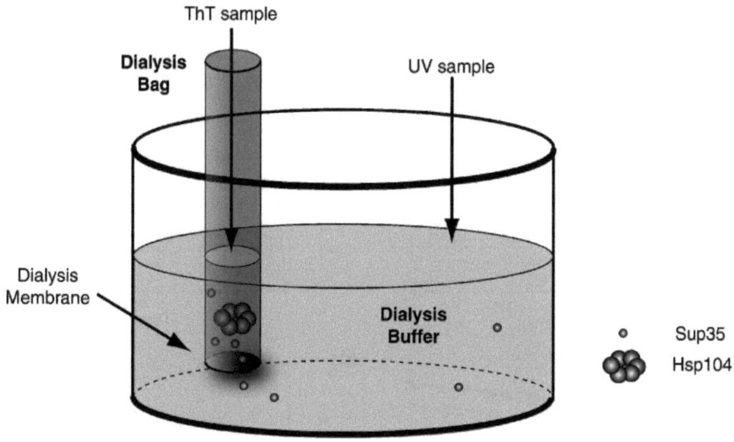

Fig. 9 Schematic representation of the dialysis experiment. A Sup35$^{5-26}$ solution is incubated in a dialysis bag in the presence and absence of Hsp104 and ATP using different MWCO membranes. ThT and tyrosine fluorescence are measured separately to quantify the number of Sup35$^{5-26}$ fibrils in the dialysis bag and the amount of peptide diffusing out of the dialysis bag, respectively

(Fig. 10a). Tyrosine fluorescence intensities are normalized such that 100% corresponds to complete equilibration between the dialysis solution and the reservoir. At the end of the experiments, 60% of Sup35$^{5-26}$ diffused through the membrane in the absence of Hsp104, whereas in the presence of Hsp104, 80% of Sup35$^{5-26}$ could escape the dialysis bag in the presence and absence of ATP, respectively (Fig. 10a).

The experiment was repeated using a 10-kDa MWCO membrane, corresponding to the size of tetrameric Sup35$^{5-26}$. At the end of this experiment, approximately 80% of Sup35$^{5-26}$ diffused out of the membrane in the absence of Hsp104. A similar level of equilibration is achieved in the presence of Hsp104 and ATP. However, in the latter case the initial rate of diffusion is increased by a factor of 2 (Fig. 10c). No amyloidogenic intermediates are observed in the ThT fluorescence assay using the 10-kDa MWCO membrane, irrespective of the presence of Hsp104 and ATP (Fig. 10d). These observations suggest that, prior to aggregation, Sup35$^{5-26}$ forms small oligomeric species, which can pass through the 10-kDa membrane but are retained in the case of the 3.5-kDa-cutoff membrane.

2.7
Analysis of the Sup35$^{5-26}$ Disaggregation Process

The temperature dependence of the intensities of the 1D ^{1}H NMR spectra in Fig. 2 indicate that Sup35$^{5-26}$ molecules in fibrillar structures are in chemical exchange with Sup35$^{5-26}$ molecules in solution. This raised the question as to

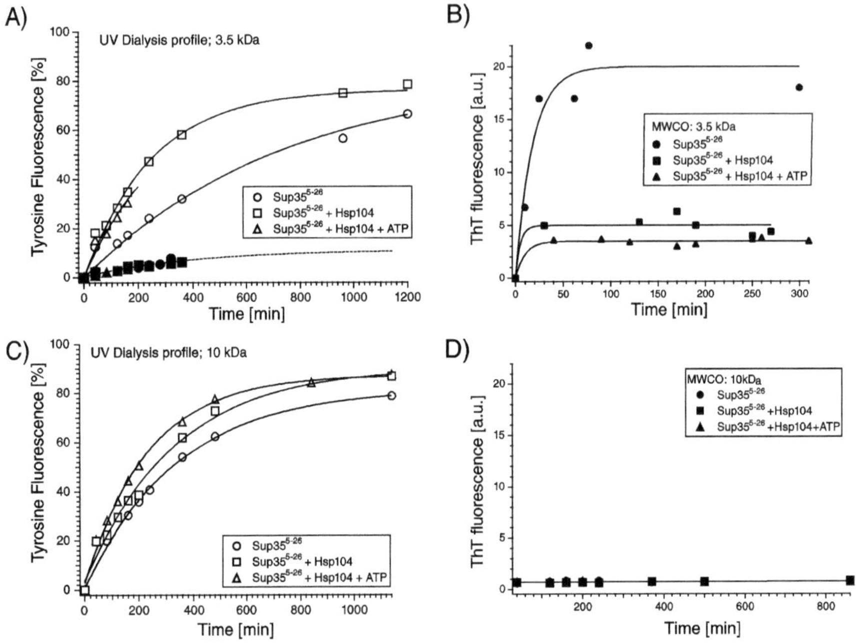

Fig. 10 Experimental results of the dialysis experiment. After solubilization of Sup35$^{5-26}$, the amount of peptide diffusing out of the dialysis bag is quantified via the tyrosine fluorescence (**a** MWCO = 3.5 kDa; **c** MWCO = 10 kDa). Addition of Hsp104 to the Sup35$^{5-26}$ solution increases significantly the rate of diffusion through the membrane. A 100% tyrosine fluorescence intensity corresponds to equilibration of the peptide solution between the dialysis bag and the dialysis buffer. The *solid symbols* in **a** represent the intensities in tyrosine fluorescence measurements using preaggregated Sup35$^{5-26}$. The amount of aggregated Sup35$^{5-26}$ is quantified using a ThT fluorescence assay (**b** MWCO = 3.5 kDa; **d** MWCO = 10 kDa). Using a 3.5-kDa MWCO membrane, addition of Hsp104 strongly reduces the buildup of the ThT fluorescence signal

whether aggregation and disaggregation follow distinguishable pathways. In order to follow the disaggregation of preformed aggregates and to better understand the specific role of Hsp104 in this process, aggregated Sup35$^{5-26}$ was injected into the dialysis bag. Using a 3.5-kDa MWCO membrane, we found that only 9% of Sup35$^{5-26}$ diffused out of the membrane irrespective of the presence of Hsp104 (Fig. 10a, filled symbols). Using a 10-kDa MWCO membrane, which allows the passage of tetrameric Sup35$^{5-26}$, we found approximately 40% of the Sup35$^{5-26}$ peptide in the dialysis reservoir after 300 min (Fig. 11). At the same time, ThT fluorescence was not affected. After changing the membrane cutoff value to 25 kDa (at 400 min), we observed that ThT fluorescence rapidly decreased. In the absence of Hsp104, ThT fluorescence was reduced to 75% after 900 min. In the presence of Hsp104 and ATP, the decrease was enhanced and ThT fluorescence was reduced to 50%. Irrespective

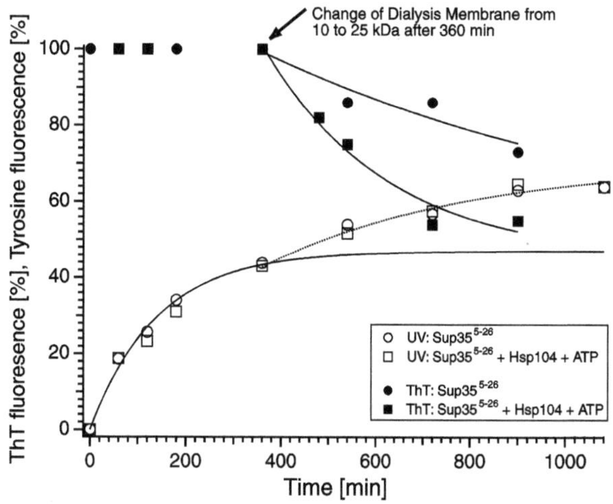

Fig. 11 Dialysis experiment employing preaggregated Sup35$^{5-26}$. Tyrosine and ThT fluorescence are indicated by *open* and *filled symbols*, respectively. Change of the MWCO of the dialysis membrane from 10 kDa to 25 kDa after approximately 360 min induces a significant decrease in the ThT fluorescence intensity, indicating a loss of fibrillar Sup35$^{5-26}$

of the presence of Hsp104, 60% of the total amount of Sup35$^{5-26}$ is diffusing through the dialysis membrane. A MWCO of 25 kDa corresponds to approximately decameric Sup35$^{5-26}$. The results indicate that disaggregation does not occur by substraction of monomers, but rather by removal of oligomers consisting of eight to ten Sup35$^{5-26}$ monomer units.

2.8
Interaction Model for Sup35 and Hsp104

We could demonstrate that the presence of catalytic amounts of the chaperone Hsp104 significantly reduces the aggregation of Sup35$^{5-26}$. At lower temperatures ($T = 12\,^\circ$C), only the rate and the extent of conversion from the soluble into the aggregated form is reduced. At higher temperatures (27 °C), we find that Hsp104 also induces structural changes in Sup35$^{5-26}$. It has been shown previously that temperature can affect the equilibrium between different oligomeric states of Sup35 [61]. Our observations suggest that Hsp104 may facilitate this process by either lowering the activation barrier or energetically favoring an alternatively folded state of Sup35$^{5-26}$.

A model of the different equilibria between monomeric and oligomeric species of Sup35$^{5-26}$, and their interaction with Hsp104, is depicted in Fig. 12. After dissolution, Sup35$^{5-26}$ quickly forms oligomers. These oligomers may be related to nuclei that precede fibril formation or aggregation [62]. Ultimately, aggregation of Sup35$^{5-26}$ leads to the formation of high oligomeric structures that bind the amyloid-specific dye Thioflavin T (ThT). Their mo-

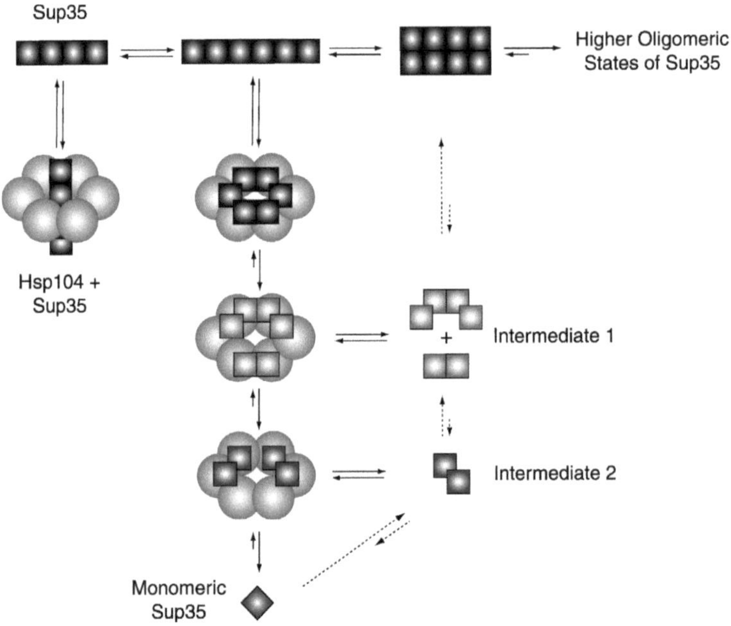

Fig. 12 Interaction model for Hsp104 and Sup35. $Sup35^{5-26}$ can exist in various oligomeric states. At 12 °C, Hsp104 induces an increase in population of smaller oligomeric states of $Sup35^{5-26}$. Higher oligomeric states ($n > 6$) of $Sup35^{5-26}$ cannot interact with Hsp104. At 27 °C, refolding of an initially hexameric form of $Sup35^{5-26}$ to a monomeric form via two intermediate species is observed. This is reflected by a characteristic chemical shift change of the resonances of $Sup35^{5-26}$. The initial, hexameric species of $Sup35^{5-26}$ display the largest interaction with Hsp104, whereas the lower oligomeric states show a reduced interaction with the chaperone. The implicated oligomeric states of $Sup35^{5-26}$ and Hsp104 are verified experimentally by DOSY NMR and DLS experiments

lecular weight is too large to be detected by solution-state NMR spectroscopy. This process is responsible for the loss of signal shown in Fig. 1.

2.9 Substoichiometric Amounts of Hsp104 Suppress the Aggregation of $Sup35^{5-26}$

How can we explain the observation that the addition of small amounts of Hsp104 to a solution of $Sup35^{5-26}$ strongly reduces the aggregation of the peptide? Our data suggest that Hsp104 may simply bind to the nuclei and thus block further polymerization. Alternatively, the chaperone may be able to dissolve nuclei and/or aggregated peptide, even in the absence of ATP. A third model, in which Hsp104 binds to monomeric peptide and prevents its incorporation into aggregates, is less probable since, according to our data, binding of monomeric $Sup35^{5-26}$ to Hsp104 is relatively weak and Hsp104 is active in substoichiometric amounts.

2.10 Oligomeric Species of $Sup35^{5-26}$ Dissociate in the Presence of Hsp104

Our results obtained with DOSY NMR suggest that at 27 °C, the freshly dissolved peptide is mainly hexameric (Fig. 12). It is not known whether these molecules or some other species in the initial oligomer distribution represent the nuclei for the subsequent polymerization process. However, the concentration of nuclei must be high in order to account for the high rate of aggregation.

After addition of Hsp104 to $Sup35^{5-26}$, the hexameric state is slowly converted to smaller species in a multistep process. The different intermediates are characterized by their specific chemical shifts. A characterization of the exact oligomeric state of $Sup35^{5-26}$ is difficult. First, multiple oligomeric states are likely to be in fast exchange. Second, we cannot assume that $Sup35^{5-26}$ is a spherical molecule. Therefore, only weighted averages of diffusion constants and approximated molecular weights can be determined.

Eventually, all soluble peptide becomes monomeric. This observation indicates that, under the conditions of the experiment, the monomeric peptide may be more stable than the oligomeric species. The 1D NMR spectra thus monitor the establishment of the equilibrium distribution, and Hsp104 may serve as a catalyst in this process. This is difficult to assess, since the reaction can only be observed in the presence of the chaperone. According to this notion, the initial distribution of $Sup35^{5-26}$ does not reflect the equilibrium, but is highly enriched in oligomeric species that serve as seeds for rapid aggregation. After its synthesis, the peptide was purified by reversed-phase HPLC and subsequently lyophilized. It is possible that organic solvents stabilize the oligomeric states of $Sup35^{5-26}$ and are thus responsible for the over-representation of these species in the starting material. Furthermore, the dissolution process will generate high local concentrations of peptide, which may in turn lead to the formation of metastable oligomers. Thus, one would expect that the "equilibrated" peptide solution would show a decreased tendency to aggregate, because it no longer contains nuclei. This is indeed what we observe: after dialysis of the $Sup35^{5-26}$/Hsp104 solution, $Sup35^{5-26}$ no longer aggregates, even in the absence of Hsp104 (data not shown).

Like many other chaperones, Hsp104 requires the hydrolysis of ATP to carry out its biological function [35, 56]. It would thus be obvious to investigate whether the presence of ATP changes the way in which Hsp104 interacts with $Sup35^{5-26}$. However, this turned out to be very difficult. The turnover rate of Hsp104 for ATP is in the order of $k_{cat} \approx 70\,\text{min}^{-1}$ [53]. Thus, the rate of ATP hydrolysis would be $\sim 4\,\text{mM}\,\text{min}^{-1}$ in cases where $\text{Hsp104} : \text{Sup35}^{5-26} = 1 : 25$. The required concentration of ATP is not amenable for an NMR experiment, which takes several minutes to hours. Therefore, we focused on the intrinsic binding properties of Sup35 to Hsp104 in the present study. It has been shown for a number of ATP-dependent molecular chaperones that some features of their mechanism, e.g., substrate specificity, can be studied in the absence of nucleotide [63, 64]. Experiments have been carried out using a nonhydrolyzable ATP analogon (AMP-PNP) to study the influence of ATP on binding of Sup35 to Hsp104 (data not shown). These experiments show quantitatively the same behavior as in the case without AMP-PNP.

2.11 Interactions Between Sup35$^{5-26}$ and Hsp104

The chemical groups on the peptide that are involved in the interaction with Hsp104 could be identified using STD NMR. STD signals were observed for the side chains of Asn/Gln and Tyr. Since binding kinetics are presumably fast compared to the time required for recording an STD experiment, a change in the STD signal over time can only be caused by changes in the concentration or in the affinity of the ligand. Our data show that Sup35$^{5-26}$ remains soluble throughout the experiment, and thus its total concentration is constant. We can conclude from DOSY experiments and 1D NMR spectra that both the degree of oligomerization and the structure of the peptide change over time. Therefore, the time dependence of the STD signal can be attributed to the interconversion of the various Sup35$^{5-26}$ species. Moreover, conclusions can be drawn about the relative affinities of these species toward Hsp104. The data recorded at 27 °C suggest that Hsp104 preferentially binds to oligomeric peptide species. Although we have currently no structural information on the peptide binding site(s) of the Hsp104 hexamer, this is a rather intriguing result. It indicates that peptide binding may be cooperative, i.e., the peptide oligomer behaves like a multivalent ligand that binds to several Hsp104 subunits simultaneously (Fig. 12). This idea is in agreement with recent results of Lindquist and coworkers, who showed that polylysine stimulates ATP hydrolysis by Hsp104 in a cooperative manner [65]. Multivalent peptide binding has also been observed for the molecular chaperone GroEL from *E. coli*, which consists of two rings of seven subunits each [66]. Another interesting observation in this context is that the octameric species, which dominates the initial distribution at 12 °C, does not bind to Hsp104. One can speculate whether this molecule is not recognized because of its size or because of its inappropriate symmetry.

2.12 Equilibrium Between Fibrillar and Soluble Sup35$^{5-26}$

An increase of the temperature causes a shift of the aggregation equilibrium toward the soluble form (Fig. 13). This is in agreement with an earlier report on Sup35-NM (N-terminal (residues 1–123) and center domain (residues 124–253) of Sup35), which shows that temperature influences the

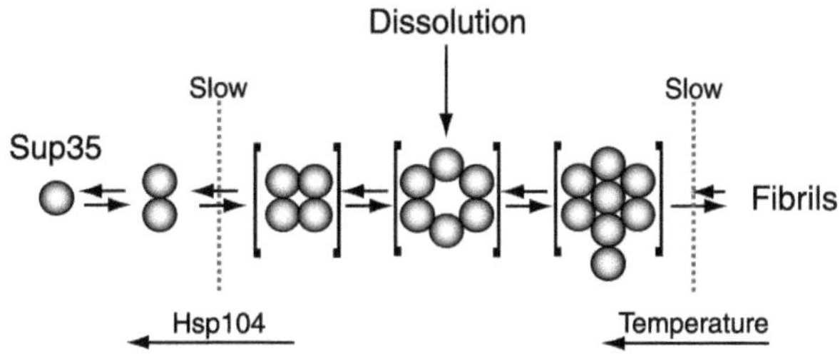

Fig. 13 Schematic representation of oligomeric intermediate states of Sup35 aggregation upon interaction with Hsp104

equilibrium between the soluble and the aggregated form of Sup35 [61]. The intensities of the ^{1}H NMR resonances represented in Fig. 2 show a hysteresis behavior upon heating and cooling. The establishment of the equilibrium between soluble and aggregated Sup35$^{5-26}$ must therefore be slow, since the temperature was set 20 min before each individual spectrum was recorded. This is in agreement with STD experiments. An STD signal would be expected if Sup35$^{5-26}$ exchanged rapidly between its soluble and aggregated forms. Fast exchange between the soluble and the aggregated state of a protein was observed previously for Alzheimer's disease β-amyloid peptide [67]. The observation that the ^{1}H chemical shift of Sup35$^{5-26}$ recorded at different temperatures does not vary significantly argues for a model in which the oligomeric states of Sup35$^{5-26}$ are rather stable. The finding that the diffusion constant—as measured in DOSY experiments—is changing as a function of temperature might therefore be indicative of a change in equilibrium between critical oligomers which are different in size ($4 \leq n \leq 8$). Chemical exchange between fibrillar and soluble states can, however, be ruled out.

2.13 Influence of Hsp104 on Sup35$^{5-26}$ Fibril Assembly

The aggregation kinetics of solubilized Sup35$^{5-26}$ in the dialysis experiment (Fig. 9) can be characterized as competition between an aggregation and a dilution process. The former is driven by fibril assembly and the latter is governed by diffusion of Sup35$^{5-26}$ out of the dialysis bag. In the case of the 3.5-kDa MWCO membrane and in the absence of Hsp104, less time is required for aggregation (15 min) than for dilution (300, 120, and 80 min for Sup35$^{5-26}$, Sup35$^{5-26}$ + Hsp104, and Sup35$^{5-26}$ + Hsp104 + ATP, respectively). Within 15 min, the concentration of Sup35$^{5-26}$ inside the dialysis bag is only reduced by 3 and 8% in the absence and presence of Hsp104, respectively (Fig. 9a). The rate of release of monomer is increased by a factor of 2.5 upon incubation of Sup35$^{5-26}$ together with Hsp104. ThT fluorescence is quenched at the same time. Addition of ATP has only a slight effect on both the dilution process and in preventing Sup35$^{5-26}$ aggregation. The increase of the rate of diffusion in the presence of Hsp104 and the quenching of the ThT fluorescence can only be accounted for if one assumes that Hsp104 shifts the equilibrium between different Sup35$^{5-26}$ aggregation states toward monomeric Sup35$^{5-26}$.

Using a membrane with a 10-kDa MWCO, we observe almost equal concentrations (up to 90%) of the peptide inside and outside of the dialysis bag after 1000 min, irrespective of the presence of Hsp104 (starting with solubilized Sup35$^{5-26}$). The time constants for the dilution process are—with the exception of the Sup35$^{5-26}$ sample in the absence of Hsp104—comparable to the experiment in which a 3.5-kDa MWCO membrane was employed (160, 120, and 80 min for Sup35$^{5-26}$, Sup35$^{5-26}$ + Hsp104, and Sup35$^{5-26}$ + Hsp104 + ATP, respectively). No fibrillar structures could be detected in the ThT fluorescence experiments.

Comparing the 3.5- and 10-kDa dialysis data, we find that the system is almost completely diffusion controlled in the presence of Hsp104 and ATP, since the differences in the rate of increase of the ThT fluorescence intensities are only very modest. This implies that Hsp104 must very efficiently convert oligomeric Sup35$^{5-26}$ into monomers. The fact that the rate of diffusion for Sup35$^{5-26}$ alone is different using 3.5- and 10-kDa MWCO membranes is in agreement with the observation that Sup35$^{5-26}$ is not monomeric after dissolution [47]. The exchange kinetics from oligomeric to monomeric Sup35$^{5-26}$ must be slow, as otherwise similar dialysis profiles would be expected for the 3.5- and 10-kDa MWCO membranes. Fibril assembly depends critically on the concentration of Sup35$^{5-26}$. However, it is not expected that a decrease in concentration by 3–8% within 15 min (the characteristic time required for aggregation) results in the abolition of fibril formation. The quenching of ThT fluorescence observed in the presence of Hsp104 using a 3.5-kDa MWCO membrane must therefore be due to a destabilization of oligomeric Sup35$^{5-26}$ molecules. This implies, however, that fibril assembly occurs by agglomeration of $(\text{Sup35}^{5-26})_n$ oligomers, with $n \approx 4$. This seems to be in disagreement with results obtained by Weissman and coworkers who demonstrated that fibril growth occurs via monomer addition [68]. However, these authors were investigating the fibril elongation process, not the formation of critical oligomers which precedes fibril formation. In addition, we focus only on the N-terminal part of the N-domain of Sup35 in our studies. It is conceivable that the repeat sequence in the N-domain (residues 40–114) of Sup35 and the M-domain modulate the aggregation behavior.

2.14 Influence of Hsp104 on Sup35$^{5-26}$ Fibril Disaggregation

Equilibrium dialysis experiments on preaggregated Sup35$^{5-26}$ show that 9% of the Sup35$^{5-26}$ monomers are released from the dialysis bag using a 3.5-kDa

MWCO membrane (Fig. 9a). This demonstrates that fibrils are not static structures, but that $Sup35^{5-26}$ molecules in fibrillar structures are undergoing chemical exchange between the fibrillar state and the soluble state. Using dialysis bags with a 10-kDa MWCO membrane, we observe a release of 40% of the total amount of $Sup35^{5-26}$ present in the dialysis bag. This release is not accompanied by a decrease in the population of amyloidogenic intermediates and fibrils, as detected by ThT fluorescence (Fig. 10). Upon changing the membrane to a MWCO of 25 kDa, a fast release of $Sup35^{5-26}$ out of the dialysis bag and simultaneously a decrease in ThT fluorescence was observed. This demonstrates that $(Sup35^{5-26})_{6-10}$ molecules can be subtracted from fibrillar structures which induces a breakdown of ThT fluorescence. We therefore speculate that critical oligomers important for fibril disassembly must consist of six to ten $Sup35^{5-26}$ molecules. $Sup35^{5-26}$ oligomers consisting of four or fewer molecules are not directly linked to fibril disassembly, since a MWCO of 10 kDa does not influence the ThT fluorescence intensity. The experiments also indicate that these $(Sup35^{5-26})_n$ oligomers ($6 \leq n \leq 10$) must be sufficiently stable, since an interconversion to $(Sup35^{5-26})_{n\leq 4}$ oligomers would have resulted in a loss of ThT fluorescence at the beginning of the experiment ($0 < t < 360$ min).

Irrespective of the presence of Hsp104, we find that the same amount of $Sup35^{5-26}$ (60%) diffuses out of the membrane, though ThT fluorescence showed a differential effect (faster decrease in fluorescence for $Sup35^{5-26}$ fibrils in the presence of Hsp104) (Fig. 10). This behavior is expected if we assume that Hsp104 interacts only with low oligomeric weight $(Sup35)_n$ oligomers ($n \leq 6$–8). A shift of equilibrium among $(Sup35^{5-26})_n$ oligomers ($1 \leq n \leq 8$) would not result in a change of the rate of diffusion out of the dialysis bag. However, dissolution of $(Sup35^{5-26})_n$ oligomers ($n = 6$–8) by Hsp104 would imply the recruitment of new $(Sup35^{5-26})_{6-8}$ oligomers out of $Sup35^{5-26}$ fibrils, and therefore induce a destabilization of the fibril structure.

2.15
Structural Changes in $Sup35^{5-26}$ Upon Interaction with Hsp104

The monotonously decaying STD intensity of intermediate 2 indicates that the monomeric peptide does not interact with Hsp104. The observation that signals arising from exchangeable protons disappear in the course of the 1D ^{1}H NMR spectra of $Sup35^{5-26}$, whereas the intensity of protons from nonexchangeable sites remains constant, suggests that hydrogen bonds, in which amide protons are protected from exchange through the formation of hydrogen bonds, are lost. This conclusion is supported by the NMR diffusion data. Moreover, the chemical shift dispersion for the two tyrosine aromatic resonance lines becomes degenerate, indicating a loss of structure in both intermediate 2 and the release state of $Sup35^{5-26}$.

2.16 Implications for Fibril Assembly and Prion Propagation

The proposed model for the interaction between Sup35 and Hsp104 also has possible implications for the assembly process of amyloid fibrils, as indicated by dashed lines in Fig. 11. It is plausible to assume that monomeric Sup35 associates to small oligomers and eventually larger species that may serve as nuclei for amyloid fibril formation or aggregation. In the absence of Hsp104, these intermediates are short-lived and cannot be detected by NMR. However, in the presence of the chaperone, they are stabilized and therefore amenable to a structural investigation [61].

With respect to prion propagation, our observations favor a model in which Hsp104 interacts only with low oligomeric state ($4 \leq n \leq 6$) species of Sup35$^{5-26}$, as has been postulated by Lindquist and coworkers [37, 38]. No interaction between Hsp104 and Sup35$^{5-26}$ ($n > 6$) could be detected. The latter observation comes from experiments that were carried out at 12 °C, since species with $n > 6$ are only stable at lower temperature. We cannot, however, rule out the possibility that Hsp104 also interacts with fibrillized Sup35$^{5-26}$. At 27 °C, fibrils are not formed: the total intensity (Fig. 5) is approximately constant during the experiment. At 12 °C, however, a significant amount of Sup35$^{5-26}$ is aggregating and is presumably forming fibrils, which are too large to be detected by solution-state NMR. The state-of-the-art detection limit is in the order of $M < 900$ kDa [69]. This would correspond to an oligomer comprising ~ 350 Sup35$^{5-26}$ subunits. Scheibel and Lindquist reported that the nucleus which precedes fibril elongation consists of 10–30 subunits of the N + M domain of Sup35 [61]. Therefore, we exclude the possibility that interactions between Hsp104 and a nucleus of this size are not detected. To characterize interactions between fibrillized Sup35$^{5-26}$ and Hsp104, solid-state NMR experiments are envisaged. Such interactions have recently been reported by Yoshida and coworkers [70]. However, their importance with respect to prion propagation is so far unclear.

3 Alzheimer's Disease β-Amyloid

Alzheimer's disease is the most abundant age-related neurodegenerative disease. The β-amyloid peptide, which exists in different lengths (39–43 amino acids), is generated after processing of a transmembrane protein, APP [71, 72]. The disease is characterized by two fundamental events: the accumulation of insoluble fibrillar aggregates of Aβ, and the degeneration and death of neurons in the brain regions that are concerned with learning and memory processes. Abnormal protein deposition is also a shared characteristic of other age-related neurodegenerative diseases, such as Parkinson's disease,

Huntington's disease, and the prion diseases. There is increasing evidence that the mechanism of this aggregation may be similar in each of these diseases [73]. At the same time, it is found that probably not the fibrillar state, but a protofibrillar state which is in fast equilibrium with a so-called LMW (low molecular weight) state, is responsible for its neurotoxicitiy [74–76]. Most interestingly, it was found recently that antibodies can specifically recognize these soluble oligomers [77]. So far, the molecular basis for the neurotoxicity of Aβ could not be identified. It is speculated that $A\beta^{1-40}$ forms a pore in the membrane and leads to an unregulated flux of Ca^{2+} into and out of the cell [76]. Alternatively, $A\beta^{1-40}$ may generate radicals that are neurotoxic themselves [78–80]. However, it is not understood why only certain aggregation states of $A\beta^{1-40}$ should be associated with neurotoxicity given the identical primary structure of $A\beta^{1-40}$. So far, NMR structural studies of $A\beta^{1-40}$ show that the peptide is mostly unstructured in aqueous solution [81, 82]. It was shown as well that $A\beta^{1-40}$ adopts a helix-turn-helix structure upon addition of trifluoroethanol (TFE) [83, 84] or sodium dodecyl sulfate (SDS) [85, 86]. Recently, solid-state NMR structural data have also become available for $A\beta^{1-40}$ in the fibrillized state [8]. Interestingly, a similar loop at approximately the same position in the primary structure was found around residues S^{26}NKG, which was already observed in the SDS solution-state structure [85]. In the solid-state structure, the only charged side chains in the core are those of D23 and K28 which form a salt bridge [8]. The C-terminus of the peptide folds back onto the hydrophobic core, so that hydrophobic core and the aliphatic side chains of the C-terminus come into close contact.

3.1 Solution Conditions Change the Aggregation State of Aβ

In order to study the β-amyloid aggregation process, we searched for (physiological) solution-state conditions which would possibly allow shifting of the chemical equilibrium between different aggregation states of $A\beta^{1-40}$ [67]. Goto and Aimoto [87] showed by CD spectroscopy that addition of anions to the buffer solution can induce an α-helical structure in otherwise unstructured amphiphilic polypeptides. The stabilization is proportional to the strength of the anion added in the order of triphosphate^{4-} > diphosphate^{3-} > phosphate^{2-} > SO_4^{2-} > ClO_4^- > Cl^- > F^-. An increase in propensity for the α-helical structure in $A\beta^{1-40}$ should—in analogy to the results obtained by Goto and Aimoto—decrease the propensity for amyloid aggregate formation. We added 100 mM of F^-, Cl^-, NO_3^-, ClO_4^-, and SO_4^{2-} to the buffer solution and observed an increase in solubility for $A\beta^{1-40}$ in the order indicated (data not shown). In all NMR experiments, $A\beta^{1-40}$ was directly dissolved in 450 μL of buffer and 50 μL of D_2O yielding a final $A\beta^{1-40}$ concentration of 0.5 mM. No attempt was made to dissolve possible aggregates before starting the experiment. In all samples, β-mercaptoethanol was added at a concentration of

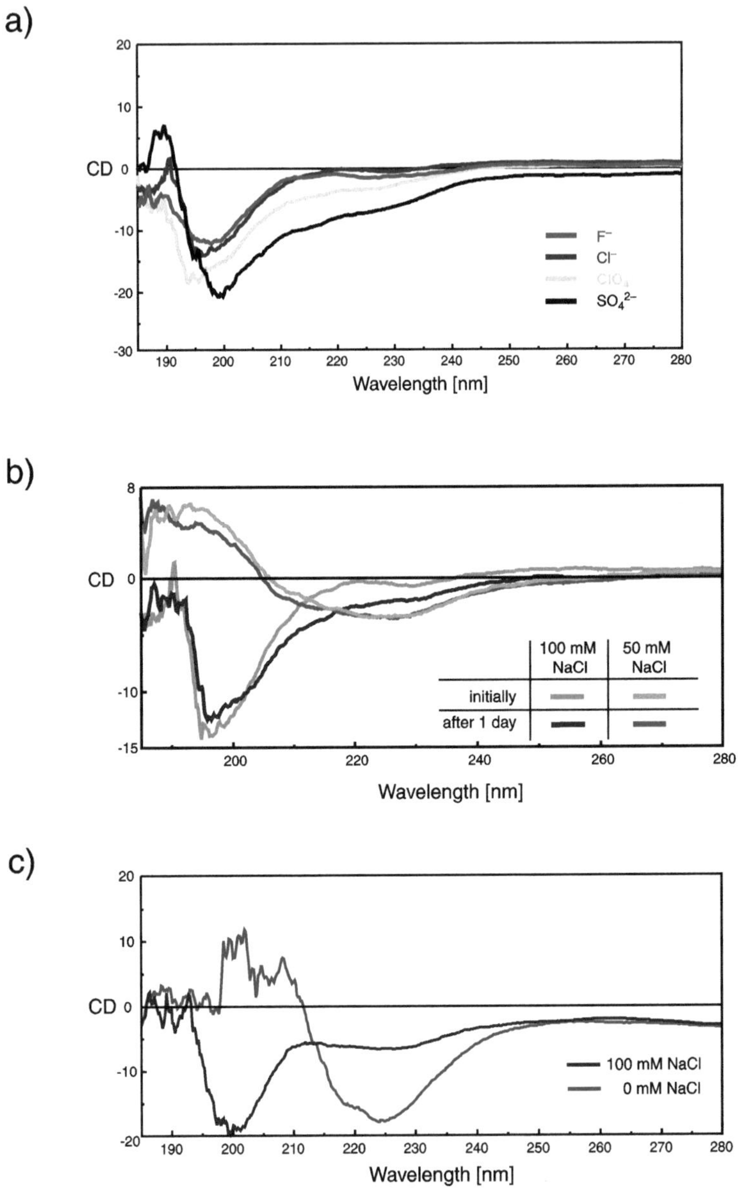

Fig. 14 **a** CD spectra of 0.15 mM Aβ^{1-40} dissolved in buffer solution with additional 100 mM salt: NaF (*red*), NaCl (*dark blue*), $NaClO_4$ (*light blue*), and Na_2SO_4 (*black*). The sample was filtered through a membrane with a 0.2-µm pore diameter before transfer into the CD cuvette. **b** CD spectra of a 0.15 mM Aβ^{1-40} solution employing a salt concentration of 50 mM NaCl (*red*) and 100 mM NaCl (*blue*). In contrast to **a**, the spectra were obtained without filtering the sample prior to the measurement. This preparation is analogous to the NMR sample preparation. **c** CD spectra of a 0.50 mM Aβ^{1-40} solution employing a salt concentration of 0 mM NaCl (*red*) and 100 mM NaCl (*blue*)

200 μM as an internal standard for the DOSY experiments. Care was taken to reproduce the same peptide concentration in all NMR experiments. The error in balancing $A\beta^{1-40}$ is estimated to be in the order of 5%.

The CD spectrum displays a decrease of contributions due to random coils upon increase of the anionic strength (Fig. 14a). At the same time, the level of contributions of α-helical secondary structure elements is increased, as could be shown by a quantitative analysis using the programs CDSSTR and CONTINLL [88]. Before the measurements, all samples are sterile filtered (membrane pore diameter: 0.2 μm) to eliminate large aggregates. We can therefore exclude that the spectrum is due to fibrillar structures in the CD sample. A similar effect is observed for decreasing concentrations of, e.g., Cl^-. In this case the sample was not filtered prior to the measurement. Soluble and aggregated components are therefore both present in the sample. The CD spectrum is stable in the order of 24 h, which is a prerequisite for the solution-state NMR experiments presented below. A reduction of the salt concentration from 100 to 50 mM (at an $A\beta^{1-40}$ concentration of 0.15 mM) results in a decrease of unstructured $A\beta^{1-40}$ in the CD spectrum (Fig. 14b). Quantitative fitting of the data yields a distribution for α-helical, β-sheet, turn, and random coil secondary structure elements of 24.1, 37.9, 19.5, and 22.1% for 100 mM NaCl, and of 58.5, 15.9, 14.3, and 12.4% for 50 mM NaCl, respectively. A high degree of α-helical structure for an intermediate oligomeric state during Aβ aggregation was observed previously [89]. At higher peptide concentrations (0.5 mM), a similar effect is visible (Fig. 14c). For the NMR experiments, conditions are employed, where $A\beta^{1-40}$ is soluble for several days (0.5 mM $A\beta^{1-40}$, 50 mM phosphate, 50 mM sodium sulfate or chloride, pH 6.9).

Figure 15 shows a comparison of the 1D ^1H NMR spectra that are obtained of $A\beta^{1-40}$ dissolved in buffer containing 50 mM NaCl and 50 mM Na_2SO_4. We observe that the intensity in the spectrum of the sample containing Cl^- as the anion is reduced by a factor of 0.77. Care was taken in order to assure the same amount of peptide in both samples. Since the error due to balancing errors can be estimated to be smaller than 5%, we conclude that the equilibrium between soluble and insoluble $A\beta^{1-40}$ molecules must be shifted toward the fibrillar species in the case of Cl^-. This observation is corroborated by the spectra presented in Fig. 15b. Increasing concentrations of Cl^- yield an increase in the observed signal-to-noise ratio in the 1D ^1H NMR spectra, again indicating that the equilibrium must be shifted toward a soluble monomeric structure of $A\beta^{1-40}$ at higher salt concentration. Interestingly, using Cl^- in the buffer we observe a very broad peak at around 0 ppm. The intensity of this peak is increasing with decreasing salt concentration. This resonance is not observed if we use $SO_4{}^{2-}$ as anion in the buffer (Fig. 15a). We show below that the molecule giving rise to this resonance has a diffusion coefficient corresponding to a molecular weight of > 100 kDa. We hypothesize that this resonance belongs to an oligomeric $A\beta^{1-40}$ molecule which could be

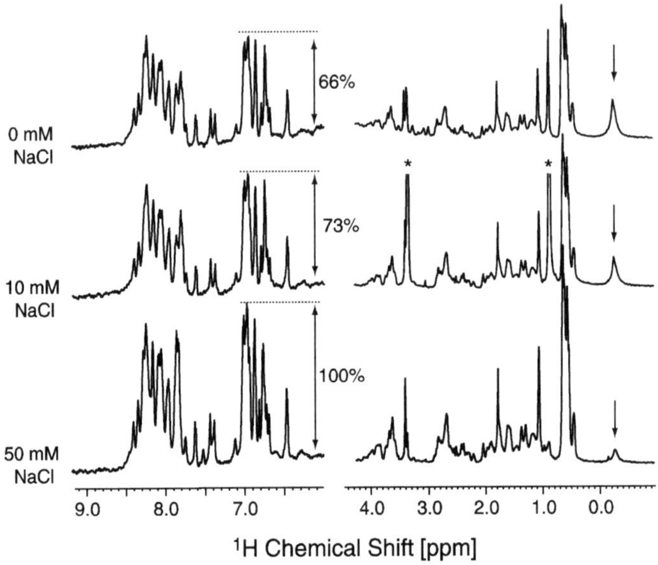

Fig. 15 1D ^{1}H NMR spectra of Aβ^{1-40} at various NaCl concentrations (*top*: 0 mM; *middle*: 10 mM; *bottom*: 50 mM). The intensity of the resonances is increasing as the salt concentration is increased. Note that the resonance upfield of 0 ppm is increased as the NaCl concentration is decreased. The amount of peptide is the same in all samples. The error of balancing Aβ^{1-40} can be estimated to be smaller than 5%

identical to a critical oligomer which was postulated to be essential for fibril formation.

In order to estimate the molecular weight of Aβ^{1-40} under these various salt conditions, we recorded DOSY experiments. The molecular weight that can be fitted to the magnetization decay curve corresponds to an average molecular weight of approximately 20.8 kDa (for SO_4^{2-}), assuming a spherical shaped Aβ^{1-40} molecule (Fig. 16). This molecular weight is larger than that expected for a monomeric Aβ^{1-40} molecule ($M = 4330$ Da). The deviation can be explained either by assuming that Aβ exists as an aggregated oligomer (on average tetramer) or by assuming that Aβ undergoes chemical exchange between a soluble and a fibrillar form, yielding on average a higher molecular weight. DOSY experiments recorded for samples using 100 mM Cl^- instead of 100 mM SO_4^{2-} as anion in the buffer solution yield a systematically higher average molecular weight in the order of ca. 24 kDa for Aβ^{1-40}, supporting the hypothesis that different oligomeric states of Aβ undergo chemical exchange. Interestingly, the broad resonance at around 0 ppm diffuses with a very small diffusion constant corresponding to a molecule with an approximate molecular weight of > 100 kDa.

In order to confirm the hypothesis that soluble Aβ^{1-40} undergoes chemical exchange with higher oligomeric state aggregates, and to show that the

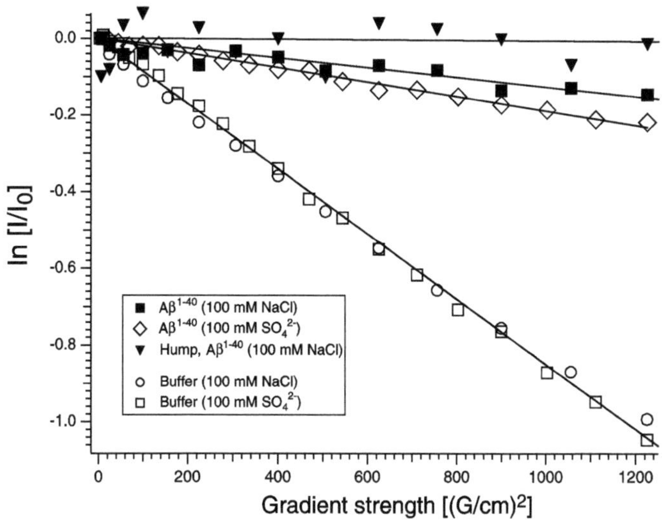

Fig. 16 DOSY NMR experimental data of $A\beta^{1-40}$. The natural logarithm of the relative intensity is represented as a function of the square of the gradient strength. Under the solution conditions used in the experiments, $A\beta^{1-40}$ ($M^{mono} = 4329.8$ Da) diffuses with a molecular weight of approximately 23 900 and 20 781 Da using 100 mM NaCl (*filled squares*) and 100 mM Na_2SO_4 (*open diamonds*) in the buffer solution, respectively. As a molecular weight standard, we use a resonance of oxidized β-mercaptoethanol ($M = 156$ Da) which is also present in the buffer of the sample (*open circles* and *squares*). The decay of the broad resonance line at around 0 ppm, which is present in samples containing Cl^-, can be fitted assuming a molecular weight of approxiamtely 14 MDa

broad resonance at around 0 ppm is related to $A\beta^{1-40}$, we have carried out STD NMR experiments (Fig. 17). The STD experiments are performed with an on-resonance saturation at around 0 ppm for 1.5 s. Significant STD intensities are observed for several resonances. As shown in Fig. 17a, the STD intensity is dependent on the employed salt conditions in the sample. In the case of Cl^-, we observe a higher intensity in the STD experiment, even though the effective number of nuclei is smaller (reduced intensity in a reference 1D ^1H experiment). Especially, aromatic resonances at around 6.8 ppm are observed which are missing in the spectrum obtained from the SO_4^{2-}-containing sample. Before performing experiments with $A\beta^{1-40}$, the STD experiments were carefully calibrated using two peptide samples, $Sup35^{6-25}$ and LPFFD. In both cases, the signal-to-noise ratio of the reference experiment (performed in the absence of a protein with a peptide binding affinity) is negligible using a comparable number of scans. We assume therefore that we observe in this experiment a soluble $A\beta^{1-40}$ molecule in chemical exchange with an insoluble, aggregated $A\beta^{1-40}$ molecule. The fibrillar state itself cannot be observed directly by solution-state NMR methods. The NMR resonances

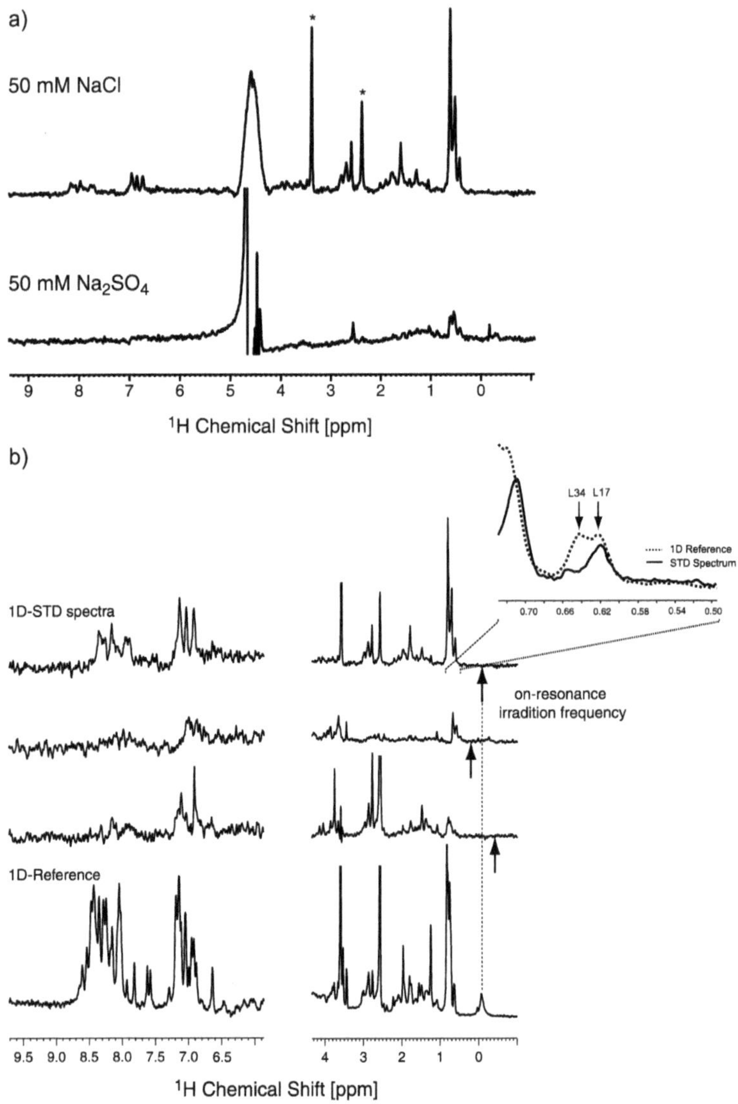

Fig. 17 **a** 1D ^{1}H STD spectra recorded for Aβ^{1-40} employing 50 mM NaCl (*top*) and Na_2SO_4 (*bottom*) in the buffer solution. In the case of NaCl, a higher STD signal intensity can be observed, suggesting that higher oligomeric states of Aβ^{1-40} (which are not visible by solution-state NMR) undergo chemical exchange with the soluble monomeric form of Aβ^{1-40}. **b** 1D ^{1}H STD spectra recorded for Aβ^{1-40} employing 50 mM NaCl in the buffer. The spectrum on the *top* is obtained after saturating the broad resonance line at 0 ppm. The spectra below are recorded after changing the on-resonance radiofrequency as indicated by the *arrow*. The highest STD intensity is obtained using a saturation frequency corresponding to 0 ppm, suggesting that the molecules giving rise to this resonance are related to the soluble form of Aβ^{1-40}. For reference, a 1D ^{1}H spectrum of Aβ^{1-40} is represented on the *bottom*

of these molecules are broadened beyond detection due to the long correlation time of an amyloid aggregate. Figure 17b represents the spectra obtained upon irradiation of the broad resonance at around 0 ppm which we hypothetically assigned to an oligomeric state of $A\beta^{1-40}$. On-resonance irradiation of the broad peak induces a strong difference signal on amide, aromatic, and aliphatic resonances of $A\beta^{1-40}$, whereas a much weaker effect is observed in the case of off-resonance irradiation. We conclude that the broad resonance must be due to an Aβ conformer which is in chemical equilibrium with soluble $A\beta^{1-40}$.

In order to identify the chemical groups involved in $A\beta^{1-40}$–$A\beta^{1-40}$ contacts, we carried out a 2D STD total correlation spectroscopy (TOCSY) experiment. The spectral assignments that have been obtained previously [82, 90] could be reproduced. Figure 18 represents the aliphatic region of the 2D STD-TOCSY spectrum (red), superimposed with a standard TOCSY for reference (black). We find that mostly the methyl groups of leucine, isoleucine, valine, and alanine side chains contribute to the interaction. We observe a high STD peak intensity for the aromatic resonances of tyrosine (Y10). Careful analysis of the 1D STD spectrum indicates that Leu-17Hδ, but not Leu-34Hδ, contributes to Aβ–Aβ interactions (Fig. 17b). Except for Tyr-10 (H$\tilde{\varepsilon}\delta$), Val-18 (H$\tilde{\beta}$Hγ), Ile-31/Ile-32 (Hγ2/Hγ1; Hγ2/Hβ), and the buffer resonances, no STD

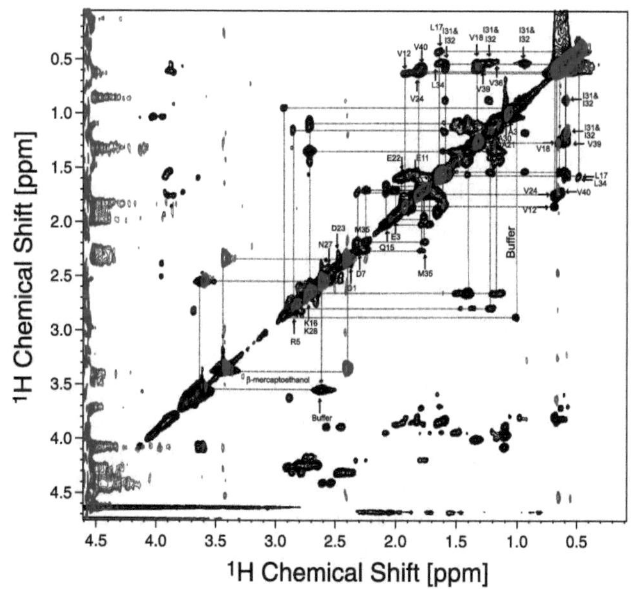

Fig. 18 Superimposition of a 2D STD-TOCSY of $A\beta^{1-40}$ (*red*) and a standard 2D TOCSY (*black*), displaying the (a) aromatic and (b) aliphatic region of the spectrum. The STD experiment yields highest intensities for those chemical groups which are involved in interactions or in chemical exchange between soluble and aggregated $A\beta^{1-40}$ molecules

intensity can be detected for cross peaks. An unambiguous assignment of all chemical groups that are involved in Aβ–Aβ contacts is therefore difficult and requires a uniformly ^{15}N, ^{13}C- labeled Aβ$^{1-40}$ sample. Work in this direction is currently in progress in our laboratory. All residues that contribute to Aβ$^{1-40}$–Aβ$^{1-40}$ interactions are represented in the scheme below:

D^{1} – **AEFR**HDSG**Y**^{10} – EVH*HQ***KLVFF**^{20} – **AED***V*G*SN***KGA**^{30} –
IIGLMVGGVV^{40}

Residues which can be unambiguously resolved are underlined and at the same time highlighted in bold letters. Alanine Hβ resonances are not resolved in the TOCSY correlation experiment. Therefore, they are represented only in bold, but are not underlined. At the current resolution, it is not possible to differentiate if Glu-22Hβ or Val-12Hβ, and Glu-11Hβ or Val-24Hβ contribute to the diagonal peak intensity in the STD-TOCSY experiment. At the same time, it is not possible to unambiguously assign a STD diagonal peak to lysine Hε (Lys-16/Lys-28) or to Phe-Hβ. A similar ambiguity exists for the chemical shift of Arg-5Hδ and His-Hβ. Therefore, these amino acids are represented all in bold. However, we find that the region around the hydrophobic core (residues 15–24) contributes most to Aβ$^{1-40}$ · Aβ$^{1-40}$ contacts.

In contrast to previous NMR studies [90–93], we find that Aβ exists in equilibrium with higher oligomeric state structures and that the peptide appears to adopt on average a tetrameric oligomeric state. Recently, DOSY studies were carried out on the fragment Aβ$^{12-28}$ at acidic pH (pH 2.9) [91]. In this early study, the diffusion constant was measured as a function of peptide concentration, in order to follow a shift of equilibrium from a monomeric to a dimeric Aβ$^{12-28}$. Gräslund and coworkers [93] investigated the same fragment (Aβ$^{12-28}$) at pH 5.0 and found that the peptide exists mostly in a monomeric, random coil conformation. Studies by Maggio and coworkers [92] on the full-length peptide Aβ$^{1-40}$ show that the peptide is mostly monomeric under NMR conditions. Size exclusion chromatography indicates, however, a population of dimeric Aβ$^{1-40}$. In this case, the peptide (0.23 mM) was directly dissolved in phosphate-buffered saline (PBS; 10 mM sodium phosphate, 100 mM NaCl, pH 7.5). We explain the different results compared to our studies by the different concentration of phosphate used in the buffer and the different peptide concentration. Similarly, Zagorski and coworkers [90] demonstrated that the peptide is predominantly monomeric at neutral pH. There, the sample was prepared using an explicit disaggregation protocol [94] to dissolve preformed fibrils and to obtain monomeric Aβ in solution.

These DOSY NMR results are in contrast to observations made in biochemical experiments which indicate the presence of oligomeric structures in vitro. Beyreuther and coworkers observe dimeric Aβ$^{10-43}$ using size exclusion chromatography and water as a solvent [95]. If salt is present in the sample (e.g., 500 mM NaF), only monomeric $A\beta^{10-43}$ is observed. Variation of the anionic strength and the concentration of the salt results in a shift between lower and higher order oligomeric states. The results that we observe in our experiments follow the same trend. Further evidence that Aβ is not monomeric in solution comes from cross-linking experiments [96]. Photoinduced cross-linking of unmodified proteins (PICUP) allowed the identification of short-lived metastable assemblies. In addition, Teplow and coworkers showed that the Aβ fibril assembly proceeds in vitro via α-helical intermediates [89]. The structural intermediates have an approximate molecular weight larger than 100 kDa, as estimated from dialysis experiments. Interestingly, upon addition of 50 mM NaCl we observe a CD spectrum which contains a high degree of α-helical structure. At the same time, we observe a very broad resonance at around 0 ppm in the ^{1}H spectrum that is related to monomeric $A\beta^{1-40}$ by chemical exchange, and for that we determined a molecular weight of > 100 kDa. We speculate that the addition of anions can stabilize this transient α-helical intermediate that was observed in the process of fibril formation.

SDS-stable Aβ oligomers are also obtained when Aβ is either incubated in the presence of apolipoprotein J under physiological conditions [97] or in cell culture medium where the oligomeric assemblies were referred to as ADDLs (amyloid-derived diffusible ligands) [98, 99]. LaDu and coworkers showed that addition of 150 mM NaCl to the culture medium significantly reduces the amount of larger oligomers and aggregates [100]. At the same time, a larger number of protofibrillar structures and oligomeric aggregates are observed by atomic force microscopy (AFM). A similar trend is observed for in vitro studies. AFM studies indicate that the presence of salt yields a faster growth of fibrillar structures [101]. We speculate that addition of salt reduces the population of oligomeric intermediate structures and thus yields larger amounts of fibrillar aggregates, but also increases the population of monomeric Aβ.

In contrast to the NMR studies referred to above, we directly dissolved Aβ in buffer. In addition, we added various salts to the buffer in order to allow for a shift of equilibrium between different oligomeric states. To our knowledge, NMR experiments reported so far were carried out without the addition of salt to the sample. The observation that we detect on average a higher oligomeric state compared to other groups is corroborated by the observation of a differential CD spectrum for differentially prepared samples. In the absence of salt, the CD spectrum displays characteristic spectral features which do not originate from random coil structures.

At the same time, we do not observe a significant change in the H^N,H^α chemical shift pattern in the presence and absence of salt (Fig. 19). We can only explain this finding by assuming that higher oligomeric states are being populated which are not detectable by solution-state NMR. This also explains the reduced signal intensity in the 1D NMR spectra. Comparing a TOCSY correlation spectrum recorded without and with 50 mM NaCl (Fig. 19), we find

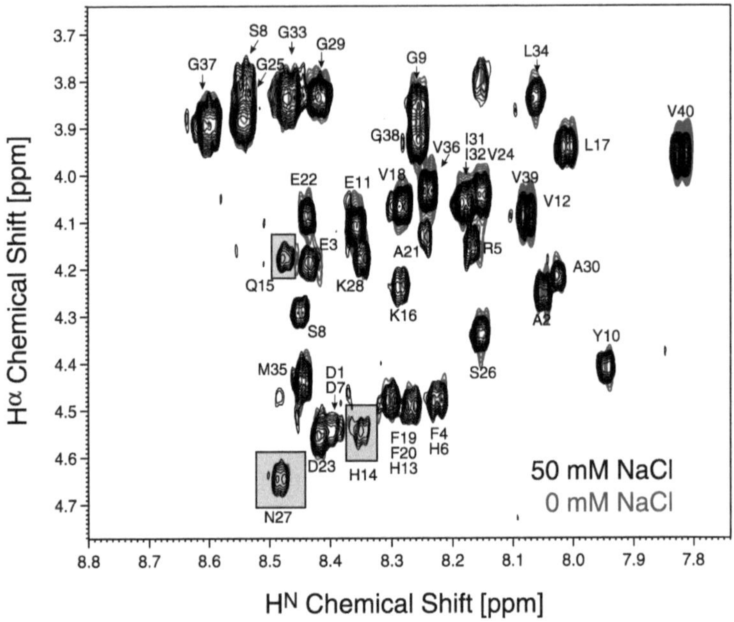

Fig. 19 Superposition of $A\beta^{1-40}$ 2D TOCSY spectra employing 50 mM NaCl (*black*) and 0 mM NaCl (*red*) in the buffer. The H^N,H^α spectral region is displayed. Assignments are given in the figure

that the chemical shifts of Gln-15, Asn-27, and His-14 are mostly affected. Removal of the salt induces a broadening of the respective H^N and H^α resonances resulting in the disappearance of the respective signal. We therefore speculate that the anion binds to positively charged side chains and, this way, prevents peptide–peptide interactions. A very similar result is obtained if we compare TOCSY correlation spectra that were recorded with 50 mM NaCl and 50 mM Na_2SO_4 (data not shown). Again, the resonance frequency of the H^N,H^α correlation peaks of His-13, His-14, Gln-15, Val-24, Ser-26, and Asn-27 are affected. Residues which are influenced upon addition or variation of salt are represented in italics in the scheme above displaying the $A\beta^{1-40}$ primary structure. At the same time, using Cl^- as the anion in the buffer, the aromatic resonances of His-13 and His-14 are broadened and only the cross peak involving His-6 in the aromatic part of the spectrum can be detected. The respective cross peaks are clearly visible in the SO_4^{2-} sample. This effect again might be due to the differential chemical environment of a SO_4^{2-} anion in contrast to a Cl^- anion. Interestingly, Ser-26 and Asn-27 are also affected. In the structure that was obtained on fibrillar $A\beta^{1-40}$ [8], S26 is involved in a loop (SNKG) which is stabilized by a salt bridge between Asp-23 and Lys-28. We observe a correlation between the binding of anions to this loop and the aggregation characteristics of $A\beta^{1-40}$. We therefore speculate that the folding and stability of this loop are influenced by the presence of anions.

Solution-state NMR structural studies revealed a collapsed coil structure of Aβ in an aqueous environment [81, 82, 102]. Characteristic restraints in this structure are a nuclear Overhauser effect (NOE) contact between the S-methyl group of Met35 and the side-chain and backbone atoms of Phe-19, Ala-21, and Lys-28. Furthermore, medium-range contacts between Ile-31/Ile-32 and the H^N atom of Ser-26 are observed. The side chain of Val-18 was determined to be buried within the central hydrophobic core of Aβ. Dynamic studies indicate that the central region of Aβ (residues 5–35) must adopt a relatively compact kind of molten globular state [82, 103]. The N- and C-terminal residues show increased mobility. In contrast, a helix-turn-helix structure was observed in aqueous SDS micellar environment [85], with the turn located around residues Ser-26, Asn-27, Lys-28, and Gly-29. If $A\beta^{1-40}$ undergoes chemical exchange between a monomeric form and an oligomeric state, the NOE spectroscopy (NOESY) correlation peaks would in fact be tr-NOE [104] correlations and should reflect the structure in the oligomeric state due to its high correlation time. Also, the STD signals reflect these NOE contacts and correspond to the tertiary contacts observed in aggregated $A\beta^{1-40}$. In the correlation experiments that we have carried out so far, we do not observe a change in the chemical shift pattern as a function of time. Such a change would be indicative of a structural rearrangement in the course of time. At the same time, the diffusion constant measured for $A\beta^{1-40}$ in DOSY experiments stays constant over a period of approximately 1 week (data not shown). We can therefore exclude the influence of a kinetic effect on our data.

The STD NMR experiments give information about the aggregation mechanism of $A\beta^{1-40}$. We observe the largest STD peak intensities for methyl and aromatic resonances in the neighborhood of the hydrophobic core of $A\beta^{1-40}$. The flanking region around this hydrophobic core shows reduced STD intensities. For example, we observe only weak intensity for Lys-Hε, and no STD signal for Lys-Hγ/β. Since the STD peak intensity is correlated to the proximity of the respective chemical groups to the amyloid aggregates, we conclude that the region around the hydrophobic core (residues 15–24) contributes most to $A\beta^{1-40}$–$A\beta^{1-40}$ intermolecular interactions. Therefore, hydrophobic interactions determine the association of Aβ monomers, before electrostatic interactions, like formation of salt bridges, can occur. This is in agreement with thermodynamic data, such as those from isothermal titration calorimetry (ITC) experiments, which show that aggregation is initially entropy driven [105]. The release of structured water is the driving force for self-association. Chemical shift changes upon addition or variation of salt occur predominantly in the flanking region around the hydrophobic core residues. We observe that residues His-13, His-14, Gln-15, Val-24, Ser-26, and Asn-27 are mostly affected. We therefore conclude that anions can bind to these positively charged side chains and attenuate protein interactions by preventing attractive electrostatic interactions.

The intensities in the 1D STD experiment recorded for Aβ dissolved in NaCl containing buffer are significantly higher compared to the intensities which are obtained if Na_2SO_4 is present in the solution. The increased STD intensity in the case of Cl^- can be explained by a decreased dissociation constant which would allow a more efficient transfer of saturation. An exact quantitative analysis of the STD amplification factor with respect to the dissociation constant is difficult, since the saturation difference signal depends on many parameters like temperature and proton density, and requires knowledge about the intermediate aggregation states of Aβ. A detailed theoretical characterization of the STD amplification factor as a function of various experimental conditions is given by Krishna and coworkers [106]. In theory, a comparison of STD interactions in Cl^- and SO_4^{2-} would allow identification of the chemical groups which become increasingly important for aggregation, as the equilibrium is shifted toward the aggregated state.

4
β-Amyloid and αB-Crystallin

Recently, molecular chaperones were identified as binding partners of Aβ using a transgenic *Caenorhabditis elegans* model to search for intracellular factors that can contribute to the neurotoxicity and metabolism of Aβ [107]. Importantly, a deranged expression profile of molecular chaperones was observed in the brain of Alzheimer's disease (AD) patients [108]. Especially, a major lenticular, non-tissue-specific, small heat shock protein (sHsp), αB-crystallin, was found to coprecipitate with Aβ [108, 109]. Further evidence for a relation between AD and αB-crystallin stems from clinical studies which show that nearly all cases of AD patients with trisomy-21 (Down's syndrome) display abnormal expression of αB-crystallin [110]. Notably, all Down's syndrome (DS) patients show AD pathology above 40 years of age [111]. Additionally, the observation of frequent equatorial supramolecular cataracts in lenses of AD patients suggests a relation between crystallins and Aβ [112]. αB-crystallin also plays a major role in several neurological [113, 114] and neuromuscular diseases [115, 116]. The molecular mechanistic interactions of αB-crystallin with Aβ are, however, far from clear. In vitro, αB-crystallin was shown to interact physically with amyloid peptides and was reported to inhibit fibril formation [117, 118]. At the same time, it was demonstrated that preincubation of Aβ with αB-crystallin yields increased neurotoxicity despite reduced fibril formation [119].

The studies described below were motivated to obtain a better understanding of the mechanism due to which αB-crystallin increases the neurotoxicity of Aβ. We have carried out STD NMR experiments to identify the chemical groups of $A\beta^{1-40}$ which contribute to the interaction of αB-crystallin (unpublished results). As we showed previously, even in the absence of the chaperone, STD intensities are observed for several $A\beta^{1-40}$ resonances [67]. As shown above, this effect is attributed to chemical exchange between $A\beta^{1-40}$ molecules in the aggregated state and the soluble state. In the presence of substoichiometric amounts of αB-crystallin, STD intensities are significantly increased, indicating that $A\beta^{1-40}$ and αB-crystallin interact.

Figure 20 represents a proton 1D STD spectrum recorded for $A\beta^{1-40}$ in the presence (top) and in the absence (middle) of αB-crystallin. In addition, a 1D reference spectrum is displayed at the bottom of the figure. The 1D STD experiment recorded in the presence of αB-crystallin yields a much stronger STD signal, indicating that $A\beta^{1-40}$ and αB-crystallin do indeed interact. This interaction must represent the interaction between the soluble form of $A\beta^{1-40}$ and αB-crystallin, since no increase in the STD signal would be expected if αB-crystallin exclusively interacted with the fibrillar form of $A\beta^{1-40}$.

Figure 21 represents the aliphatic region of a 2D STD TOCSY spectrum, superimposed with a standard 2D TOCSY spectrum for reference. The inter-

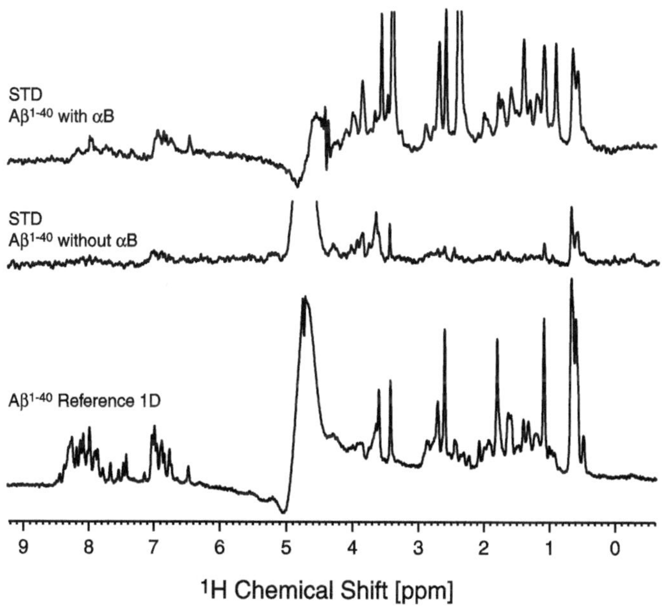

Fig. 20 1D ^{1}H reference spectrum for $A\beta^{1-40}$ (*bottom*) and 1D ^{1}H STD spectra recorded for $A\beta^{1-40}$ in the absence (*middle*) and presence (*top*) of αB-crystallin at a molar ratio of $[\alpha B] : [A\beta^{1-40}] = 1 : 25$. STD signals reflect the chemical groups of $A\beta^{1-40}$ interacting with αB-crystallin. Even in the absence of the chaperone, significant STD signals can be observed which are due to chemical exchange between the fibrillized state and the soluble state of $A\beta^{1-40}$. In the presence of αB-crystallin, the intensities of these STD signals are strongly enhanced, indicating an interaction between αB-crystallin and $A\beta^{1-40}$

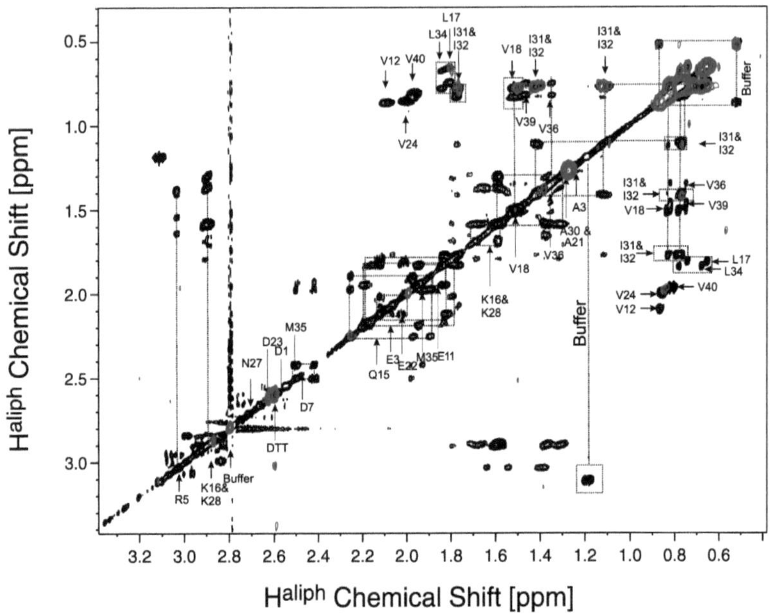

Fig. 21 Superimposition of a STD-TOCSY of Aβ$^{1-40}$ (*red*) and a standard TOCSY (*black*), recorded in the presence of αB-crystallin (molar ratio 1 : 25). The STD experiment shows only the chemical groups that are in close proximity to the chaperone. The spectra display the methyl and parts of the Hβ spectral regions only. Amino acids from the hydrophobic core region of Aβ$^{1-40}$ show the strongest STD correlations

acting residues of Aβ$^{1-40}$ with αB-crystallin are indicated in bold letters in the Aβ$^{1-40}$ sequence:

D^{1} – EAFRHDSGY^{10} – EVHHQK**LVFF**^{20} –
AED**V**GSNKG**A**^{30} – **II**GLMVGGVV^{40} .

We find that the region around the hydrophobic core (residues 17–21) contributes most to the interaction with αB-crystallin. In addition, STD signals are observed for the side-chain resonances of the two isoleucine residues I31 and I32. CD spectra recorded in the presence and absence of αB-crystallin indicate a structural change of Aβ$^{1-40}$ induced by αB-crystallin (data not shown). At the same time, no major time-dependent changes of the NMR chemical shifts of Aβ$^{1-40}$ in the presence of αB-crystallin are occurring. We therefore interpret our finding by assuming that interactions between Aβ$^{1-40}$ and αB-crystallin involve higher oligomeric states of Aβ$^{1-40}$ that are not observable using solution-state NMR spectroscopy.

Careful inspection of the NMR spectra indicates the accumulation of Aβ$^{1-40}$ molecules in which Met35 is populated in its oxidized state (Fig. 22). The assignment of the chemical shifts is corroborated by a recent study which

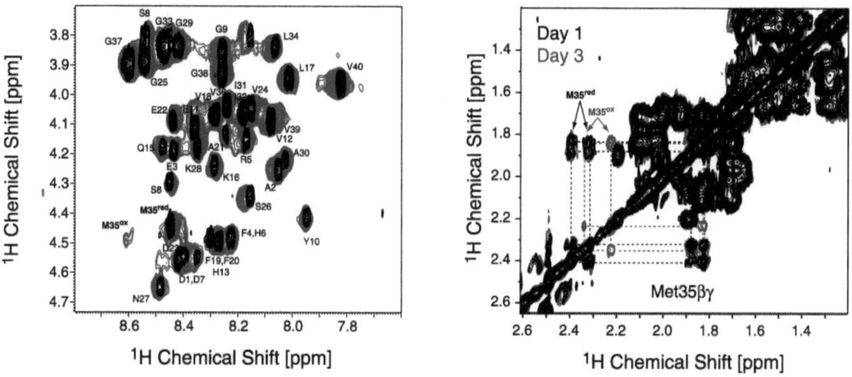

Fig. 22 TOCSY spectra of $A\beta^{1-40}$ in the presence of αB-crystallin displaying the H^N-H region (*left*) and the aliphatic region (*right*) of the spectrum as a function time. The spectra recorded on the first and third days after incubation are colored *black* and *red*, respectively. Newly appearing resonances can be assigned to Met35 in its oxidized state

has been carried out by Zagorski and coworkers [90]. In addition, we find that the rate of auto-oxidation of β-mercaptoethanol is increased by a factor of 2 at low temperature (7 °C) and by a factor of 5 at higher temperature (25 °C) in the presence of αB-crystallin (data not shown). β-Mercaptoethanol was added to all samples as a reducing agent to prevent oxidation. This led us to the investigation of the role of the redox characteristics of αB-crystallin and the implications for its interactions with $A\beta^{1-40}$.

In order to study the redox characteristics of the system $A\beta^{1-40}$/αB-crystallin, we suggest following the auto-oxidation of monomeric glutathione (GSH) to its dimeric form (GSSG) in the absence and presence of αB-crystallin and various cosolutes by solution-state NMR spectroscopy. This is a simple method which allows qualitative determination of the redox properties of proteins with respect to reduced glutathione. If the protein which is added to the GSH solution possesses reducing activity compared to GSH, GSSG dimer formation occurs more slowly. Auto-oxidation of GSH is mostly due to solubilized oxygen in the sample buffer, but is affected by the presence of other proteins in the sample. The monomeric/reduced (GSH) and the dimeric/oxidized (GSSG) form of glutathione can be differentiated by their different $^1H^N$ chemical shifts. GSH monomer and GSSG dimer can be unambiguously identified by DOSY NMR experiments. The ratio of GSSG to GSH is obtained from 1D NMR experiments at each point in time. As a reference, the rate of auto-oxidation of GSH was determined in the absence of any added substance. To determine the redox behavior of $A\beta^{1-40}$, a 1.0 mM solution of GSH was incubated together with 50 M of $A\beta^{1-40}$ (oligomer/fibril) and/or 10 μM of αB-crystallin. We tested this method first using ascorbic acid as coadded substance. As expected, a retardation of the rate of auto-oxidation of GSH is observed in the presence of ascorbic acid (Fig. 23). The differences of the initial rates of the GSSG dimer formation can be used to quantify the electrochemical potential of a given molecule added to the GSH solution. Figure 23a (triangles down) represents the effect of αB-crystallin on the auto-oxidation of GSH. Addition of αB-crystallin significantly slows down the auto-oxidation process, whereas $A\beta^{1-40}$ has almost no effect (Fig. 23a, squares). A mixed sample of $A\beta^{1-40}$ and αB-crystallin displays an intermediate effect. If $A\beta^{1-40}$ and αB-crystallin act as independent redox chemicals in the reaction, the same curve would be expected as in the case where αB-crystallin was added alone to the GSH solution. The fact that an intermediate effect was observed can only be explained by a cooperative effect of the complex of $A\beta^{1-40}$ and αB-crystallin. More interestingly, copper has a strong effect on the redox characteristics of αB-crystallin (Fig. 23b). Addition of Cu(II) to GSH yields a faster oxidation of GSH in the presence of αB-crystallin. Since Cu(II) alone does not change the rate of auto-oxidation of GSH, we conclude that Cu(II) binds tightly to αB-crystallin and alters its redox potential.

In order to validate the NMR data, we carried out colorimetric experiments using Uptiblue (Uptima) as a redox indicator. Uptiblue is a resazurin-based sensitive dye, originally designed to sense the electron transport of mitochondria in living systems. The dye changes its color from blue to red upon re-

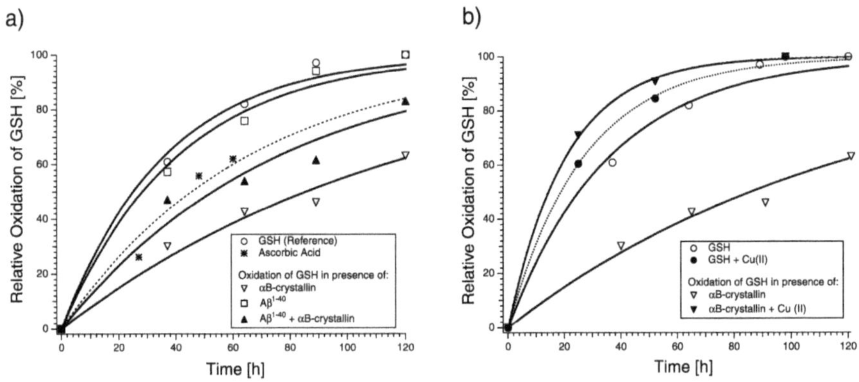

Fig. 23 Auto-oxidation of GSH upon addition of $A\beta^{1-40}$ and αB-crystallin as monitored by NMR. The figure displays the relative intensity of oxidized to reduced glutathione I(GSSG)/I(GSH) as a function of time. As reference, we record the auto-oxidation of GSH due to solubilized oxygen in the NMR tube (*circles*). αB-crystallin has a reducing effect on GSH (*triangle down*). $A\beta^{1-40}$ alone (*squares*) has almost no effect on the auto-oxidation of GSH, whereas $A\beta^{1-40}$ in combination with αB-crystallin (*filled triangles up*) shows an intermediate effect, indicating an altered redox potential for the complex of $A\beta^{1-40}$ and αB-crystallin in contrast to the individual components. As reference, the effect of ascorbic acid on the auto-oxidation process of GSH was recorded (*stars*). The concentration of GSH was 1.0 mM. Ascorbic acid, $A\beta^{1-40}$, and αB-crystallin were added using a concentration of 1.0 mM, 50.0 and 10.0 μM, respectively

duction. The end product is very stable and allows an easy comparison of the redox activity of various proteins. Each well of a 96-well plate was filled with 10 μL of Uptiblue. Then, 0.1 mL of an $A\beta^{1-40}$ solution (concentration varying from 80 to 400 μM) was added to the wells. In the case of αB-crystallin, concentrations of 8 to 40 μM were used, varying the concentration in steps of 8 μM. After incubation for 48 h, the samples were centrifuged and 50 μL of the supernatant was transferred into a new 96-well plate. The colorimetric change was observed directly by scanning the samples. Increasing amounts of αB-crystallin induce a reduction of Uptiblue (Fig. 24, lane 1). This reaction is abolished in the presence of Cu(II) (lane 2). A less pronounced colorimetric effect is observed as a function of $A\beta^{1-40}$ concentration (lane 3). Note the much higher concentration of $A\beta^{1-40}$ needed in this case. This behavior is in agreement with the GSH auto-oxidation data from the NMR experiment. $A\beta^{1-40}$ reduces Uptiblue only at much higher concentrations (concentration range from 0 to 400 μM). In the presence of αB-crystallin (at a molar ratio

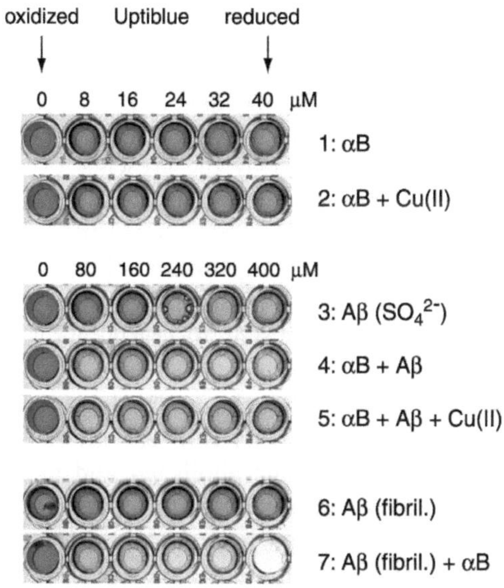

Fig. 24 Colorimetric assay to probe $A\beta^{1-40}$/αB-crystallin redox properties. Reduction of the dye Uptiblue due to an added compound causes a change from the oxidized (non-fluorescent, *blue*) form to the reduced (fluorescent, *red*) form. The oxidation–reduction potential of Uptiblue corresponds to $E_0 = +380$ mV at pH 7.0, 25 °C. Increasing amounts of an added compound yield an increased redshift, depending on the electrochemical potential of the compound. Strong reducing agents should induce a rapid redshift. In all experiments, the concentration of αB-crystallin was always kept tenfold lower with respect to the concentration of $A\beta^{1-40}$. Uptiblue is always used in excess. Lane 1: αB-crystallin; lane 2: αB-crystallin + Cu(II); lane 3: $A\beta^{1-40}$; lane 4: $A\beta^{1-40}$+αB-crystallin; lane 5: $A\beta^{1-40}$ and αB-crystallin + Cu(II); lane 6: fibrillized $A\beta^{1-40}$; lane 7: fibrillized $A\beta^{1-40}$ + αB-crystallin

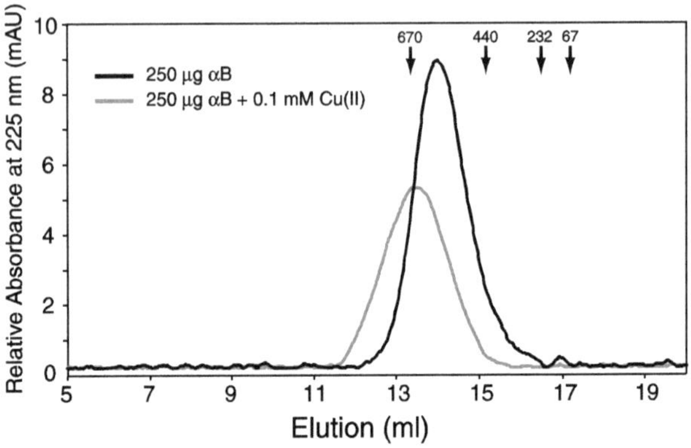

Fig. 25 Size exclusion chromatography of αB-crystallin in the presence and absence of copper. It is found that the average molecular weight of αB-crystallin in the presence of 0.1 mM copper increases. Molecular weight standards corresponding to 67 kDa (albumin), 232 kDa (catalase), 440 kDa (ferritin), and 670 kDa (thyroglobulin) are indicated by *arrows*

of 1 : 10 with respect to $A\beta^{1-40}$) (lane 4), the observed effect is larger compared to the case when $A\beta^{1-40}$ is incubated alone. The redshift induced by the reaction occurs at lower concentrations of $A\beta^{1-40}$ and αB-crystallin. This colorimetric effect is not influenced by the presence of Cu(II) (lane 5). Interestingly, fibrillized $A\beta^{1-40}$ seems to have an electrochemical potential that is larger than + 380 mV, since it does not reduce Uptiblue even at very high concentrations. This situation is changed once αB-crystallin is added. Now, the colorimetric effect is similar to the case in which nonfibrillized $A\beta^{1-40}$ is mixed with αB-crystallin, indicating that αB-crystallin dissociates preformed fibrils.

Gel filtration studies indicate that the molecular weight of αB-crystallin is increased in the presence of Cu(II) (Fig. 25). At the same time, binding of $A\beta^{1-40}$ to αB-crystallin seems abolished and association of $A\beta^{1-40}$ molecules into aggregates is promoted. Addition of EDTA restores the endothermic behavior in the ITC experiment, indicating that copper ions become complexed and that the interaction between $A\beta^{1-40}$ and αB-crystallin can be recovered.

4.1
Interaction Mechanism of αB-Crystallin and $A\beta^{1-40}$

Analysis of the 2D STD-TOCSY experiment allows the identification of the chemical groups of $A\beta^{1-40}$ that are involved in interactions with respect to αB-crystallin. We find that the amino acids Gln^{15}–Asp^{23} and Ile^{31}–Ile^{32} of $A\beta^{1-40}$ show especially strong interactions. These residues are located in the hydrophobic core of $A\beta^{1-40}$. ITC experiments are in agreement with this observation. We observe a positive binding enthalpy and conclude that the interactions are hydrophobic in nature. Figure 20 demonstrates that STD resonances can be observed even in the absence of the chaperone. The observed signals are not subtraction artifacts, as the same experiment using a peptide sample which is not prone to aggregation yields a spectrum without signals. We have shown above that this effect is due to chemical exchange: $A\beta^{1-40}$ monomers are in equilibrium with $A\beta^{1-40}$ molecules in the aggregated state. The observed resonances reflect the chemical groups which are involved in $A\beta^{1-40}$–$A\beta^{1-40}$ interactions [67]. Since the intensities of the 1D STD spectrum are much increased upon addition of αB-crystallin, we conclude that αB-crystallin effectively competes for $A\beta^{1-40}$ monomer–monomer interactions. This way, αB-crystallin might destabilize $A\beta^{1-40}$ fibril structures. This is in agreement with fluorescence studies which have shown that $A\beta^{1-40}$ participates in subunit exchange of αB-crystallin which promotes $A\beta^{1-40}$ protofibril formation [117]. DOSY experiments show that the average molecular weight of soluble $A\beta^{1-40}$ does not change upon interaction with αB-crystallin. Given the fact that we observe strong STD signals, we conclude that the binding of monomeric $A\beta^{1-40}$ to αB-crystallin is weak and short-lived. The helical propensity of $A\beta^{1-40}$ observed in the CD experiments (data not shown) slowly disappears upon incubation with αB-crystallin. At the same time, the CD spectrum of $A\beta^{1-40}$ alone does not change as a function of time. Dramatic changes in ThT fluorescence were observed upon incubation of $A\beta^{1-40}$ with αB-crystallin [119], indicating a structural reorganization. We interpret this finding by suggesting that a conformational change is induced for the fibrillar or the protofibrillar state of $A\beta^{1-40}$, but not for the soluble state. By competing for binding to the hydrophobic core region of $A\beta^{1-40}$, αB-crystallin might shift the equilibrium between fibrillar and oligomeric/protofibrillar $A\beta^{1-40}$ more toward the oligomeric state.

4.2
Redox Activity of αB-Crystallin

In redox reactions proton and electron transfers are coupled. Many enzymes act as electron or hydrogen atom carriers. We conclude from our NMR as well as from the colorimetric redox experiments (Figs. 23 and 24) that αB-crystallin possesses a biological redox activity. It was shown previously that the protective activity of human αB-crystallin against TNFα-mediated cell death results from its ability to raise the intracellular concentration of glutathione [120]. In addition, many redox-active proteins are differentially up-/downregulated due to overexpression of αB-crystallin in HeLa cells (unpublished data). Especially, on downregulation of the cellular quality control protein disulfide isomerase, Erp57, and the cellular redox-state balancing protein, PDX1 is observed (data not shown). These findings indicate a relation of αB-crystallin to cellular redox regulation. Figure 26 summarizes qualita-

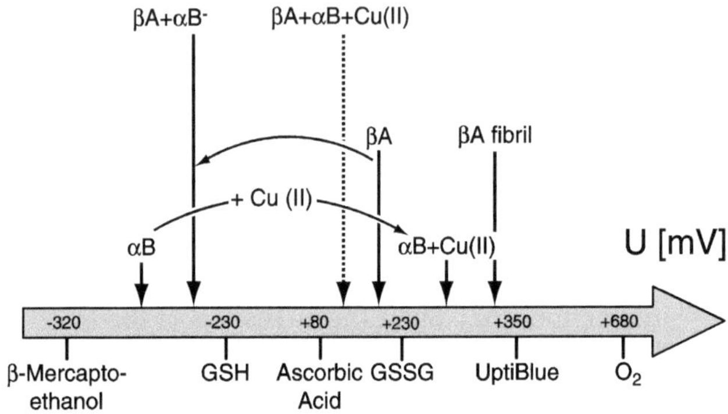

Fig. 26 Proposed redox electrochemical potentials for $A\beta^{1-40}$ and αB-crystallin in the presence and absence of Cu(II)

tively the findings from the NMR GSH titration experiments and the Uptiblue colorimetric analysis in terms of electrochemical potentials for the different mixtures of $A\beta^{1-40}$ and αB-crystallin. As reference, the electrochemical potentials of β-mercaptoethanol, GSH, ascorbic acid, GSSG, Uptiblue, and molecular oxygen are indicated. Arrows that are placed on the left-hand side of the diagram represent αB/Aβ complexes that tend to give electrons to the molecules that are placed on the right-hand side. Experimentally, we find that GSH oxidation is retarded in the presence of αB-crystallin (Fig. 23, open triangle down). Therefore, the electrochemical potential of αB-crystallin is positioned left with respect to GSH. Similarly, we observed that GSH oxidation is unaffected in the presence of $A\beta^{1-40}$ (Fig. 23a, squares). The electrochemical potential of $A\beta^{1-40}$ should therefore be similar to the potential of GSSG. Upon addition of $A\beta^{1-40}$ together with catalytic amounts of αB-crystallin to the GSH solution, the rate of auto-oxidation of GSH lies in between the two rates observed for the pure compounds. This is represented in the diagram by a horizontal arrow pointing in the direction of smaller electrochemical potentials. Upon addition of αB-crystallin and copper (Fig. 23b, filled triangle down), αB-crystallin does not have a reducing activity on GSH any more, but an oxidative one. The diagram accounts for this observation by a horizontal arrow pointing to larger electrochemical potentials for the complex of αB-crystallin and Cu(II). Surprisingly, we find that the electrochemical potential of aggregated $A\beta^{1-40}$ appears to be very close to the potential of molecular oxygen, since $A\beta^{1-40}$ fibrils are not able to reduce Uptiblue (Fig. 24, lane 6). Coaddition of $A\beta^{1-40}$ fibrils and αB-crystallin to the GSH solution (Fig. 24, lane 7) induces fast reduction of Uptiblue, which indicates that the electrochemical potential of the complex is largely changed.

4.3 Oxidation of $A\beta^{1-40}$ (Met35) Might Explain Reduced Aβ Fibril Formation and Increased Neurotoxicity in Presence of αB-Crystallin

Figure 26 shows that αB-crystallin possesses reducing activity with respect to $A\beta^{1-40}$. At the same time, we observe that Met35 in $A\beta^{1-40}$ changes its oxidation state upon interaction with $A\beta^{1-40}$ (Fig. 26). Two explanations can account for these—at first sight contradictory—observations. First, because of its large size, αB-crystallin might change the redox potential of $A\beta^{1-40}$ oligomers as a whole. As shown by the 2D STD correlation experiment, there is no detectable interaction between $A\beta^{1-40}$(Met35) and αB-crystallin (Fig. 21b). This residue was shown to be not surface accessible under nonoxidative conditions (PDB code: 1BA4). It might be that the increased difference in the redox potential between the complex $A\beta^{1-40}$–αB-crystallin and $A\beta^{1-40}$ alone with respect to molecular oxygen is sufficient to facilitate the oxidation of Met35. Second, the observed 2D STD correlations reflect interactions between monomeric $A\beta^{1-40}$ and αB-crystallin. If αB-crystallin interacts with fibrillar $A\beta^{1-40}$, this interaction is not easily observable by solution-state NMR. This interpretation is also supported by the fact that the NMR chemical shift and the intensities are not changing over time, whereas the CD spectrum is changing. The oxidized form of $A\beta^{1-40}$ can then only be observed as a consequence of the chemical exchange between the fibrillar state of $A\beta^{1-40}$ and a soluble state. Oxidation of methionine induces an increased polarity by changing its dipole moment from 1.60 to 5.28 debye [121]. This increased polarity yields an increased solubility of $A\beta^{1-40}$ which destabilizes fibrillar structures [121]. This is in agreement with previous observations which show that αB-crystallin does indeed dissolve preformed $A\beta^{1-40}$ fibrils [117, 119]. Butterfield and coworkers showed that Met35(ox) in $A\beta^{1-40}$ induces an increased neurotoxicity, since the oxidized form can propagate more easily free radicals to adjacent residues [122]. Recent experiments suggest that probably not the fibrillar state, but a high oligomeric state of $A\beta^{1-40}$ is responsible for its neurotoxicity [123–125]. Taken together, we conclude that oxidation of Met35 in $A\beta^{1-40}$ is a consequence of the redox properties induced by αB-crystallin, and that αB-crystallin induces increased neurotoxicity due to the population of oligomeric $A\beta^{1-40}$ assemblies.

4.4 Regulation of αB-Crystallin Substrate Binding by Cu(II) In Vitro

STD and ITC experiments show that $A\beta^{1-40}$ specifically binds to the hydrophobic region of αB-crystallin in the absence of copper. In the presence of copper, the binding affinity is drastically reduced. This can be attributed either to a change in the conformation of $A\beta^{1-40}$ or to αB-crystallin or to both. Titrating EDTA into the $A\beta^{1-40}$–αB-crystallin–Cu(II) solution reconstitutes the endothermic effect and, thus, binding of $A\beta^{1-40}$ to αB-crystallin. Gel filtration studies are in agreement with this interpretation and show that copper alters the oligomeric state of αB-crystallin (Fig. 25). Combining the results from the redox assay, ITC, and gel filtration experiments, we conclude that αB-crystallin uses copper as a switch to regulate binding to $A\beta^{1-40}$.

4.5 Regulation of αB-Crystallin Redox Properties by Cu(II)

Many neurodegenerative diseases including Alzheimer's disease seem to be dependent on the concentration of copper in the cell [126, 127]. Familial amyotrophic lateral sclerosis (ALS) is a motor neuron disease which has its molecular origin in an inactivation of the radical scavenging protein superoxide dismutase 1 (SOD1), which transports copper across mitochondria [128]. In ALS patients, expression of the small heat shock proteins Hsp25 and αB-crystallin is upregulated. These proteins are found specifically cofractioned with insoluble SOD1 [129]. Involvement of copper and crystallins in cataractogenesis [130] and modulation of chaperone activity of crystallin in the presence of transition metal ions [131] confirms the hypothesis that copper might be a factor in regulating the function of αB-crystallin. So far, nothing is known about the involvement of copper in the modulation of αB-crystallin function and its implication with respect to substrate binding. Since no cystein residue is present in the sequence, we speculate that coordination of copper occurs via binding of the metal to the histidine imidazole ring. It was recently postulated that APP is important for maintaining neuronal copper homeostasis [132]. In addition, it is found that a mutant copper transporter CuATPase7b induces elevated levels of intracellular copper, reducing the amyloid burden in the brain at the same time [133]. It could be shown that the elevation of the intracellular copper concentration stabilizes the SOD1 activity and decreases the production of $A\beta^{1-40}$ [127]. Importantly, regulation of SOD1 activity is not only restricted to the copper chaperone CCS, but can also be regulated by cellular redox conditions: an increased level of cellular (reduced) glutathione (GSH) provides an alternative pathway to activate SOD1 [134]. At the same time, the authors find that high levels of intracellular glutathione result in a lower level of SOD1, suggesting that GSH can affect the posttranslational stability of SOD1 [134]. We therefore propose that αB-crystallin has a regulation function with respect to cellular redox conditions. We find that the redox activity of αB-crystallin is modulated by copper. Addition of copper abolishes binding of $A\beta^{1-40}$ to αB-crystallin. These findings indicate a relationship between copper transport, redox activity, and the mechanism of the disease.

References

1. Pepys MB (1996) In: Weatherall DJ, Ledingham JGG, Warell DA (eds) The Oxford textbook of medicine. Oxford University Press, Oxford, p 1512
2. Kelly JW (1998) Curr Opin Struct Biol 8:101
3. Selkoe D (1991) Neuron 6:487
4. Prusiner SB, Scott MR, DeArmond SJ, Cohen FE (1998) Cell 93:337
5. Guijarro JI, Sunde M, Jones JA, Campbell ID, Dobson CM (1998) Proc Natl Acad Sci USA 95:4224
6. Jimenez JL, Guijarro JI, Orlova E, Zurdo J, Dobson CM, Sunde M, Saibil HR (1999) EMBO J 18:815
7. Fandrich M, Fletcher MA, Dobson CM (2001) Nature 410:165
8. Petkova AT, Ishii Y, Balbach JJ, Antzutkin ON, Leapman RD, Delaglio F, Tycko R (2002) Proc Natl Acad Sci USA 99:16742
9. Tycko R (2003) Biochemistry 42:3151
10. Petkova AT, Leapman RD, Guo ZH, Yau WM, Mattson MP, Tycko R (2005) Science 307:262
11. Jaroniec CP, MacPhee CE, Astrof NS, Dobson CM, Griffin RG (2002) Proc Natl Acad Sci USA 99:16748–16753
12. Jaroniec CP, MacPhee CE, Bajaj VS, McMahon MT, Dobson CM, Griffin RG (2004) Proc Natl Acad Sci USA 101:711
13. Ritter C, Maddelein M-L, Siemer AB, Lührs T, Ernst M, Meier BH, Saupe SJ, Riek R (2005) Nature 435:844
14. Calzolai L, Lysek DA, Güntert P, Schroetter CV, Riek R, Zahn R, Wüthrich K (2000) Proc Natl Acad Sci USA 97:8340
15. Ventura S, Zurdo J, Narayanan S, Parreño M, Mangues R, Reif B, Chiti F, Giannoni E, Dobson CM, Aviles FX, Serrano L (2004) Proc Natl Acad Sci USA 101:7258
16. Hilbich C, Kisters-Woike B, Reed J, Masters CL, Beyreuther K (1992) J Mol Biol 228:460
17. Schubert U, Anton LC, Gibbs J, Norbury CC, Yewdell JW, Bennink JR (2000) Nature 404:770
18. Seckler R, Jaenicke R (1992) Faseb J 6:2545
19. Dobson CM (2003) Nature 426:884
20. Bukau B, Horwich AL (1998) Cell 92:351
21. Hartl FU, Hayer-Hartl M (2002) Science 295:1852
22. Braig K, Otwinowski Z, Hegde R, Boisvert DC, Joachimiak A, Horwich AL, Sigler PB (1994) Nature 371:614
23. Boisvert DC, Wang J, Otwinowski Z, Horwich AL, Sigler PB (1996) Nat Struct Biol 3:170
24. Landry SJ, Zeilstra-Ryalls J, Fayet O, Georgopoulos C, Gierasch LM (1993) Nature 364:255
25. Zahn R, Spitzfaden C, Ottiger M, Wüthrich K, Plückthun A (1994) Nature 368:261
26. Kobayashi N, Freund SMV, Chatellier J, Zahn R, Fersht AR (1999) J Mol Biol 292:181
27. Horst R, Bertelsen EB, Fiaux J, Wider G, Horwich AL, Wuthrich K (2005) Proc Natl Acad Sci USA 102:12748
28. Riek R, Fiaux J, Bertelsen EB, Horwich AL, Wuthrich K (2002) J Am Chem Soc 124:12144
29. Morshauser RC, Wang H, Flynn GC, Zuiderweg ERP (1995) Biochemistry 34:6261
30. Pellecchia M, Montgomery DL, Stevens SY, Vander Kooi CW, Feng HP, Gierasch LM, Zuiderweg ERP (2000) Nat Struct Biol 7:298

31. Revington M, Zhang YB, Yip GNB, Kurochkin AV, Zuiderweg ERP (2005) J Mol Biol 349:163
32. Dedmon DD, Christodoulou J, Wilson MR, Dobson CM (2005) J Biol Chem 280:14733
33. Sanchez Y, Lindquist SL (1996) Science 248:1112
34. Lindquist S, Kim G (1996) Proc Natl Acad Sci USA 93:5301
35. Parsell DA, Kowal AS, Singer MA, Lindquist S (1994) Nature 372:475
36. Kim KI, Cheong GW, Park SC, Ha JS, Woo KM, Choi SJ, Chung CH (2000) J Mol Biol 303:655
37. Chernoff YO, Lindquist SL, Ono B, Inge-Vechtomov SG, Liebman SW (1995) Science 268:880
38. Patino MM, Liu JJ, Glover JR, Lindquist S (1996) Science 273:622
39. DePace AH, Santoso A, Hillner P, Weissman JS (1998) Cell 93:1241
40. Reddy PS, Houseman DE (1997) Curr Opin Cell Biol 9:364
41. Ross CA (1997) Neuron 19:1147
42. Krobitsch S, Lindquist S (2000) Proc Natl Acad Sci USA 97:1589
43. Cohen FE, Prusiner SB (1998) Annu Rev Biochem 67:793
44. Paushkin SV, Kushnirov VV, Smirnov VN, Ter-Avanesyan MD (1996) EMBO J 15:3127
45. Kushnirov VV, Ter-Avanesyan MD (1998) Cell 94:13
46. Schirmer EC, Lindquist S (1997) Proc Natl Acad Sci USA 94:13932
47. Narayanan S, Bösl B, Walter S, Reif B (2003) Proc Natl Acad Sci USA 100:9286
48. Borchsenius AS, Wegrzyn RD, Newnam GP, Inge-Vechtomov SG, Chernoff YO (2001) EMBO J 20:6683
49. Liu J-J, Sondheimer N, Lindquist SL (2002) Proc Natl Acad Sci USA 99:16446
50. Stejskal EO, Tanner JE (1965) J Chem Phys 42:288
51. Mayer M, Meyer B (1999) Angew Chem Int Ed 38:1784
52. Takahashi H, Nakanishi T, Kami K, Arata Y, Shimada I (2000) Nat Struct Biol 7:220
53. Hattendorf DA, Lindquist SL (2002) EMBO J 21:12
54. Wilkins DK, Grimshaw SB, Receveur V, Dobson CM, Jones JA, Smith LJ (1999) Biochemistry 38:16424
55. Goloubinoff P, Mogk A, Zvi APB, Tomoyasu T, Bukau B (1999) Proc Natl Acad Sci USA 96:13732
56. Glover JR, Lindquist S (1998) Cell 94:73
57. Schlee S, Beinker P, Akhrymuk A, Reinstein J (2004) J Mol Biol 336:275
58. Kryndushkin DS, Alexandrov IM, Ter-Avanesyan MD, Kushnirov VV (2003) J Biol Chem 278:49636
59. Yoshida M (2003) Euresco conference and FEBS advanced course, Tomar, Portugal
60. Narayanan S, Walter S, Reif B (2006) Chem Bio Chem 7:757
61. Scheibel T, Lindquist SL (2001) Nat Struct Biol 8:958
62. Podlisny MB, Walsh DM, Amarante P, Ostaszewski BL, Stimson ER, Maggio JE, Teplow DB, Selkoe DJ (1998) Biochemistry 37:3602
63. Rudiger S, Germeroth L, Schneider-Mergener J, Bukau B (1997) EMBO J 16:1501
64. Walter S, Lorimer GH, Schmid FX (1996) Proc Natl Acad Sci USA 93:9425
65. Cashikar AG, Schirmer EC, Hattendorf DA, Glover R, Ramakrishnan MS, Ware DM, Lindquist SL (2002) Mol Cell 9:751
66. Chen L, Sigler PB (1999) Cell 99:757
67. Narayanan S, Reif B (2005) Biochemistry 44:1444
68. Collins SR, Douglass A, Vale RD, Weissman JS (2004) PLoS Biol 2:1582
69. Fiaux J, Bertelsen EB, Horwich AL, Wüthrich K (2002) Nature 418:207

70. Yoshida M (2002) FASEB summer research conference on protein folding in the cell, Vermont, USA
71. Baumeister R, Eimer S (1998) Angew Chem Int Ed Engl 37:2978
72. Selkoe DJ (1999) Nature 399:A23
73. Lansbury PT Jr (1997) Neuron 19:1151
74. Walsh DM, Lomakin A, Benedek GB, Condron MM, Teplow DB (1997) J Biol Chem 272:22364
75. Walsh DM, Hartley DM, Kusumoto Y, Fezoui Y, Condron MM, Lomakin A, Benedek GB, Selkoe DJ, Teplow DB (1999) J Biol Chem 274:25945
76. Hartley DM, Walsh DM, Ye CP, Diehl T, Vasquez S, Vassilev PM, Teplow DB, Selkoe DJ (1999) J Neurosci 19:8876
77. Kayed R, Head E, Thompson JL, McIntire TM, Milton SC, Cotman CW, Glabe CG (2003) Science 300:486
78. Stadtman E (1992) Science 257:1220
79. Butterfield DA, Yatin SM, Varadarajan S, Koppal T (1999) Methods Enzymol 309:746
80. Varadarajan S, Yatin S, Aksenova M, Butterfield DA (2000) J Struct Biol 130:184
81. Zhang S, Iwata K, Lachenmann MJ, Peng JW, Li S, Stimson ER, Lu Y-A, Felix AM, Maggio JE, Lee JP (2000) J Struct Biol 130:130
82. Riek R, Günthert P, Döbeli H, Wipf B, Wüthrich K (2001) Eur J Biochem 268:5930
83. Barrow CJ, Zagorski MG (1991) Science 253:179
84. Sticht H, Bayer P, Willbold D, Dames S, Hilbich C, Beyreuther K, Frank RW, Rösch P (1995) Eur J Biochem 233:293
85. Coles M, Bicknell W, Watson AA, Fairlie DP, Craik DJ (1998) Biochemistry 37:11064
86. Shao H, Jao S, Ma K, Zagorski MG (1999) J Mol Biol 285:755
87. Goto Y, Aimoto S (1991) J Mol Biol 218:387
88. Sreerama N, Woody R (2000) Anal Biochem 287:252
89. Kirkitadze MD, Condron MM, Teplow DB (2001) J Mol Biol 312:1103
90. Hou L, Shao H, Zhang Y, Li H, Menon NK, Neuhaus EB, Brewer JM, Byeon I-JL, Ray DG, Vitek MP, Iwashita T, Makula RA, Przybyla AB, Zagorski MG (2004) J Am Chem Soc 126:1992
91. Mansfield SL, Jayawickrama DA, Timmons JS, Larive CK (1998) Biochim Biophys Acta 1382:257
92. Tseng BP, Esler WP, Clish CB, Stimson ER, Ghilardi JR, Vinters HV, Mantyh PW, Lee JP, Maggio JE (1999) Biochem 38:10424
93. Jarvet J, Damberg P, Bodell K, Eriksson LEG, Gräslund A (2000) J Am Chem Soc 122:4261
94. Zagorski MG, Yang J, Shao H, Ma K, Zeng H, Hong A (1999) Methods Enzymol 309:189
95. Hilbich C, Kisters-Woike B, Reed J, Masters CL, Beyreuther K (1991) J Mol Biol 218:149
96. Bitan G, Lomakin A, Teplow DB (2001) J Biol Chem 276:35176
97. Oda T, Wals P, Osterburg HH, Johnson SA, Pasinetti GM, Morgan TE, Rozovsky I, Stine WB, Snyder SW, Holzman TF, Krafft GA, Finch CE (1995) Exp Neurol 136:22
98. Lambert MP, Barlow AK, Chromy BA, Edwards C, Freed R, Liosatos M, Morgan TE, Rozovsky I, Trommer B, Viola KL, Wals P, Zhang C, Finch CE, Krafft GA, Klein WL (1998) Proc Natl Acad Sci USA 95:6448
99. Gong Y, Chang L, Viola KL, Lacor PN, Lambert MP, Finch CE, Krafft GA, Klein WL (2003) Proc Natl Acad Sci USA 100:10417
100. Stine BW, Dahlgren KN, Krafft GA, LaDu MJ (2003) J Biol Chem 278:11612

101. Harper JD, Wong SS, Lieber CM, Lansbury PT Jr (1999) Biochemistry 38:8972
102. Zhang S, Casey N, Lee JP (1998) Fold Des 3:413
103. Massi F, Peng JW, Lee JP, Straub JE (2001) Biophys J 80:31
104. Ni F (1994) Prog NMR Spectrosc 26:517
105. Terzi E, Holzemann G, Seelig J (1994) Biochemistry 33:1345
106. Jayalakshmi V, Krishna NR (2002) J Magn Reson 155:106
107. Fonte V, Kapulkin V, Taft A, Fluet A, Friedman D, Link CD (2002) Proc Natl Acad Sci USA 99:9439
108. Yoo BC, Kim SH, Cairns N, Fountoulakis M, Lubec G (2001) Biochem Biophys Res Commun 280:249
109. Shinohara H, Inaguma Y, Goto S, Inagaki T, Kato K (1993) J Neurol Sci 119:203
110. Renkawek K, van Workum FAP, de Jong WW (1995) Dev Brain Dysfunct 8:35
111. Siegel GJ, Agranoff BW, Fisher SK, Albers RW, Uhler MD (eds) (1999) Basic neurochemistry, vol 6. Lippincott-Raven, Philadelphia
112. Goldstein LE, Muffat JA, Cherny RA, Moir RD, Ericsson MH, Huang X, Mavros C, Coccia JA, Faget KY, Fitch KA, Masters CL, Tanz RE, Chylack LT Jr, Bush AI (2003) Lancet 361:1258
113. van Rijk AF, Bloemendal H (2000) Ophthalmologica 214:7
114. Lowe J, Landon M, Pike I, Spendlove I, Mc Dermott H, Mayer RJ (1990) Lancet 336:515
115. Vicart P, Caron A, Guicheney P, Li ZL, Prevost MC, Faure A, Chateau D, Chapo F, Tome F, Dupret JM, Paulin D, Fardeau M (1998) Nat Genet 20:92
116. Selcen D, Engel AG (2003) Ann Neurol 54:804
117. Liang JJ (2000) FEBS Lett 484:98
118. Kudva YC, Hiddinga HJ, Butler PC, Mueske CS, Eberhardt NL (1997) FEBS Lett 416:117
119. Stege GJ, Renkawek K, Overkamp PS, Verschuure P, van Rijk AF, Reijnen-Aalbers A, Boelens WC, Bosman GJ, de Jong WW (1999) Biochem Biophys Res Commun 262:152
120. Mehlen P, Kretz-Remy C, Preville X, Arrigo AP (1996) EMBO J 15:2695
121. Bitan G, Tarus B, Vollers SS, Lashuel HA, Condron MM, Straub JE, Teplow DB (2003) J Am Chem Soc 125:15359
122. Kanski J, Varadarajan S, Aksenova M, Butterfield DA (2001) Biochim Biophys Acta 1586:190
123. Haass C (2001) Nat Neurosci 4:859
124. Caughey B, Lansbury PTJ (2003) Annu Rev Neurosci 26:267
125. Kim HJ, Chae SC, Lee DK, Chromy B, Lee SC, Park YC, Klein WL, Krafft GA, Hong ST (2003) Faseb J 17:118
126. Bush AI, Masters CL, Tanzi RE (2003) Proc Natl Acad Sci USA 100:11193
127. Bayer TA, Schäfer S, Simons A, Kemmling A, Kamer T, Tepest R, Eckert A, Schüssel K, Eikenberg O, Sturchler-Pierrat C, Abramowski D, Staufenbiel M, Multhaup G (2003) Proc Natl Acad Sci USA 100:14187
128. Gurney ME, Pu H, Chiu AY, Dal Canto MC, Polchow CY, Alexander DD, Caliendo J, Hentati A, Kwon YW, Deng HX, Chen WJ, Zhai P, Sufit RL, Siddique T (1994) Science 264:1772
129. Wang J, Slunt H, Gonzales V, Fromholt D, Coonfield M, Copeland NG, Jenkins NA, Borchelt DR (2003) Hum Mol Genet 12:2753
130. Goldstein LE, Leopold MC, Huang X, Atwood CS, Saunders AJ, Hartshorn M, Lim JT, Faget KY, Muffat JA, Scarpa RC, Chylack LT Jr, Bowden EF, Tanzi RE, Bush AI (2000) Biochemistry 39:7266

131. Ganadu ML, Aru M, Mura GM, Coi A, Mlynarz P, Kozlowski H (2004) J Inorg Biochem 98:1103
132. Barnham KJ, McKinstry WJ, Multhaup G, Galatis D, Morton CJ, Curtain CC, Williamson NA, White AR, Hinds MG, Norton RS, Beyreuther K, Masters CL, Parker MW, Cappai R (2003) J Biol Chem 278:17401
133. Phinney A, Drisaldi B, Schmidt SD, Lugowski S, Coronado V, Liang Y, Horne P, Yang J, Sekoulidis J, Commaraswamy J, Chishti MA, Cox DW, Mathews PM, Nixon RA, Carlson GA, St George-Hyslop P, Westaway D (2003) Proc Natl Acad Sci USA 100:14193
134. Carroll MC, Girouard JB, Ulloa JL, Subramaniam JR, Wong PC, Valentine JS, Culotta VC (2004) Proc Natl Acad Sci USA 101:5964

Top Curr Chem (2007) 272: 169–185
DOI 10.1007/128_062

Published online: 11 August 2006

NMR Analysis of Bioprotective Sugars: Sucrose and Oligomeric (1→2)-α-D-glucopyranosyl-(1→2)-β-D-fructofuranosides

Daniela Fischer · Armin Geyer (✉)

Fachbereich Chemie, Philipps-Universität Marburg, Hans-Meerwein-Strasse,
35032 Marburg, Germany
geyer@staff.uni-marburg.de

Abstract NMR spectroscopy yields structural information on a new class of bioprotectants with the general composition oligo(1 → 2)-α-D-glucopyranosyl-(1 → 2)-β-D-fructofuranoside. Their smallest representative is sucrose for which conflicting models try to explain its cryoprotective properties. Starting from sucrose, we characterize the influence of the growing chain of oligo(1 → 2)-α-D-glucopyranoses. An analytical approach will be presented that can identify ultraweak membrane affinities, a property often discussed in the context of cryoprotectants. The most prominent feature of the vicinally connected (1 → 2)-α-D-glucopyranose oligomers is their well-defined solution structure with a stand-alone sugar helix. The unfolding barrier of this helix is large enough to become observable as line-broadening in the ^{1}H NMR.

Keywords NMR Spectroscopy · Conformation analysis · Bioprotectants · Glucans · Helical structures

1 Introduction

The disaccharides sucrose and trehalose are important bioprotectants but the essential structural features that distinguish them from other sugars remain elusive. Membrane binding or the disturbance of the water structure are the two main lines of thought about how cryoprotectants act on a molecular level and are distinguished from other sugars [1–3]. A complex dynamic behavior complicates the identification of a bioactive conformation that interferes with the cellular membrane or its solvation sphere. Homooligomeric saccharides like fructans (inulins and levans) are a further class of commercially

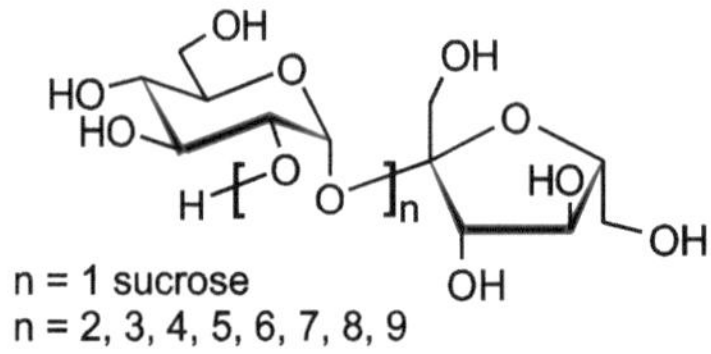

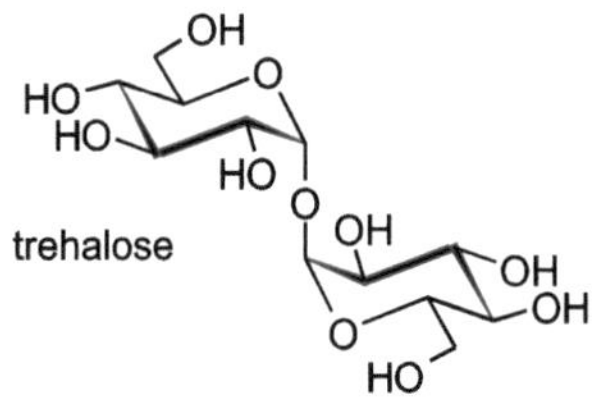

important bioprotectants [4]. The oligo(1 → 2)-α-D-glucopyranosyl-(1 → 2)-β-D-fructofuranosides described here are a new class of cyanobacterial glucans, which bridge the gap between sucrose and the larger homooligomeric saccharides. Their unparalleled signal dispersion in the ^{1}H NMR allows a detailed structural analysis with the aim of identifying the essential structural features.

2 Disaccharides

Furanose-ring puckering and glycosidic bond rotations make sucrose a flexible molecule at room temperature. This dynamic picture evolved from the early hard-sphere studies [5] to more recent investigations [6, 7]. NMR spectroscopy played a dominant role in the development of this view [8]. Intramolecular hydrogen bonds, which are detected in the solid state and in modeling studies [9], have a mere transient character in solution. Aqueous solutions require supercooling [10–12] to detect the solvent-accessible hydroxyl protons, and the excellent proton acceptor DMSO interferes with hydrogen bonds to blur possible biologically relevant conformations. Solvent mixtures like DMSO/water or acetone/water (so-called cryo-mixtures) [13, 14] face the same problems [15–17] and therefore alternative solvent conditions are welcome. The polarity of methanol is sufficiently increased by small amounts of calcium chloride to be able to dissolve sucrose or many other saccharides. Even though it is a protic and polar solvent, the hydroxyl chemical exchange rates are effectively slowed down to make them observable at 300 K. The proton chemical exchange rates in methanol are below 10 s^{-1}, thus giving rise to separated signals with coupling pattern in 1D ^{1}H NMR and to cross signals in 2D COSY and NOESY spectra, respectively, as long as acid or base catalysis is excluded (which otherwise would

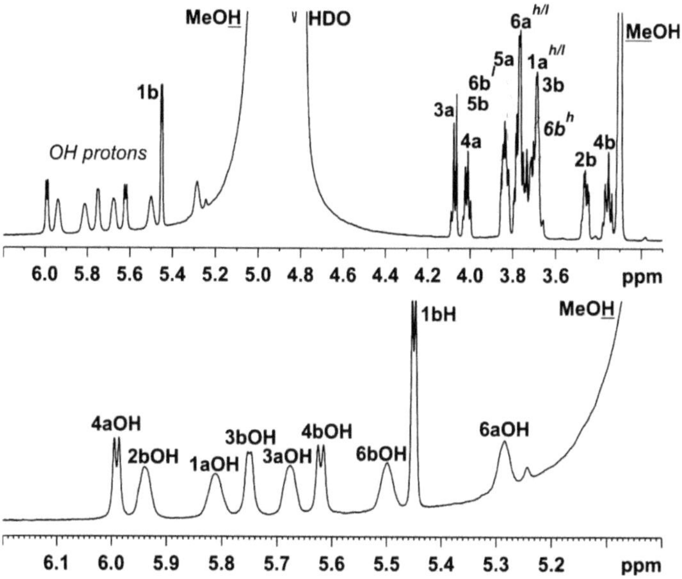

Fig. 1 ^{1}H NMR of sucrose (CD_3OD/CD_3OH 1 : 1, 4 eq. $CaCl_2$, 600 MHz, 292 K, h = high-field, l = low-field proton). The *upper spectrum* shows the well-resolved hydroxyls that are assigned in the lower expansion. Gauss transformation identifies ^3J coupling constants for 4aOH, 3bOH, and 4bOH. 2bOH and 3aOH are broadened from isotope effects. Multiplet structures are not resolved for primary hydroxyls

speed up chemical exchange by several orders of magnitude). The 1 : 1 mixture of D3- and D4-methanol visualizes coupled hydrogen bonds on the basis of their isotope effect [18] (Fig. 1). This provides a simple setup to identify the transient intramolecular contacts at ambient temperature. Sucrose 2bOH and 1aOH show the only chemical exchange cross peak in the ROESY, which can be interpreted as a hydrogen bond similar to the one observed in the solid state. The smallest temperature coefficient of all eight protons (– 6.3 ppb/K between 280 and 310 K) was measured for 6aOH, which can coordinate to the ring oxygen of glucose. These data add to the picture of a dynamic sucrose with an average solution structure close to the crystal structure.

It has been stated that sugars that protect membranes from desiccation, heat, or salt stress bind to biological membranes to perform their membrane protecting action [1–3]. NMR spectroscopy can detect membrane affinities that are mediated by hydrophobic effects and/or by salt bridges [19–21] but the specific binding of neutral polyols onto a lipid bilayer has not been observed yet. Such affinities can be only weak and are out of range of traditional NMR titrations although they play an important role on cellular membranes. On the cell surface, individual ligand–receptor affinities cooperatively multiply to reflect strong polyvalent interactions where selectivity and not affinity is the key issue in each individual binding event [22–25]. An NMR method is developed here to scrutinize the properties of the plain membrane bilayer as a structure-selective receptor for several polyols, including cryo-protective sugars [26].

Solution NMR methods like DOSY [27–29], STD [30–32] or trNOE [33, 34] extend the affinity range of classical titration methods towards weaker molecular recognition processes with the advantage that the spectra are confined to the resonance signals of the ligand. Residual dipolar couplings [35–38] (RDCs) can identify bound ligand conformations by transferred dipolar couplings (trDCs) [39, 40] and provide information on the conformation of the ligand in the bound state and on its relative orientation to the receptor [41–43]. RDCs are applied here for the analysis of transient binding events between phosphatidylcholine bilayers and polyols. Large signal responses are expected even for low percentages of "ligand–receptor complexes" making RDCs a sensitive tool for the analysis of molecular recognition events. A small fraction of receptor-bound ligand can be detected as an additional signal splitting (Fig. 2).

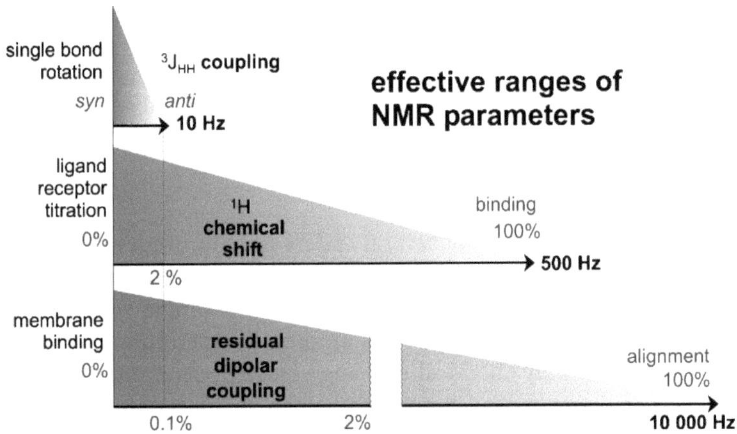

Fig. 2 Accessible measuring windows (dynamic ranges) of three different ^{1}H NMR parameters. The effective range of RDCs is compared to other ^{1}H NMR parameters like three-bond couplings $^3J_{H,H}$ and chemical shifts δ_H. A typical NMR titration with a chemical shift variation $\Delta\delta_H$ upon ligand binding of 1 ppm covers an effective range of 500 Hz on a 500 MHz spectrometer. 10 Hz variation (2% binding) compares to the maximal response detectable with $^3J_{H,H}$ couplings yet to only 0.1% signal response observed for RDCs. The strong dipolar coupling in the solid state ($^1D_{CH} \sim 10$ kHz) is averaged to zero in solution as a result of the fast random reorientation of the molecules, whereas the alignment of on average only one in a thousand molecules in a liquid-crystalline medium is sufficient to measure RDCs of 10 Hz with the precision of high-resolution NMR methods. They can be followed up to approximately 200 Hz in high resolution NMR, which corresponds to 2% binding. In spite of this restriction, RDCs yield exceptionally large signal responses for weak binding affinities. Thus, RDCs are a sensitive tool for the analysis of relative binding specificities of different ligands

A liquid crystalline phase that is suitable for the analysis of membrane binding is provided by phospholipid bicelles, which consist of a mixture of long-chain phospholipid (e.g., DMPC, C_{14}) and a short-chain detergent (e.g., DHPC, C_6) [44–47]. The bicelles are the receptors in the present study. Bicelles can be either regarded as disc-shaped bilayered phospholipid membrane fragments or as a "spongelike system, in which DMPC bilayers are disrupted by DHPC pores" [48]. With variation of the composition of the bicelle solution, i.e., different total amount of lipid (c) or different molar ratios (q) of DMPC : DHPC it is possible to tune the degree of alignment for dissolved molecules, which depends linearly on the surface area of the liquid crystal [49]. Sucrose and other disaccharides exhibit RDCs not exceeding ± 3 Hz for heteronuclear couplings [50]. Even at high lipid concentrations ($q = 2.7$, $c = 12\%$) $^1D_{CH}$ do not exceed 10 Hz. The anisotropic tumbling is caused by steric interaction with the bicelles, not by a specific molecular recognition of the phospholipid bilayers. Only strongly anisotropic linear oligosaccharides resembling a cylindrical (prolate) shape with a ratio of inertial axes around $\sim 1:5$ approach higher values of + / – 20 Hz [51–54].

Structure-dependent interactions of small, polar molecules with bicelles have been discussed by Bax et al. for threonine [47]. But how does a weak membrane affinity of a cryo-sugar look like in the ^{1}H NMR? Polyols of type **1** combine a rigid bicyclic ring system with defined stereochemistry and a globular shape [55, 56]. Their polarity and charge can be diversified to exclude contributions from non-specific membrane interactions. Variation of c, q and the ratio of **1** : DMPC reveals a strong dependence of the measured RDC values on the composition of the bicelle solution (Fig. 3). The higher the amount of **1**, the lower the RDCs measured, as expected for a binding event when the ratio of ligand to receptor is increased. As RDCs also depend on the total amount of lipid (c) we can confirm that the degree of alignment is related to the surface area of liquid crystal available, as proposed by Gaemers and Bax [47]. First-order spectra are obtained only with relatively high concentrations of the ligand **1** in dilute solutions of the bicelle receptor ($c \sim 5\%$). Still, the quantity of the $^1D_{CH}$ coupling constants are in the range of the $^1J_{CH}$. $^1D_{CH}$ can approach several hundred Hertz even in diluted bicelle solutions ($c \sim 5\%$, $q = 3$). Individual ^{1}H NMR resonance signals become intractable above signal splittings of 200 Hz for $|D_{HH} + J_{HH}|$.

The explanation for the strong alignment is the molecule's affinity towards the membrane surface. This affinity cannot be explained by a non-specific hydrophobic binding nor by a simple charge effect since it displays the characteristics of a specific molecular recognition process with the ligand's stereochemistry, conformation, and electrostatic properties as the crucial parameters. More examples of polar membrane-binding molecules are given in [26]. All studied cryo-sugars are different. Their RDCs are in the range of expected values for small molecules without membrane binding that are independent of the type of medium used. This is in contrast to **1**, which

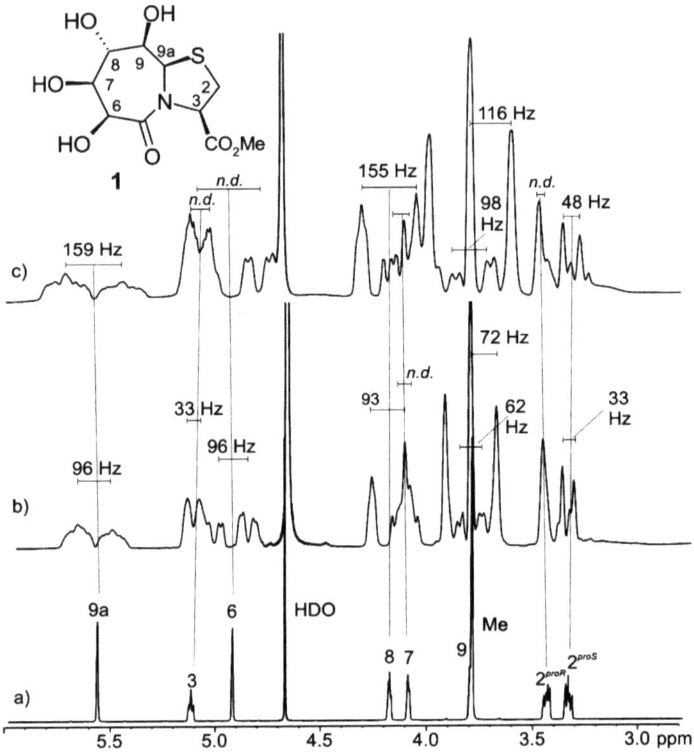

Fig. 3 $|D_{HH} + J_{HH}|$ of **1** for different q and c. **a** ^{1}H NMR of **1** (58 mM) in D_2O (600 MHz, 308 K). The $^3J_{HH}$ couplings of H6–H7 and H9–H9a are small (< 2 Hz) and therefore H6 and H9a appear as singlets. **b** ^{1}H T_1-inversion-recovery NMR of **1** (55 mM, $q = 2.9$, $c = 5\%$, 3% CTAB), 700 µL buffered D_2O (600 MHz, 308 K, $\Delta = 420$ ms). Signal splitting due to the alignment of **1** between the bicelles is given in Hz. **c** ^{1}H T_1-inversion-recovery NMR of **1** (64 mM, 55 mM NaCl, $q = 2.7$, $c = 13\%$, 4% CTAB), 700 µL D_2O (600 MHz, 308 K, $\Delta = 430$ ms)

shows a dependence of its RDCs on the medium. In media that orient the solute only due to steric effects (e.g., PEG/*n*-hexanol/water) the RDCs are distinctively smaller than for media with additional electrostatic effects (e.g., bicelles). The GDO value (a scalar parameter that mirrors the overall strength of the molecule's orientation) and Euler angles, which relate the alignment of orientation to any molecular coordinate system (e.g., crystal structure), were calculated for both the bicelle solution and PEG/*n*-hexanol/water. The relative strengths of alignment of the polyol **1** and sucrose differ by a factor of 25.

The calculated GDO values for **1** under different sample conditions are in accordance with the idea that the binding constant of a ligand–receptor system is independent of sample conditions but specific for every ligand–receptor combination.

The affinity between **1** and the DMPC/DHPC model membrane is strong enough to become observable by alternative methods. Saturation transfer difference (STD) measurements can visualize the transfer of saturation between the bicelles and **1** but not between bicelles and sucrose. The STD spectrum exhibits intense signals for protons that are in close vicinity to the membrane while more distant protons are mediated to zero. The saturation transfer from saturated DMPC-/DHPC-groups onto specific protons of **1** confirm the assumption of a weak but selective binding between **1** and the bicelles.

Polar compounds like **1** or the example investigated in [26] can exhibit considerable membrane affinities with a strong dependence on the electrostatic properties of the investigated molecules. The degree of alignment, as quantity of the ligand–receptor binding strength, is in accordance with the assumption that the overall charge and the hydrogen bonding pattern are crucial for the membrane affinity. Hence, RDCs expand the sensitivity range of NMR spectroscopy to characterize the structure-dependence of ultraweak molecular recognition processes between small molecules and the phosphatidylcholine head groups of the bilayered membrane models. But, due to the fact that strong binding will lead to complete dephasing of magnetization and thus no observable RDCs, the approach presented here will be limited to ultraweak interactions, which are discussed for polar small molecules like drugs or cryoprotectants [57–60]. Sucrose is a dynamic molecule that does not interact with membranes, but what about its cryoprotective derivatives with larger molecular weights?

3
Cyanobacterial Oligosaccharides

α-D-Gluco-oligosaccharides and glucans are ubiquitous saccharides connected by either $(1 \rightarrow 3)$-, $(1 \rightarrow 4)$-, or $(1 \rightarrow 6)$-linkages. Beside their role as storage sugars, certain oligosaccharides are important bioprotectants of cellular membranes and proteins. The oligosaccharides isolated from the cyanobacteria *Nostoc spec.* and *N. ellipsosporum* extend this class of carbohydrates by the $(1 \rightarrow 2)$-linked oligomers of α-D-glucose with the highest degrees of polymerization known so far. In contrast to the widespread inulins [61], where a single pyranose ring terminates a fructan oligomer: α-D-Glc$(1 \rightarrow 2)[$-β-D-Fru$(1 \rightarrow 2)]_n$, the cyanobacterial oligosaccharides described here have an inverted composition with a single fructofuranose terminating a linear chain of glucopyranoses: $[\alpha$-D-Glc$(1 \rightarrow 2)]_n$ – β-D-Fru. The individual oligomers could be separated up to the decasaccharide ($n = 9$). They adopt right-handed helices with a tetrasaccharide repeat, a diameter of 12 Å, and a pitch of 5 Å. Although any glucan can assume helical conformations in solution, only the vicinal α-$(1 \rightarrow 2)$-glycosidic linkages of the cyanobacterial oligosaccharides induce a compact helix which – in the presence of appropriate helix-capping by fructofuranose – exhibits a relevant energy barrier against unwinding. Kojiheptaose with a fructofuranose cap [α-D-Glc(1 $\rightarrow$ 2)]$_7$-β-D-Fru is the smallest biopolymer with an NMR-observable unfolding barrier (^{1}H NMR, 300 K, 600 MHz), thus demonstrating the co-operative effect of helix capping and helix pitch of the individual building blocks. The secondary structure of the native oligosaccharide collapses upon selective hydrolysis of the helix-capping fructofuranose.

The amount and composition of cyanobacterial intracellular saccharides varies in response to environmental stress. Total intracellular carbohydrate levels of the investigated *Nostoc* species rise up to tenfold in response to salt (75 mM NaCl, 2 days) and temperature (40 °C, 1 day), respectively. Especially under heat stress, sucrose and the lower gluco-oligosaccharides in *N. ellipsosporum* are displaced by higher oligomers up to the decasaccharide. The quality of this new class of non-reducing sugars to act as bioprotectants [62] are described elsewhere [63]. Oligo-(1 $\rightarrow$ 2)-α-D-glucopyranosyl-(1 $\rightarrow$ 2)-β-D-fructofuranosides were isolated and characterized up to the decasaccharide, thus exceeding previously known koji-oligomers obtained from in vitro synthesis [64] or natural sources [65] and establishing a complete series of α(1 $\rightarrow$ 2)-glucans with a fructofuranoside capping on the reducing end terminus. Transglycosidic heteronuclear $^3J_{CH}$ couplings identify α(1 $\rightarrow$ 2)-linkages and assign the fructofuranoside on the reducing end of the α(1 $\rightarrow$ 2)-glucan chain. Well-separated anomeric resonances in the ^{1}H NMR spectra with ^3J coupling constants around 3.5 Hz are observed even for the hexasaccharide [α-D-Glc(1 $\rightarrow$ 2)]$_5$-β-D-Fru. The homo- and heteronuclear NMR assignments are obtained in a completeness that cannot be achieved for other glucans because they exhibit the minimal ^{1}H chemical shifts dispersions typical of homo-oligomers. The anomeric protons of the α-(1 $\rightarrow$ 2)-linked hexasaccharide are at least 0.1 ppm (^{1}H NMR, 600 MHz) separated from each other, while, for comparison, the anomeric protons of the three glycosidic linkages of the α-(1 $\rightarrow$ 4)-glucan maltotetraose are not separated at all in the ^{1}H NMR (Figs. 4 and 5).

The unexpectedly large chemical shift dispersion is a consequence of the limited conformational flexibility of the vicinal progression of glycosidic linkages. One helix turn is formed by four glucose rings, and hydrophobic stacking between neighboring helix turns stabilizes the overall secondary structure. The characteristic NOEs are discussed in Fig. 6 and a model is shown in Fig. 7. The glycosidic torsions of each α-Glc(1 $\rightarrow$ 2)-Glc disaccharide moiety of the right-handed helix assume values of approximately $\phi = -30°$ (ϕ = H1-C1-O-C2′) and $\psi = +30°$ (ψ = C1-O-C2′-H2′), which are well within the global energy minimum of the Ramachandran plot of an isolated kojibiose disaccharide encompassing values for ϕ between 0 and – 60° and ψ between – 60° and + 60° [66]. In contrast to an isolated kojibiose disaccharide, the ϕ/ψ plots of the longer α(1 $\rightarrow$ 2)-glucans are much more restricted because the borderline between *P* and *M* helicity divides the global energy minimum of

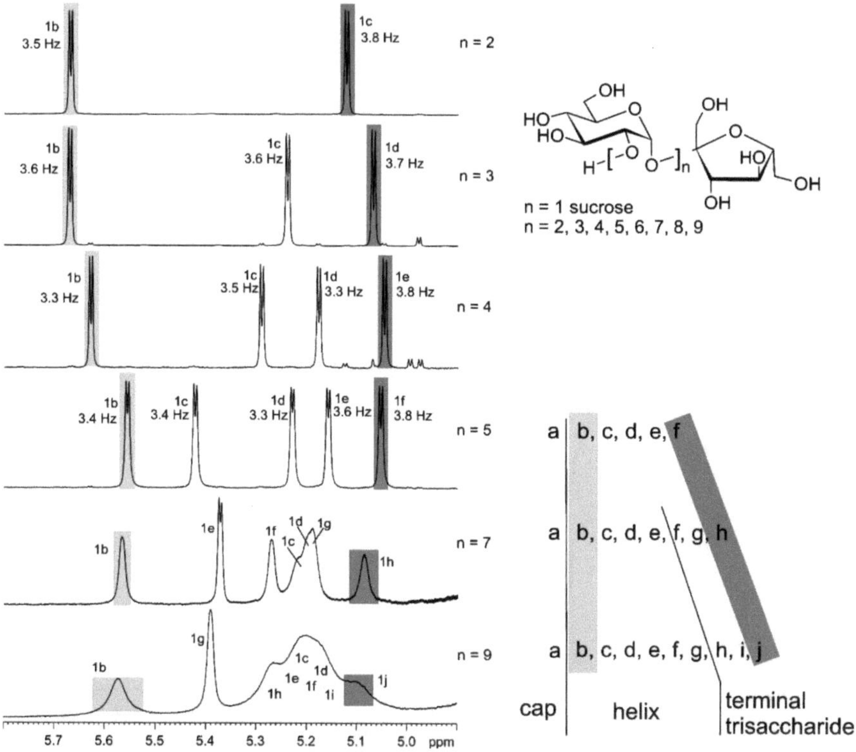

Fig. 4 Expansion of the anomeric region from the ^{1}H NMR spectra (D_2O, 300 K, 600 MHz) of the α-D-gluco-oligosaccharides $[\alpha\text{-D-Glc}(1 \rightarrow 2)]_n$-$\beta$-D-Fru isolated from *N. ellipsosporum*. The signal assignments are based on homo- (DQF-COSY) and heteronuclear (HMQC, HMBC) correlations and additionally on dipolar contacts (ROESY) in the fast-exchange regime (320 K) for the octa- and decasaccharide. In all compounds studied, the anomeric proton of pyranose b (*light gray*) is resonant at lowest field (highest ppm value) while the glucopyranose on the opposite end of the oligomer appears at highest field

the ϕ/ψ plots along a line where the sum of ϕ and ψ equals − 60°. Together with the approximately + 60° (H2-C2-C1-H1) added by each pyranose ring, the preference for an overall zero helicity is obtained.

In other words, the combination of glycosidic torsions preferred by an individual $\alpha(1 \rightarrow 2)$-glycosidic bond is not accessible for $\alpha(1 \rightarrow 2)$-glucans, which evade the zero pitch region towards the neighboring tight helical conformations. Both helix senses are accessible by only minor changes of the glycosidic torsions and the preference for either helix sense is controlled by secondary effects (Fig. 7).

Such a secondary effects are minor structural variations on the termini of the $\alpha(1 \rightarrow 2)$-glucan, which suffice to induce either *P* or *M* helix sense or an equilibrium of both. Fructofuranose acts as a helix-cap for a right-handed

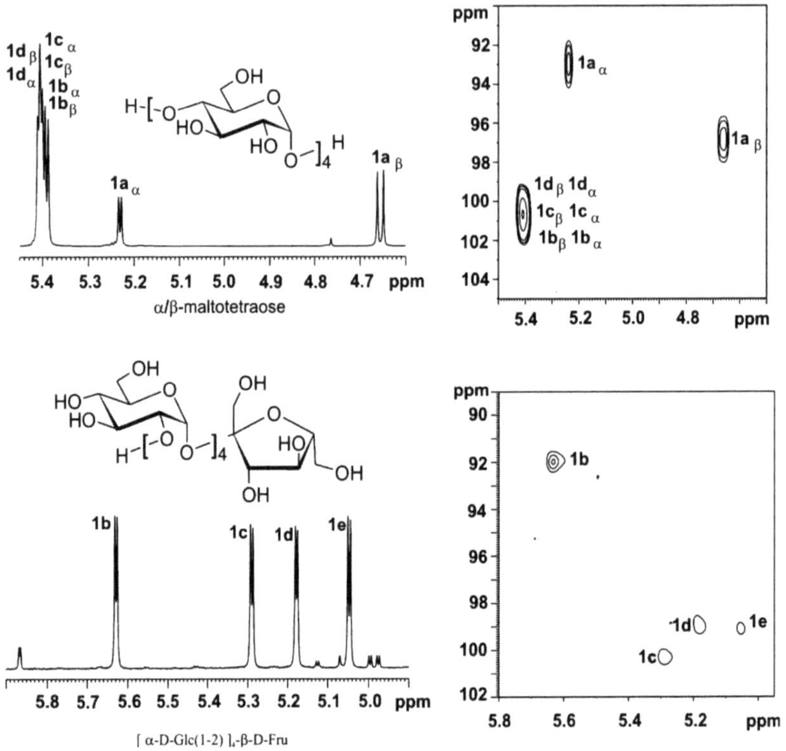

Fig. 5 Comparison of ^{1}H and ^{13}C chemical shift dispersion of tetramaltoside (α/β mixture) and the pentasaccharide [α-D-Glc(1 $\rightarrow$ 2)]$_4$-β-D-Fru (D_2O, 300 K, 600 MHz). The anomeric regions of ^{1}H NMR and 2D HMQC spectra are shown

helix by coordinating to the terminal trisaccharide moiety (Fig. 8). The dynamics of the fructose cap remain largely unaffected by the growing chain of (1 $\rightarrow$ 2)-α-linked glucopyranoses.

The conformation of the glycosidic linkage of the fructofuranose moiety is found to be similar to the solid-state structure of sucrose. Hydrophobic contacts are visible as NOEs but no permanent hydrogen bonds were identified for the end-capping fructose. The end-capping fructofuranose effectively terminates the hydrophobic ring planes of the four terminal glucopyranoses. The average furanose ring plane is close to perpendicular to the helix axis (Fig. 8). Values for the homooligomeric helix are given in Table 1.

Selective acidolysis of the single ketosidic bond of the pentasaccharide ($n = 4$) causes the collapse of the secondary structure apparent from significant overall chemical shift changes and the altered NOE pattern. The kojitetraose [α-D-Glc(1 $\rightarrow$ 2)]$_4$-OH appears as two separate ^{1}H NMR signal sets, one for the α- and one for the β-anomer of the reducing glucose. Both differ significantly from the fructose-capped α-(1 $\rightarrow$ 2)-glucan; even the anomeric

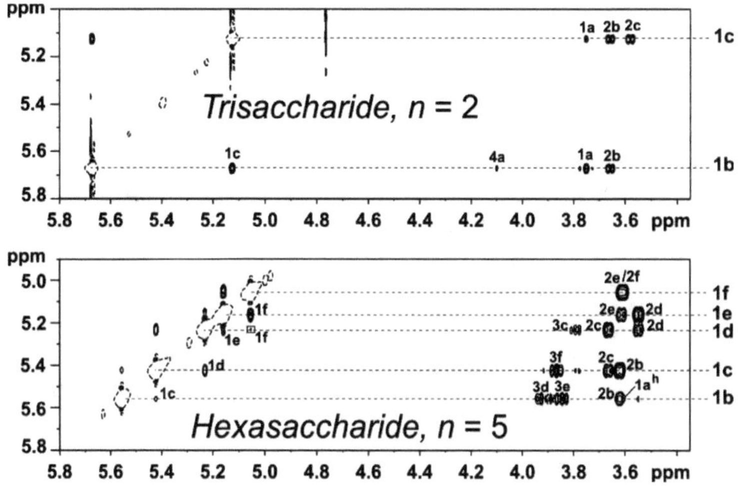

Fig. 6 *Top*: Expansion of the anomeric region from the ROESY spectrum of the hexasaccharide [α-D-Glc(1 $\rightarrow$ 2)]$_5$-β-D-Fru including the NOE assignments. The average distance for NOEs $H1_i$-$H1_{i+1}$ = 2.9–3.4 Å, $H2_i$-$H1_{i+1}$ = 2.1–2.2 Å, and the long range NOEs $H1_i$-$H3_{i+3}$ = 2.4–2.5 Å. Long-range NOEs are absent after acidolysis of the fructofuranoside bond and the average interglycosidic NOEs shorten, ($H1_i$-$H1_{i+1}$ = 2.3 Å for the α-epimer and 2.4–2.8 Å for the β-epimer) as a result of conformational flexibility

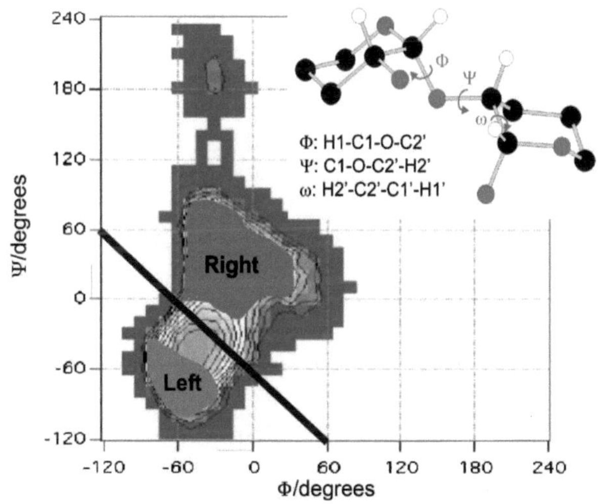

Fig. 7 Global energy minimum in the ϕ/ψ-plot of $\alpha(1 \rightarrow 2)$-glucans is divided by a line where the sum of ϕ and ψ equals – 60°. The accessible ϕ/ψ-combinations for longer $\alpha(1 \rightarrow 2)$-glucans are either right-handed or left-handed helical to both sides of this forbidden region. The basic ϕ/ψ-plot shown was taken from GlycoMaps [65] and calculated for a disaccharide unit

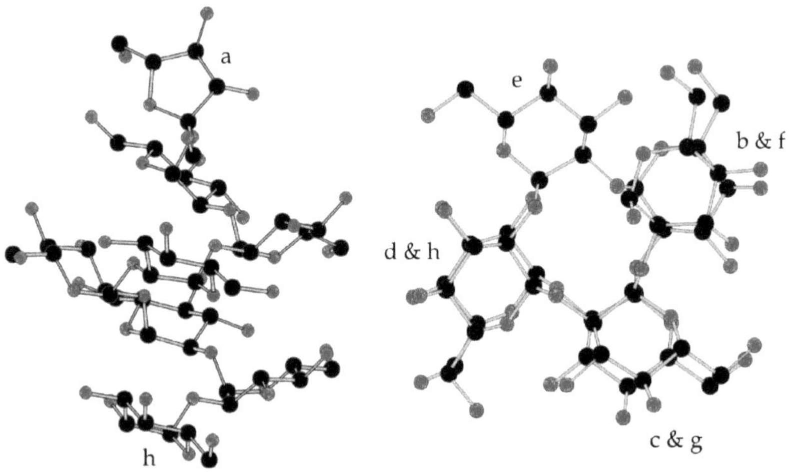

Fig. 8 *Left*: Side view of the energy-minimized average structure of the octasaccharide including NOE-derived distance restraints. The glycosidic torsions ($\phi = 91.7°$ and $\psi = -31.3°$) assume values similar to the ones observed in the solid state for sucrose ($\phi = 107.9°$ and $\psi = -44.8°$). *Right*: Top view along the helix axis showing the tetrasaccharide repeat (ring a is not shown in this representation). Molecules were constructed in HyperChem and minimized in vacuo to 0.01 kcal (mol Å)$^{-1}$. The MM2+ parameter set was used for the minimization procedure in conjunction with the Polak–Ribiere conjugate gradient method implemented in HyperChem(TM) [67]

Table 1 Average rotating frame NOE distances across the glycosidic linkage (1-2 ROE) and between the anomeric protons of sequential pyranoses (1-1 ROE) and the corresponding Φ/Ψ angle ranges of the O-glycosidic linkages that were obtained from MD simulations

Oligosaccharide	1-1 ROE (Å)	1-2 ROE (Å)	Φ/Ψ angle (defined via protons)
greenα-Defruct	Short (~ 2.4)	Short (~ 2.4)	$\Phi = -42$*to* $-69°$ $\Psi = -50$*to* $-61°$
β-Defruct	Short/medium (2.4 – 2.8)	Missing	Not calculated
Penta/hexa capped oligosaccharides	Long (2.6 – 3.4)	Short (2.1 – 2.2)	$\Phi = -30$*to* $-45°$ $\Psi = -32$*to* $-55°$
Helix model right-handed ($\Phi = 60°$, $\Psi = 120°$) (defined via heavy atoms)	Long (2.8 – 2.9)	Short (~ 2.3)	$\Phi = \sim -50°$ $\Psi = \sim 0°$
Helix model left-handed ($\Phi = 60°$, $\Psi = 60°$)	Short (2.1 – 2.2)	Long (3.0 – 3.1)	$\Phi = -55$*to* $-60°$ $\Psi = \sim 60°$

Experimentally determined proton–proton distances of the various oligosaccharides investigated were set as restraints in these calculations, whereas the calculations of the model helices are based on Φ/Ψ angle restraints

proton of the terminal glucose on the opposite end of the homo-oligomer senses a low-field shift of 0.17 ppm (α-anomer) and 0.66 ppm (β-anomer), compared to [α-D-Glc(1 $\rightarrow$ 2)]$_4$-β-D-Fru, respectively. Such large chemical shift variations document how the fructofuranose helix-cap influences the entire chain of the α(1 $\rightarrow$ 2)-glucan (Fig. 8). The cooperative collapse of the helical secondary structure upon cleavage of the fructofuranose is visible even in the NMR spectra of the highest oligomers studied here although only the shorter ones can be fully assigned in the ^{1}H NMR (Fig. 9 and Fig. 10).

The line-broadening observed for the larger octa- and decasaccharide (Fig. 4) could be nearly resolved in the fast exchange regime at 320 K but the slow-exchange regime with separate signal sets for the exchanging isomers was not accessible. At 300 K, the fourth-last anomeric proton (pyranose e of the octasaccharide and pyranose g of the decasaccharide, respectively) and the anomeric protons of the terminal pyranoses are the only resolved anomeric protons. The anomeric proton of the fourth-last pyranose is resonant at a chemical shift of approximately 5.4 ppm, baseline-separated from all other resonances. The discrete chemical environment of a single pyranose within the α-(1 $\rightarrow$ 2)-glucan chain is caused by different secondary structures of the reducing end and the non-reducing end moieties. These opposing helix senses accessible by the α(1 $\rightarrow$ 2)-glucans are bridged by the fourth-last pyranose forming the saddle required between the *P* and *M* helices. The right-handed helix induced by the helix-capping fructofuranose is opposed by the terminal trisaccharide moiety (f, g, h of the octa- and h, i, j of the decasaccharide) of the larger oligomers forcing the fourth-last pyranose into the saddle position between the two opposing helix senses. The energy barrier does not further increase for higher polymerization degrees, a phenomenon which is also observed also for racemization barriers

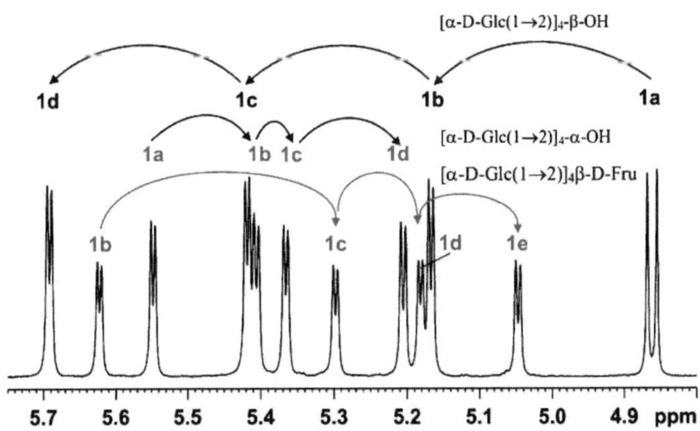

Fig. 9 Anomeric region (^{1}H NMR, D_2O, 300 K, 600 MHz) of the pentasaccharide [α-D-Glc(1 $\rightarrow$ 2)]$_4$-β-D-Fru and the defructosylated koji-tetrasaccharides [α-D-Glc(1 $\rightarrow$ 2)]$_4$-α/β-OH obtained from selective acidolysis of the ketosidic bond

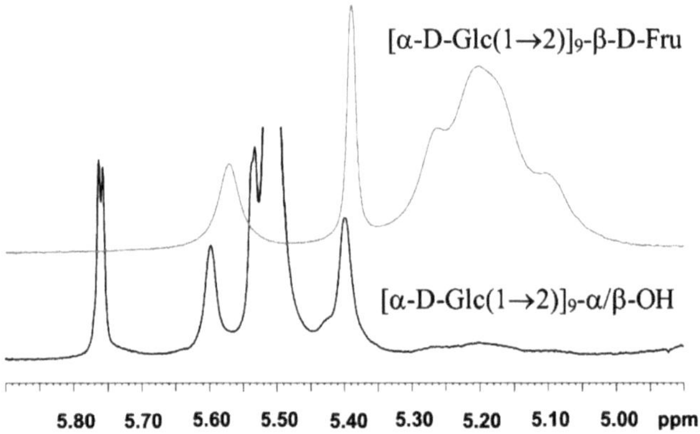

Fig. 10 Anomeric region (^{1}H NMR, D_2O, 300 K, 600 MHz) of the decasaccharide [α-D-Glc(1 → 2)]$_4\beta$-D-Fru and the defructoslylated koji-nonasaccharide [α-D-Glc(1 → 2)]$_9$-OH (*black*, not to scale). Selective acidolysis of the single ketosidic bond in the decasaccharide ($n = 9$), again, as in the case of the defructosylated [α-D-Glc(1 → 2)]$_4$-OH, results in a complete reshaping of the ^{1}H NMR. Yet, no signal assignment is feasible due to the insufficient chemical shift dispersion, as expected for a homo-oligomer with fraying ends. The collapse of the secondary structure is accompanied by a low-field shift of the anomeric resonances

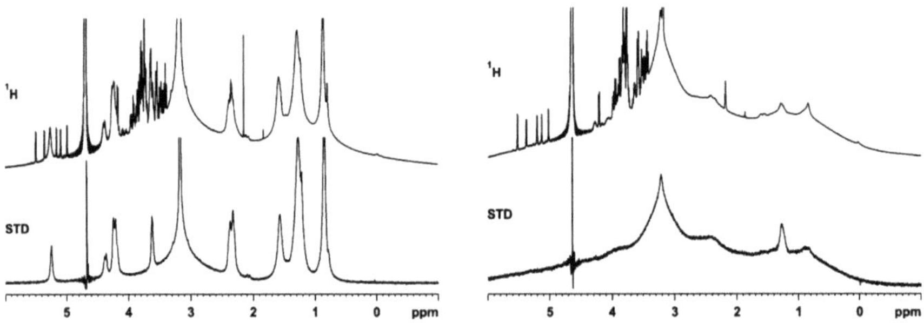

Fig. 11 *Left*: ^{1}H NMR (600 MHz) and STD spectra of [α-D-Glc(1 → 2)]$_5\beta$-D-Fru in the presence of DMPC/DHPC liposomes at 300 K (isotropic solution). *Right*: The same spectra at 308 K in bicelles (nematic liquid-crystalline medium). No saturation transfer is observed in the STD spectra, neither in the case of liposomes (*left*) nor in the case of the nematic phase (*right*). The RDCs are in the region of a few Hz as expected for a molecule that does not bind to the phosphatidylcholine bilayer. The broad underground stems from the DMPC/DHPC liposomes or bicelles, respectively

of achiral homo-oligomers [68]. Illustrative examples of macroscopic systems with saddle structures are found in twines of bryony (*Bryonia alba*) or vine tendrils that have to adopt both helix senses to relieve torsion during curling.

Based on STD NMR measurements and RDCs, membrane affinity can be excluded for the α-(1 $\rightarrow$ 2)-glucans (Fig. 11) as it was excluded for the more common cryoprotective sugars sucrose and trehalose. Taken together, α-(1 $\rightarrow$ 2)-glucans have a strong tendency to form helical secondary structures, with structural modifications on the reducing end of the α(1 $\rightarrow$ 2)-gluco-oligosaccharide exerting a long-range conformational effect. The fructofuranose cap on the reducing end terminus stabilizes a right-handed helix, which is independent from the length of the attached homo-oligomer, ending however, at the fourth-last pyranose.

4 Conclusion

Sucrose and its higher homologs show cryoprotectant properties, although varying considerably in their structural properties like molecular shape and dynamics. Oligo-(1 $\rightarrow$ 2)-α-D-glucopyranosyl-(1 $\rightarrow$ 2)-β-D-fructofuranosides challenge the general view of linear oligosaccharides as being flexible biomolecules. This new class of glucans show a slow equilibrium between a folded helix and a partially unfolded structure. How these unique conformational properties correlate with the bioprotective properties of the fructose-capped oligo-kojibioses remains to be elucidated.

Acknowledgements This work was supported by the Deutsche Forschungsgemeinschaft. The authors thank Dr. Eckhard Loos and Edelgard Herold for the isolation of the oligo-(1 $\rightarrow$ 2)-α-D-glucopyranosyl-(1 $\rightarrow$ 2)-β-D-fructofuranosides.

References

1. Bryant G, Koster KL, Wolfe J (2001) Seed Science Res 11:17
2. Vereyken IJ, Chupin V, Hoekstra FA, Smeekens SCM, Kruijff B (2003) Biophys J 84:3759
3. Furuki T (2002) Carbohydr Res 337:441
4. Hincha DK, Zuther E, Hellwege EM, Heyer AG (2002) Glycobiology 12:103
5. Bock K, Lemieux RU (1982) Carbohydr Res 100:63
6. Freedberg DI (2002) J Am Chem Soc 124:2538
7. Venable RM, Delaglio F, Norris SE, Freedberg DI (2005) Carbohydr Res 340:863
8. Poppe L, van Halbeek H (1992) J Am Chem Soc 114:1092
9. Immel S, Lichtenthaler FW (1995) Liebigs Ann Chem 1925
10. Symons MCR, Benbow JA, Harvey JM (1980) Carbohydr Res 83:9
11. Poppe L, van Halbeek H (1991) J Am Chem Soc 113:363
12. Poppe L, van Halbeek H (1994) Nature Struct Biol 4:215
13. Sandström C, Baumann H, Kenne L (1998) J Chem Soc Perkin Trans 2 4:809
14. Bekiroglu S, Kenne L, Sandström C (2003) J Org Chem 68:1671
15. Adams B, Lerner L (1992) J Am Chem Soc 114:4827

16. Leeflang BR, Vliegenthart JFG, Kroon-Batenburg LMJ, van Eijck BP, Kroon J (1992) Carbohydr Res 230:41
17. Bock K, Frejd T, Kihlberg J, Magnusson G (1988) Carbohydr Res 176:253
18. Dabrowski J, Kožár T, Grosskurth H, Nifant'ev NE (1995) J Am Chem Soc 117:5534
19. Middleton DA, Reid DG, Watts A (2004) J Pharm Sci 93:507
20. Wang J, Schnell JR, Chou J (2004) J Biochem Biophys Res Commun 324:212
21. Middleton DA, Hughes E, Madine J (2004) J Am Chem Soc 126:9478
22. Morton CJ, Campbell ID (1994) Curr Biol 4:615
23. McLendon G (1991) Struct Bond 75:159
24. Kim KH, Thomas DW (1988) J Immunol 140:2500
25. Simons K, Ikonen E (1997) Nature 387:569
26. Fischer D, Geyer A (2005) Magn Reson Chem 45:893
27. Johnson CS (1999) Prog NMR Spectrosc 34:203
28. Yan J, Kline AD, Mo H, Zartler ER, Shapiro MJ (2002) J Am Chem Soc 124:9984
29. Diaz MD, Berger S (2000) Carb Res 329:1
30. Klein J, Meinecke R, Mayer M, Meyer B (1999) J Am Chem Soc 121:5336
31. Vogtherr M, Peters T (2000) J Am Chem Soc 122:6093
32. Mayer M, Meyer B (1999) Angew Chem Int Ed 38:1784
33. Campbell AP, Sykes BD (1993) Ann Rev Biophys Biomol Struct 22:99
34. Ni F (1994) Prog NMR Spectrosc 26:517
35. Tjandra N, Bax A (1997) Science 278:1111
36. Prestegard JH, Al-Hashimi HM, Tolman JR (2000) Quart Rev Biophys 33:371
37. Tolman JR (2001) Curr Opin Struct Biol 11:532
38. Simon B, Sattler M (2002) Angew Chem Int Ed 41:437
39. Koenig BW (2002) ChemBioChem 3:975
40. Bolon PJ, Al-Hashimi HM, Prestegard JH (1999) J Mol Biol 293:107
41. Koenig BW, Mitchell DC, König S, Grzesiek S, Litman BJ, Bax A (2000) J Biomol NMR 16:121
42. Koenig BW, Kontaxis G, Mitchell DC, Louis JM, Litman BJ, Bax A (2002) J Mol Biol 322:441
43. Shimizu H, Donohue-Rolfe A, Homans SW (1999) J Am Chem Soc 121:5815
44. Sanders CR II, Schwonek JP (1992) Biochemistry 31:8898
45. Ottiger M, Bax A (1998) J Biomol NMR 12:361
46. Sanders C, Hare B, Howard K, Prestegard JH (1994) Prog NMR Spectrosc 26:421
47. Brunner E (2001) Conc Magn Reson 13:238
48. Rowe BA, Neal SL (2003) Langmuir 19:2039
49. Gaemers S, Bax A (2001) J Am Chem Soc 123:12343
50. Neubauer H, Meiler J, Peti W, Griesinger C (2001) Helv Chim Acta 84:243
51. Rundlöf T, Landersjö C, Lycknert K, Maliniak A, Widmalm G (1998) Magn Reson Chem 36:773
52. Martin-Pastor M, Bush CA (2000) Biochemistry 39:4674
53. Thompson GS, Shimizu H, Homans SW, Donohue-Rolfe A (2000) Biochemistry 39:13153
54. Tian F, Al-Hashimi HM, Craighead JL, Prestegard JH (2001) J Am Chem Soc 123:485
55. Geyer A, Bockelmann D, Weissenbach K, Fischer H (1999) Tetrahedron Lett 40:477
56. Geyer A, Moser F (2000) Eur J Org Chem 7:1113
57. Lambruschini C, Relini A, Ridi A, Cordone L, Ghiozzi A (2000) Langmuir 16:5467
58. Engel A, Bendas G, Wilhelm F, Mannova M, Ausborn M, Nuhn P (1994) Int J Pharm 107:99
59. Santarius KA (1996) Cryobiology 33:118

60. Turner S, Senaratna T, Touchell D, Bunn E, Dixon K, Tan B (2001) Plant Sci 160:489
61. Stevens CV, Meriggi A, Booten K (2001) Biomacromolecules 2:1
62. Brumfiel G (2004) Nature 428:14
63. Fischer D, Geyer A, Loos E (2006) FEBS J 273:137
64. Chaen H, Nishimoto T, Nakada T, Tetsuya F, Fukuda S, Kurimoto M, Tsujisaka Y (2001) J Biosci Bioeng 92:177
65. Watanabe T, Kamo Y, Yoshihiro M, Matsuda K, Dudman WF (1982) Carbohydr Res 110:170
66. Lütteke T, Frank M, Lieth C-W v. d. (2005) Nucleic Acid Res 33:D242
67. HyperChem. Hypercube, Gainesville, FL
68. Janke RH, Haufe G, Würthwein EU, Borkent JH (1996) J Am Chem Soc 118:6031

Top Curr Chem (2007) 272: 187–215
DOI 10.1007/128_2006_088

Published online: 25 October 2006

Residual Dipolar Couplings Report on the Active Conformation of Rhodopsin-Bound Protein Fragments

Bernd W. Koenig

Structural Biology Institute, IBI-2, Research Center Jülich, 52425 Jülich, Germany
b.koenig@fz-juelich.de

Present address:
Heinrich-Heine-Universität Düsseldorf, Universitätsstrasse 1, 40225 Düsseldorf, Germany

Abstract Residual dipolar couplings (RDCs) provide unique information on the structure and dynamics of soluble biomolecules. They are easily extracted from high-resolution liquid-state NMR spectra of molecules that undergo slightly anisotropic tumbling. Options for solute alignment are discussed, followed by a summary of NMR strategies for detection of RDCs. The use of RDCs in protein structure determination and for establishing domain orientation is reviewed. Transient binding of ligand to an aligned target molecule can give rise to transferred dipolar couplings that characterize the target-bound form but are easily detected on the free form of the ligand by liquid-state NMR. There is no size limit on the target molecule. RDCs have been used to study the interaction of bound peptides with the G protein-coupled receptor rhodopsin in its natural membrane environment, with amyloid fibrils, and with lipid model membranes. Further, transient binding of small carbohydrates to soluble proteins was characterized using RDCs. A brief literature review is followed by a detailed discussion of the RDC-based structure determination of a rhodopsin-bound transducin peptide.

Keywords Ligand · Membrane protein · NMR · Residual dipolar couplings · Transient binding

Abbreviations

A	molecular alignment tensor
A_a	magnitude of the alignment tensor
CCR	cross-correlated relaxation
CT-HSQC	constant time version of HSQC
D_a	magnitude of dipolar coupling tensor
D_B^{PQ}	bound state dipolar coupling between nuclei P and Q
D_F^{PQ}	free state dipolar coupling between nuclei P and Q
D_{obs}^{PQ}	observed dipolar coupling between nuclei P and Q
DMPC	1,2-dimyristoyl-*sn*-glycero-3-phosphocholine
DHPC	1,2-dihexanoyl-*sn*-glycero-3-phosphocholine
$G_t\alpha$	α-subunit of transducin
$G_t\gamma$	γ-subunit of transducin
HSQC	heteronuclear single quantum correlation
K_D	dissociation constant
MII	metarhodopsin II state of rhodopsin
NOE	nuclear Overhauser effect
NOESY	nuclear Overhauser effect spectroscopy
PALES	prediction of alignment from structure
PAS	principal axis system
Pf1	filamentous bacteriophage
R	rhombicity of dipolar coupling tensor
RDC	residual dipolar coupling
REDCAT	residual dipolar coupling analysis tool
S	generalized order parameter
SA	simulated annealing
STD	saturation transfer difference spectroscopy
SUV	small unilamellar lipid vesicles
SVD	single value decomposition
S2	peptide with sequence IRENLKDSGLF
S^3E	spin-state selective excitation
TrDC	transferred dipolar coupling
TrNOE	transferred NOE
TROSY	transverse relaxation optimized spectroscopy
2D	two-dimensional
3D	three-dimensional

1 Introduction

Protein structure determination by high-resolution liquid-state NMR has been traditionally based on numerous short-range proton-proton distances (up to ~ 5 Å) derived from Nuclear Overhauser enhancement (NOE) experiments and to a lesser extent on torsion angles from scalar couplings and correlations between chemical shift values and secondary structure [1]. These structure restraints are local by nature. Over the last decade residual dipolar couplings (RDCs) have emerged as a new powerful structure determinant that provides both short- and long-range structure information [2, 3]. RDCs depend both on the distance between two interacting nuclei and the average orientation of the internuclear vector relative to a molecule-fixed reference frame. This orientational dependence allows specification of the mutual orientation of internuclear vectors no matter where in the structure these vectors are located, which makes RDCs truly global restraints.

The advent of transverse relaxation optimized spectroscopy (TROSY) based NMR experiments [4] has dramatically extended the size limit for soluble biomacromolecules that are amenable to liquid-state NMR structure determination. However, the TROSY principal works best in concert with highly deuterated biopolymers severely limiting the extraction of $^1H - {}^1H$ distances from NOESY experiments. Reliability of NOE-based distance restraints may also be hampered by potential spin diffusion which becomes problematic in large molecules. Therefore, RDCs take center stage as a major structure determinant in large biopolymers while also providing unique information on structure and orientation of small and medium-size biomolecules.

Measurement of RDCs by high-resolution NMR requires partial alignment of the solute under study. The introduction of widely applicable and practicable concepts that afford a tunable degree of solute alignment was a key step in the development of RDC methodology. The most popular alignment concepts that are available to the NMR spectroscopist today will be summarized. One important prerequisite of straightforward RDC measurements is the preservation of the high-resolution character of liquid-state NMR spectra without generating undue spectral complexity by the reintroduction of dipolar coupling information into the spectra. Generating just the right degree of alignment is the key to this central requirement. NMR experiments that are used for measurement of RDCs are often based on concepts originally developed for detection of scalar couplings. A reference-like overview will be given on the most widely used experimental schemes employed in gathering RDC data.

Interpretation of RDCs in terms of structure, orientation, or dynamic features of the biomolecule under study requires knowledge of the order tensor that mathematically describes partial alignment. Different strategies have been developed for determination of this or other closely related tensors. Software tools are available in the public domain that can aid in the derivation of tensor components from experimental RDCs. Determination of the order tensor is often embedded in the structure determination process. There are several RDC-based approaches that address different aspects of protein structure and biomolecular interaction. The major strategies will be introduced in order to aid the interested reader in choosing the most promising approach for solving a problem at hand. For example, RDCs are most commonly used as supplemental but highly valuable restraints in structure refinement protocols. Alternatively, they have been used to determine the relative orientation of sub-molecular domains in proteins or of interaction partners in complexes. RDCs are very sensitive to structural changes that may occur as a result of complex formation. Concepts for de novo determination of protein structure from RDCs have been developed recently. These approaches often integrate other sources of structural information like data base-derived structures of molecular fragments, structure of homologous proteins, or structure prediction schemes.

The second part of this review is dedicated to the use of RDCs in studies of weakly binding complexes. RDCs of strongly aligned molecules can be very large while RDCs of a freely moving small solute are typically small or average to zero in the case of isotropic motion. High-resolution liquid-state NMR can provide a time-weighted average of the RDCs that apply to bound and free ligand, respectively, if the exchange between target-bound and free ligand is fast on the NMR time scale. These transferred dipolar couplings (TrDCs) are very sensitive to the bound conformation of the ligand. The TrDC method allows extraction of the bound ligand structure from easily recorded spectra of the free ligand. Under fortunate conditions, the bound ligand can be characterized even if the size of the target molecule is way above the size limit that applies to liquid-state NMR. In particular, the TrDC method can provide the bound structure of ligands and other interaction partners of membrane proteins. Requirements for the application of the TrDC method will be discussed. RDCs have been used in a few studies for characterization of weakly bound complexes. These recent applications will be briefly reviewed followed by a detailed account on the use of TrDCs for determination of the receptor-bound conformation of a peptide that mimics a major rhodopsin-binding site of the G protein transducin.

2 Residual Dipolar Couplings in High-Resolution NMR

2.1 What are Residual Dipolar Couplings?

Dipolar couplings result from the direct through-space interaction between magnetically active nuclei. In the interest of simplicity we will ignore here any spin-spin interactions other than dipolar couplings that could further modify the energy levels of the spins and the NMR spectrum (e.g. scalar couplings). NMR magnets provide a very strong and homogeneous magnetic field B_0 which is parallel to the z-axis of the laboratory frame by convention. A spin 1/2 nucleus, P, (e.g., ^{1}H, ^{13}C, ^{15}N) that is placed in the B_0 field acts as a magnetic moment that is either parallel or antiparallel to B_0 and becomes the source of a magnetic dipolar field. The local field at the position of a second spin 1/2 nucleus, Q, will then be the sum of B_0 and the dipolar field arising from P. In practice, only the component of the dipolar field parallel to the much stronger B_0 field needs to be considered (secular approximation). Assuming a static and isolated spin pair P, Q, the effective field strength at Q can have two different values depending on the spin state of P. Since the resonance frequency of a nuclear spin is proportional to the local magnetic field strength this gives rise to two possible resonance frequencies for each of the two interacting nuclei. The NMR spectrum of an ensemble of spin pairs P, Q shows a symmetric doublet for each nucleus. The doublet splitting is a measure of the dipolar coupling, it depends on the distance r_{PQ} and the angle ξ between the internuclear vector **PQ** and the z-axis. In the absence of motion the dipolar coupling of nearby spins can be as large as several kHz ($\sim$ 41 kHz for ^{1}H - ^{1}H, $\sim$ 22 kHz for ^{15}N - ^{1}H, and $\sim$ 48 kHz for ^{13}C - ^{1}H interactions based on Eq. 1b and lower distance limits of 1.80 Å, 1.04 Å, and 1.08 Å, respectively). Heavy overlap of multiplet pattern may result in complete loss of spectral resolution for multispin systems. However, if the vector **PQ** undergoes motion then only a reduced dipolar coupling, D^{PQ}, will be observed which is a time and ensemble average over all orientations of **PQ**:

$$D^{PQ} = D_0^{PQ} \left\langle \frac{3\cos^2\xi - 1}{2} \right\rangle \tag{1a}$$

$$D_0^{PQ} = -\mu_0 \frac{h}{2\pi} \frac{\gamma_P \gamma_Q}{4\pi^2 r_{PQ}^3} \, . \tag{1b}$$

Equation 1 is an approximation that is valid for weakly coupled spins. Here D_0^{PQ} is the static or maximum dipolar coupling that applies if **PQ** is invariably parallel to the direction of B_0. The definition of D_0^{PQ} in Eq. 1b differs by a factor of 2 from the static dipolar coupling constant used in solid-state NMR. Further, μ_0 is the magnetic permittivity of vacuum, h is Planck's constant, and γ_X the gyromagnetic ratio of nucleus X. Isotropic Brownian motion of a solute molecule in buffer and/or internal motion that is fast compared to the inverse of the static dipolar coupling will average the dipolar coupling to zero. This situation applies to most small organic molecules in aqueous solution. Narrow NMR lines are obtained and spectral crowding can be dealt with by using 2D or 3D NMR experiments resulting in good signal dispersion. Virtually complete motional averaging of dipolar couplings is the basis of high resolution in solution-type NMR spectra of small to moderate size biomolecules. Increasing the size and thus the rotational correlation time of the solute renders the averaging incomplete with concomitant line broadening and rapid transverse relaxation which eventually prevents NMR structure determination. However, if the incomplete averaging is the result of an alignment process that applies equally to all solute molecules and imparts a uniform motional anisotropy to them then a non-vanishing residual dipolar coupling (RDC) will normally be retained. The effective scaling factor resulting from the averaging in Eq. 1a depends on the degree of anisotropy. A minute degree of alignment that corresponds to a scaling factor of about 10^{-3} is most appropriate for the RDC measurements discussed in this work.

2.2 Residual Dipolar Couplings of Aligned Molecules

Let's first consider a rigid molecule that undergoes only whole body motions. In a molecule-fixed coordinate frame all internuclear vectors are uniquely defined and time-independent. In contrast, the orientation of the external magnetic field B_0 is time-dependent in the molecule-fixed frame. Anisotropic averaging over the mutual orientations of B_0 and the molecular frame can be mathematically described in terms of a 3×3 Saupe order matrix [5,6]. The elements of this traceless and symmetric second-rank order tensor are time averages of terms that contain direction cosines of the instantaneous angles between B_0 and the three axes of the chosen molecular frame [7]. Five independent matrix elements are sufficient for a complete description of the order tensor in any given coordinate frame. In its principal axis system (PAS) the order tensor is closely related to the dimensionless diagonalized molecular alignment tensor **A** with $|A_{zz}| > |A_{yy}| \geq |A_{xx}|$ [3] and to the cognate dipolar coupling tensor with magnitude $D_a = \frac{3}{4} D_0^{PQ} A_{zz}$ and rhombicity $R = \frac{2}{3} \frac{A_{xx} - A_{yy}}{A_{zz}}$ [8,9]. In addition to D_a and R one also has to specify the orientation of the molecular frame with respect to the tensor PAS by the remaining three variables, e.g., the Euler angles. Once the alignment tensor has been determined one can predict the RDC for spin pair P, Q:

$$D^{PQ}(\varphi, \theta) = D_a \left[\left(3\cos^2\theta - 1\right) + \frac{3}{2} R \sin^2\theta \cos 2\varphi \right] . \quad (2)$$

Polar angles ϕ and θ indicate the orientation of vector **PQ** in the tensor PAS. Apparently, a measured value of D^{PQ} does not uniquely define the orientation of vector **PQ** but merely restricts it to the surface of two distorted cones. This strong angular degeneracy of RDCs can be reduced either by determining several RDCs in molecules of known structure or by repeating the measurements under conditions that provide qualitatively different alignment, i.e., different alignment tensors are obtained [10, 11].

Internal motions will complicate analysis of RDCs. Vibrational fluctuations of bond length can be accounted for by replacing the term r_{PQ}^{-3} in Eq. 1b by the time average $\left\langle r_{PQ}^{-3} \right\rangle$. A molecule or a molecular fragment can be described as a quasi-rigid entity with a single alignment tensor if only axially symmetric internuclear vector motions about their mean positions occur. In principle, a residue-specific generalized order parameter S that accounts for such angular bond vector fluctuations [12] must be introduced to scale the predicted RDCs in Eq. 2. However, S values tend to be highly homogeneous for backbone bond vectors in structured regions of proteins [13]. Assuming a uniform order parameter S in this case introduces only a negligible error in the calculated RDCs. A uniform S is most conveniently adsorbed in the components of the alignment tensor **A**. However, indiscriminate use of the simplified Eq. 2 for prediction of RDCs can cause substantial errors if the motional properties of the corresponding internuclear vectors are very different from one another, e.g. in motionally decoupled domains of a large protein.

In general RDCs are a sensitive probe of local molecular dynamics and report on motional averages up to the timescale of milliseconds where the upper limit corresponds to the inverse of the measured RDCs. In combination with spin relaxation measurements that are sensitive to motions on the pico- and nanosecond timescale, previously unavailable insight into slower motions can be derived from RDCs. However, it is very challenging to separate the determination of an "underlying" average structure from a detailed characterization of the local dynamics based on the same RDC data. Emerging concepts and first promising results in this very active research area have been discussed in detail in a recent review [14]. The current overview is focused on structural information that can be derived from RDCs.

2.3 Lining Up Molecules

A suitable degree of alignment of the solubilized protein is fundamentally important for the measurement of RDCs. Motional averaging of dipolar couplings down to $\sim 10^{-3}$ of their static values is desirable in most cases. Alignment that is too strong will cause line broadening due to dipolar coupling of distant spins and will increase spectral complexity. If alignment is too weak precise measurement of the small RDCs will be difficult and eventually impossible.

Many molecules possess an intrinsic anisotropy of their magnetic susceptibility, i.e., the magnetic field will induce natural alignment of such molecules and the degree of alignment scales with the square of the field strength. However, in most cases this natural alignment is too weak for RDC measurements. Nevertheless, Bothner-By and colleagues demonstrated that anisotropy of the magnetic susceptibility of small molecules can be sufficient to induce very small but observable alignment [15]. Prestegard et al. reported the first measurement of RDCs in a solubilized protein based on field-induced alignment of paramagnetic myoglobin [2]. Metal-binding sites in proteins or nucleic acids have been used to accommodate paramagnetic lanthanides resulting in complexes with substantial magnetic susceptibility anisotropy providing the desired degree of field-induced alignment [16–19]. A more generally applicable approach relies on engineered fusions of target proteins with small lanthanide-binding tags [20–23]. Field-induced alignment of diamagnetic molecules is usually much weaker [24]. However, the diamagnetic susceptibility tensors of individual subunits may coherently add up in favorable cases [25]. Substantial net alignment was observed due to parallel stacking of nucleic acid bases with coplanar arrangement of aromatic rings [13, 26].

A different concept for aligning solubilized molecules does not rely on the intrinsic magnetic susceptibility anisotropy of the solute. Instead, the molecule is exposed to an anisotropic environment. Weak interaction with this environment induces a non-random orientational distribution of water-soluble molecules. Implementation of this broadly applicable concept using a diverse array of alignment media has made the measurement of RDCs a standard method in high-resolution liquid-state NMR [9]. The first medium that proved the feasibility of this concept for detection of RDCs in proteins was a dilute liquid crystalline phase [3]. It is commonly referred to as bicelles and consists of a dilute binary mixture of zwitterionic phospholipids DMPC and DHPC [27]. Most likely they form stacks of perforated lipid bilayers that orient with the membrane normal perpendicular to the B_0 field or small disk-shaped membrane fragments depending on concentration and molar ratio of the lipids and on temperature [28–30]. Observation of RDCs requires conditions that support field-induced cooperative alignment of the bicelles [31]. Partial alignment of water-soluble molecules by uncharged bicelles results primarily from steric interactions upon collision with the aligned membranes and requires an asymmetric shape of the solute [32]. Doping the bicelles with charged molecules may induce electrostatic interactions that can strongly modify the solute alignment tensor [10]. Mixtures of *n*-alkyl poly(ethylene glycol) and *n*-alkyl alcohols form very stable, uncharged bicelle-like liquid crystalline alignment media [33].

Solutions of filamentous bacteriophage, Pf1, are another very popular alignment medium [34]. Pf1 carries a negative net charge and electrostatic interactions usually dominate cosolute alignment. The magnetically oriented liquid crystalline phase formed by Pf1 in buffer is stable over wide ranges of temperature and concentration [35]. The degree of solute alignment can be easily tuned by varying Pf1 concentration and/or ionic strength which allows access to a wide range of homo- and heteronuclear RDCs.

Mechanically strained polyacrylamide gels provide a qualitatively different means of solute alignment which is referred to as strain-induced alignment [36, 37]. The solubilized molecule is confined to asymmetrically shaped aqueous pores that are obtained by compressing or stretching the gel, i.e., the alignment is independent of the magnetic field. The inertness of the polyacrylamide makes this medium compatible with most biomolecular solutes and buffer conditions. It is very simple to recover the biopolymer from the gel after the NMR measurement. Solute alignment in the electrically neutral polyacrylamide gel is steric in nature. Using charged gels decreases line broadening and allows greater variability of alignment features [38, 39].

The above selection covers only the most common alignment media. The list of alignment options is still growing, a more comprehensive sample was discussed in recent reviews [9, 40]. Availability of diverse alignment media increases the chance of identifying a means of alignment that is compatible with the molecule under study. Important points to consider include the following. A suitable alignment medium should support conditions under which the biomolecule is stable, correctly folded, and functional. The alignment process must not compromise the structure nor cause aggregation of the biopolymer. An appropriate degree of alignment is more easily obtained in media where the alignment can be tuned over a broad range. Finally, availability of several alignment media that are compatible with the biopolymer but provide very different alignment tensors, mainly due to different alignment mechanisms, allows one to reduce the degeneracy of the orientational information encoded in the RDCs [10, 11].

2.4 Measuring Residual Dipolar Couplings

RDCs between scalar coupled spins manifest themselves as an addition to the J-couplings. Magnitude and sign of the RDC are extracted from the difference between the couplings measured for partially aligned and isotropically tumbling biopolymers. However, only the size of the RDC is obtained if scalar coupling is absent or too weak for detection. Standard high-resolution pulse schemes suitable for detection of J-couplings can be used after minor modifications. Measurement of one-bond ^{1}H - ^{15}N and ^{1}H - ^{13}C couplings in moderately sized biomolecules is often based on HSQC-like pulse sequences. One can distinguish frequency-resolved schemes, where the coupling is derived from multiplet splittings in a frequency domain and intensity-based methods, where the coupling is encoded in the NMR signal intensity via J-modulation.

Intensity-based methods are also referred to as quantitative J-correlation experiments [41]. They offer very high precision and allow decoupling in both dimensions which reduces spectral crowding and increases sensitivity.

However, they are prone to systematic errors like relaxation-induced effects, pulse imperfections, or presence of passive couplings which must be carefully considered in order to prevent loss of accuracy [42, 43]. High precision is particularly important for measurement of very small RDCs [24, 44].

The advent of alignment media that allow tuning the size of RDCs to larger values has shifted the focus toward frequency-based methods. Accuracy of RDCs measured in such experiments is very sensitive to phase errors and depends strongly on spectral resolution [45]. Pulse sequences must allow for J-coupling evolution. Often HSQC spectra are recorded without decoupling during the heteronuclear evolution period. This allows observation of J or J + D splittings, respectively, in the F1 dimension which promises better resolution due to smaller R_2 relaxation rates of ^{15}N relative to ^{1}H spins.

Measurement of splittings between doublet components in F1-coupled HSQC spectra works fine for most small molecules. However, the twofold higher number of peaks in comparison with a decoupled HSQC causes severe spectral overlap for larger biomolecules. An elegant approach to reduce spectral crowding utilizes spin-state selective elements. Subspectra are generated that contain only one component of each doublet, i.e., resonances of nuclei where the coupling partner is in either the $|\alpha\rangle$ or the $|\beta\rangle$ spin state. This can be accomplished by spin-state selective excitation (S^3E) in conjunction with 2D correlation experiments in an E.COSY-type [46] fashion for determination of heteronuclear [47–49] or homonuclear couplings [50]. Another concept for obtaining spin-state separated spectra utilizes combinations of separately stored in-phase and antiphase components of J-coupled HSQC spectra (IPAP approach) [51–53]. The components of an HSQC cross-peak can also be separated into subspectra using the generalized TROSY scheme [4], an approach referred to originally as the α/β-HSQC-α/β experiment [54–56]. Spin-state selective elements can be incorporated into 3D NMR experiments to further reduce resonance overlap [57, 58].

Triple-resonance 3D experiments recorded without decoupling in the dimension of interest have been used to measure RDCs while alleviating spectral crowding. Examples include the very sensitive 3D HNCO [45, 59], the 3D CT-(HA)CA(CO)NH [60], and the 3D HA(CACO)NNH [59]. However, extraction of precise signal splittings from the indirect dimension of a 3D experiment may require very long acquisition times to fulfill the requirement for high spectral resolution and small couplings may be insufficiently resolved for measurement in the frequency domain. The resolution and thus acquisition time requirement can be significantly relaxed by use of J-modulation elements that encode coupling in signal intensity. A J-modulated pseudo-3D version of the $^{1}H - ^{13}C$ CT-HSQC experiment is often used for measurement of $^{13}C - ^{1}H$ couplings [61]. Very recently a suite of J-modulated $^{1}H - ^{15}N$ HSQC-type experiments were introduced which encode H^{α}–C^{α}, C^{α}–C', C^{α}–C^{β}, H^{N}–N couplings in the peak intensities of H^{N}–N and H^{N}–C' correlations [62].

Measurement of RDCs in large biopolymers with a rotational correlation time exceeding 15 ns can greatly profit from exploiting the TROSY principle [4] in the pulse sequence [63, 64]. TROSY elements are most beneficial at the highest field strength currently available and in triply labeled (^{15}N, ^{13}C, ^{2}H) molecules. Slow tumbling of increasingly larger macromolecules causes severe line broadening of the upfield multiplet component due to cross-correlated relaxation effects which will eventually prohibit extraction of RDCs directly from frequency domain splittings of the multiplet components. Combination of quantitative J-correlation with TROSY selection of the narrow multiplet components provides a valid alternative [65]. The coupling may also be derived from the frequency difference between the TROSY-selected narrow ^{15}N doublet component and either the ^{1}H-decoupled ^{15}N resonance or the peak position measured in a J-scaled experiment [45]. J-scaling in TROSY-based HNCO experiments was utilized for measurement of several different types of coupling [64] in particular $^{1}D_{H\text{–}N}$ couplings in a ^{15}N, ^{13}C, ^{2}H-labeled 81 kDa enzyme [66]. A conceptually similar 2D coupling-enhanced (CE)-TROSY HSQC provided $^{1}D_{H\text{–}N}$ couplings in a ^{15}N, ^{2}H-labeled 53 kDa homotrimeric mannose-binding protein [67]. Alternatively, a J-evolution period has been used to displace the narrow cross-peak multiplet component selected in a 2D TROSY scheme along an additional dimension (JE-TROSY) [68]. The coupling is measured directly from the frequency displacement of this component relative to a zero frequency origin.

Besides one-bond $^{1}H - {}^{13}C$ and $^{1}H - {}^{15}N$ RDCs a wide range of different types of couplings have been measured, including $^{1}D_{N\text{–}C'}$, $^{1}D_{N\text{–}C\alpha}$, $^{1}D_{C\alpha\text{–}C'}$, $^{1}D_{C\alpha\text{–}C\beta}$, $^{2}D_{N\text{–}C\alpha}$, $^{2}D_{HN\text{–}C'}$, $^{2}D_{HN\text{–}C\alpha}$, $^{3}D_{HN\text{–}C\alpha}$ and $^{1}H - {}^{1}H$ couplings. Triple-resonance experiments offer the advantage to determine RDCs of different types simultaneously from a single 2D or 3D data set [63]. Such experiments often exploit the E.COSY principle [46] to resolve very weak couplings.

Sidechain RDCs like $^{1}D_{H\text{–}C}$, $^{2}D_{H\text{–}H}$, and $^{1}D_{C\text{–}C}$, are highly valuable for protein structure refinement and analysis of protein interactions. Common approaches to measure $^{1}D_{H\text{–}C}$ couplings include a J-modulated CT-HSQC [61, 69], a sophisticated E.COSY-type method (SPITZE-HSQC) [70], an IPAP-CT-HSQC [52], or intensity evaluation of a set of two or three J-modulated 3D CB-(CA)CONH spectra [71].

There is a still growing number of NMR approaches for determination of RDCs. The selection of a pulse scheme for a specific task depends on several aspects such as degree of alignment, size of couplings, rotational correlation time of the biopolymer, isotope labeling, field strength etc. Available methods have been discussed in more detail [72, 73] and summarized in tabular form [40, 74] in recent reviews to highlight the elements and principles utilized along with the types of RDCs that can be addressed with a particular method.

2.5 Turning RDCs Into Protein Structure and Orientation

There are different strategies for application of RDCs in structural studies. In each case, determination of an order tensor is required. Let's consider the most simple case of a protein or a protein domain where all backbone bond vectors have approximately the same generalized order parameter *S* and bond lengths are essentially known, i.e., there is a unique order tensor that relates all measured backbone RDCs and the corresponding bond orientations according to Eq. 2. Magnitude and rhombicity of the alignment tensor are often estimated from a histogram of a large ensemble of qualitatively different, normalized one-bond backbone RDCs, assuming a uniform isotropic distribution of bond vector orientations [8, 75]. Histogram-derived tensor parameters may be fine-tuned using a grid search in the stage of structure calculation [76].

Alternatively, if at least five mutually independent RDCs have been measured and the accurate molecular structure is either known a priori or a limited set of structural models has been identified, then the order tensor of a partially aligned molecule or fragment can be determined by singular value decomposition (SVD) and Monte Carlo sampling [6]. Various SVD-based software tools are available for analysis and back-calculation of RDCs, e.g. REDCAT [77] or PALES [78]. The functionality of PALES also includes prediction of solute order tensors in both charged and neutral alignment media [79]. SVD-based order tensor solutions provide orientational information rather than structure. They allow one to specify the mutual orientation of molecular subunits in macromolecules and complexes.

Protein structure refinement is the most frequent application of RDCs. The RDCs are used along with other structure-relevant NMR data (NOE-derived distances, torsion angles, chemical shifts etc.) in molecular dynamics-based structure calculation protocols. For example, an RDC-based quadratic harmonic potential energy term has been included in the target function of the program XPLOR-NIH [80]. This potential depends on the square of the difference between the measured backbone RDCs and synthetic RDCs calculated for a trial structure with respect to a trial PAS of the alignment tensor. Both the structure of the protein and the orientation of the PAS relative to an arbitrary chosen molecule-fixed coordinate frame are varied in the course of structure calculation with the objective to reproduce the measured RDCs [13]. The large diversity of sidechain dynamics complicates interpretation of dipolar couplings involving sidechain nuclei. In practice, synthetic sidechain couplings of a trial structure are calculated using the alignment tensor derived for backbone couplings. However, dipolar couplings measured for sidechains of unknown mobility are treated as a lower bound, i.e., a potential energy penalty is accrued only if the magnitude of the measured RDC exceeds the calculated RDC [61]. Similar RDC-based potentials have been added to other popular structure calculation software like DYANA [81], CNS [82] or AMBER [83]. Inclusion of an RDC term in the total energy function yields a rugged energy landscape with multiple sharp local minima which may lead to convergence problems [73]. In order to prevent the calculated structure becoming trapped in a false minimum, the RDC term is usually turned on only at the later stages of structure refinement, i.e., after the global fold has been established from conventional restraints. Inclusion of RDC data in protein structure refinement generally improves the precision and Ramachandran score of the calculated structures [13]. It also tends to drive the NMR structure closer toward the X-ray structure, indicating an increase in accuracy [84].

A second class of applications that specifically exploits the global character of RDCs is determination of mutual orientation of subunits in large proteins and of individual components in molecular complexes. Provided interdomain rigid-body motion is negligible, then both domains share a common alignment tensor. Usually alignment tensors are determined separately for each fragment. The relative domain orientation is obtained by rotating the alignment tensors until they coincide. Prior knowledge of subunit structures, e.g., from X-ray crystallography, can be very beneficial in the process and RDCs are highly sensitive reporters of structure modifications caused by complex formation [85–87]. Orientational ambiguity of the fragments in their common PAS frame can be reduced by collecting RDCs under qualitatively different alignment conditions and/or by complementing the RDCs by at least a small set of interdomain NOEs or other distance-dependent restraints. Such additional restraints are also required to fix the translational distance between the components, which can be accomplished by rigid-body simulated annealing [88, 89]. For example, RDCs were used for characterization of protein-target recognition modes in stable calmodulin complexes [90, 91]. Very recently, the methodology was applied to study a large two-domain molecular chaperone [92] and very large protein-protein complexes [93]. However, if the dynamic properties of the individual fragments are not equivalent, e.g., due to qualitatively different interactions of the domains with the alignment medium in combination with a flexible linker between the domains, then there is a good chance that alignment tensors and RDC histograms are significantly different [94]. Comparison of the alignment features may allow conclusions regarding the interdomain motion [95]. Another complication arises for weakly bound complexes, perhaps with multiple binding modes. Observed RDCs of an exchanging component represent time-weighted averages over the individual states [96].

Concepts for de novo determination of protein backbone structure that are almost exclusively based on RDC measurements have been developed over the past few years. If large sets of backbone RDCs are available the well-defined geometry of the peptide plane can serve as a starting point to screen for torsion angles ϕ and ψ, specifying the relative orientation of consecutive peptide planes that are in agreement with the measured RDCs. Both the backbone structure of protein fragments consisting of multiple amino acids and the corresponding alignment tensors are determined in parallel using iterative algorithms. Protein tertiary structure is assembled on the premise of coinciding alignment tensors and refined against all measured RDCs [97–101]. A low-resolution initial backbone fold may also be obtained by searching a large database (e.g., the PDB) for homologous structures that allow close reproduction of the measured RDCs [102]. RDC-based molecular fragment replacement is conceptually similar but more generally applicable [103–105]. All measured one-bond backbone RDCs within a sliding frame of usually 7 to 10 residues are fitted against synthetic RDCs of all fragment structures in a comprehensive database. Torsion angles ϕ and ψ of a representative number of fragments that best fit the experimental RDC data are collected for each dipeptide junction of the target protein. Statistic evaluation of this ensemble of ϕ and ψ values allows identification of residue pairs with unambiguous torsion angles, which are then assembled into larger protein backbone fragments [103]. Another way to generate a reliable initial backbone structure is incorporation of experimental RDC restraints in homology-based protein structure prediction methods like DipoCoup [106], Rosetta [107], or RDC-PROSPECT [108]. RDCs improve the accuracy of the predicted structures. Alternatively, RDCs have been used to efficiently filter predicted protein folds [109]. Very recently, the periodic behavior of RDCs in extended secondary structure elements has been exploited to map the location of helices in a soluble protein and to subsequently refine the tertiary structure based on RDCs and a minimal number of NOE-derived distances [110]. Some of the RDC-based strategies described above are very efficient and may live up to the challenges of structural genomics. Backbone-only structures are adequate for rapid screening of fold space. However, they must be obtained with minimal amount of NMR data and isotope labeling. Data processing and structure generation should be highly automated and possibly take advantage of structure prediction schemes and data base knowledge [101].

3 Residual Dipolar Couplings in Weakly Bound Systems

3.1 The Concept of Transferred Dipolar Couplings (TrDCs)

There are several high-resolution NMR approaches that allow study of protein-ligand interactions even if the target protein is way beyond the size limit of liquid-state NMR. These experiments rely on fast exchange of a small ligand molecule between the bound form and a free form in solution. NMR data are acquired on the free form using standard liquid-state NMR pulse schemes.

Spin interactions that occur in the bound state can leave their signature in the spin system of the ligand. In fortunate cases this information can be recovered from the spectra of the free ligand and may be sufficient to reconstruct the structure and/or dynamics of the bound ligand. For example, proton-proton distances in the bound ligand can be derived from the transferred NOE (Tr-NOE) experiment based on the much higher efficiency of cross-relaxation in a slowly tumbling complex compared to the rapidly reorienting free ligand (Peters, 2006, in this volume) [111, 112]. Similarly, saturation transfer difference spectroscopy (STD) utilizes the very efficient intermolecular magnetization transfer in high molecular weight complexes to screen compound libraries for potential ligands of a given target molecule and for identification of ligand spins that are in immediate contact with the target molecule (Meyer, 2006, in this volume) [113]. Finally, cross-correlated relaxation (CCR) results from the interference between dipolar- and chemical shift-based relaxation processes and is sensitive to the projection angles between the vectors defining the two relaxation processes. CCR scales with the rotational correlation time and is thus much more rapid in a large complex than in a free ligand in solution [114–116].

Transferred dipolar couplings (TrDCs) arise from fast exchange of a weakly binding ligand between a partially aligned bound complex and the free ligand in solution. The exchange must be fast enough to ensure a single set of ligand NMR signals with chemical shifts that are weighted averages of the shifts in the two states rather than individual resonance lines of free and bound ligand, respectively. The target molecule might be part of a magnetically anisotropic large particle that aligns spontaneously in the B_0 field, e.g., certain membrane patches, bicelles, or fibrils. Alternatively, the target may get aligned by secondary interactions with an anisotropic environment. The degree of alignment should be substantially larger for the protein-ligand complex than for the free ligand in order to ensure high sensitivity of the measured RDCs to the bound conformation even if only a minor fraction of ligand is bound. The observed RDCs will be referred to as TrDCs since they carry information on the bound ligand but are detected on the free ligand using liquid-state NMR. Interpretation of TrDC data is based on the assumption of a single mode of ligand binding to the aligned target that gives rise to a unique set of RDCs $\left(D_{\mathrm{B}}^{\mathrm{PQ}}\right)$ characterizing the bound conformation of the ligand. In the absence of target binding the dipolar couplings of the ligand $\left(D_{\mathrm{F}}^{\mathrm{PQ}}\right)$ might be zero due to unrestricted isotropic molecular tumbling or might assume some finite value caused by interactions of the ligand with the alignment medium or the B_0 field. Under fast exchange conditions the measured TrDCs $\left(D_{\mathrm{obs}}^{\mathrm{PQ}}\right)$ will be a population-weighted average over the fraction of ligand molecules in the bound and free states, χ_{B} and $(1\text{-}\chi_{\mathrm{B}})$, respectively [96]:

$$D_{\mathrm{obs}}^{\mathrm{PQ}} = \chi_{\mathrm{B}} D_{\mathrm{B}}^{\mathrm{PQ}} + \left(1 - \chi_{\mathrm{B}}\right) D_{\mathrm{F}}^{\mathrm{PQ}} . \qquad (3)$$

The size of the observed couplings depends on the dissociation constant of the complex, the ligand-to-target ratio in the sample, and the degree of target alignment. Binding to a very large target with a long correlation time can cause rapid dephasing of transverse magnetization in the bound ligand. The lifetime of the bound state must be sufficiently short to avoid complete T_2 relaxation before the ligand leaves the binding site. Otherwise, detection of TrDCs on the free ligand will not be possible.

3.2 Applications of TrDCs

The fast exchange condition for observation of TrDCs is usually fulfilled in carbohydrate-protein interactions which show weak affinities ($K_D > 10^{-5}$ M). Exchange-transferred RDCs were first reported for the oligosaccharide moiety of globotriaosylceramide (Gb3-OS) that weakly binds to the B-subunit homopentamer of verotoxin 1 (VTB) [117]. RDCs of the bound ligand were derived from TrDCs and free state RDCs according to Eq. 3 and subsequently used to calculate the VTB-bound conformation of Gb3-OS. The principal axis of the molecular alignment tensor of VTB-bound Gb3-OS must coincide with the 5-fold symmetry axis of the homopentameric VTB allowing one to determine the ligand orientation with respect to the known crystal structure of VTB [118].

The orientation of α-methyl mannose (AMM) relative to its target mannose-binding protein (MBP) was determined from RDCs of the ligand observed with and without MBP present in a bicelle solution [96]. The alignment tensor of the bound ligand was derived from the known structure of AMM and five bound-state RDCs calculated with Eq. 3 from experimental data. A symmetry-based argument provides the relative orientation of AMM and MBP in the bound complex without measuring RDCs of the protein [119]. Later, the weakly bound complex of MBP and the ligand trimannoside was studied. Order tensors of both MBP and the ligand in the bound state were determined separately using bound-state RDCs and the known structures of MBP and trimannoside [67]. The relative orientation was obtained by aligning the two order tensors.

Residual dipolar couplings have been used to address weak binding of small molecules to phospholipid membranes. Binding to the membrane is used intentionally as the means of ligand alignment. Information on structure and orientation of the bound ligand relative to the membrane will be encoded in the RDCs. Examples include transient binding of peptides [120–122], polyols [123], and ethanol [124] to magnetically aligned bicelles. NMR spectra of uniaxially oriented ligands are observed in the case of rapid axial rotation of the membrane-associated molecules about the bilayer normal. In the published peptide studies the fraction of membrane-bound peptide is as high as 40 to 95%, i.e., the observed NMR signals are dominated by the bound peptide population resulting in rather broad lines and large RDCs. Solid-state NMR methods are currently developed to deal with such systems [122] but this subject is outside the scope of the current review. Under the conditions of the ethanol study only about 4% of the alcohol was membrane-associated and the TrDCs could readily be extracted from high-resolution liquid-state NMR spectra [124]. The observed RDCs were in close agreement with the known solution structure of ethanol indicating that membrane binding causes no or only very minor structural changes in this case. The ethanol study demonstrates the potential of TrDC measurements for the characterization of membrane-ligand interactions in the case of weak binding.

Formation of β-amyloid fibrils is a hallmark of Alzheimer's disease. In vitro, these fibrils align spontaneously in strong magnetic fields with the fibril axis parallel to the external field. Small peptide inhibitors can block or even reverse the growth of such fibrils by transient binding to the fibril. The conformation and backbone bond vector orientation of such a peptide inhibitor in the fibril-associated state was recently characterized using TrDCs [125].

3.3 TrDC Study on Peptide Binding to the Integral Membrane Protein Rhodopsin

The overall tumbling rate of a membrane protein is severely restricted due to the slow reorientation of membrane particles. Even artificial small unilamellar lipid vesicles (SUVs) that are obtained by sonication of lipid dispersions have diameters in excess of 250 Å and reorient on a time scale larger than 10^{-6} seconds [126]. Nevertheless, anisotropic spin-spin interactions may get partially averaged out for small membrane constituents by fast lateral diffusion across the surface of highly curved SUVs in combination with rapid axial rotation [127]. However, lateral diffusion of membrane proteins is generally too slow for efficient averaging. This is particularly true for larger membrane particles like the disk membrane from rod outer segments of the retina that have diameters on the order of several micrometers. Membrane proteins embedded in such a large particle are in general unsuitable for direct observation by solution NMR. The use of liquid-state NMR in studies of membrane protein-ligand interactions is therefore restricted to the observation of weakly binding ligands and spectra are recorded exclusively on the free form of the ligand.

TrDCs in combination with TrNOE-derived $^{1}H-^{1}H$ intra-ligand distances have been used to determine the rhodopsin-bound structure and orientation of an 11-residue peptide based on high-resolution liquid-state NMR [128]. The peptide, which is referred to as S2 peptide, constitutes a major binding epitope on the surface of the G protein transducin that is involved in binding of transducin to the metarhodopsin II (MII) state of rhodopsin [129]. The G protein-coupled receptor rhodopsin is an integral membrane protein with seven membrane spanning α-helices [130]. Observation of TrDCs on the free form of the S2 peptide requires (i) a suitable degree of alignment of rhodopsin, (ii) rapid exchange between the free and bound forms of the peptide, and (iii) a large excess of peptide over receptor in the sample.

3.3.1 Spontaneous Alignment of Disk Membranes

The outer segment of a rod cell in the retina of vertebrates consists of a stack of hundreds of disk-shaped particles with a species-specific diameter in the micrometer range [131]. These particles constitute osmotically intact but flattened vesicles of lipid and protein and are referred to as disk membranes. Rhodopsin is the major protein component of the disks and makes up about 50% of the membrane by weight. The outer surface of the disk faces the cytoplasm. The cytoplasmic loops of rhodopsin contain binding epitopes for interaction with transducin and other molecular partners during signal transduction.

The positive anisotropy of the diamagnetic susceptibility of α-helices favors alignment of the helix axis in parallel with the direction of an external magnetic field [25]. The high density and large number of almost parallel transmembrane α-helices of rhodopsin in rod outer segments induces a strong net alignment of rods in the magnetic field [132]. Spontaneous alignment of micrometer-sized individual purple membrane fragments that contain about 75 wt% bacteriorhodopsin, another integral membrane protein with seven membrane-spanning helices, has also been reported [133]. Suspension of intact individual disk membranes in hypotonic solution causes considerable swelling of the vesicles and reduces their shape asymmetry. Nevertheless, it is possible to find conditions for spontaneous alignment of individual disk membranes in the 14 Tesla field of an NMR magnet [134]. This was achieved by empirically optimizing the ionic strength of the buffer with the aim to avoid both disk aggregation, which occurs at high salt, and loss of disk shape asymmetry, which happens at low ionic strength. The residual quadrupolar splitting of the solvent ^2H_2O deuterium signal can serve as an indicator of disk alignment [31]. Analysis of ^{31}P-NMR spectra of the phospholipid component of the disks revealed that the disks are oblate spheroids with an aspect ratio of approximately 2 : 1 and the unique short axis aligned in parallel with B_0. The effective alignment of a rhodopsin-bound peptide is a weighted average over the surface of these spheroids and is reduced by a factor of 0.4 relative to the alignment that would result from binding to a perfectly flat and completely aligned disk [128].

Binding of the S2 peptide to the light-induced MII state of rhodopsin serves a dual purpose. Specific binding of the peptide to its target protein induces the biologically active peptide conformation. At the same time, binding to the partially aligned membrane-embedded protein conveys an anisotropic component to the motion of the S2 peptide which encodes the active conformation in measurable TrDCs.

3.3.2
The Critical Role of Rapid Exchange

Dipolar couplings provide a dominant mechanism for spin relaxation in diamagnetic proteins. The spins of the MII-bound S2 peptide experience much stronger dipolar couplings than the spins in the free peptide. A significant increase in the apparent transverse relaxation rate of peptide spins is observed upon transient binding of the peptide to MII. Another consequence of strong dipolar couplings in the bound complex is a pronounced doublet asymmetry in proton coupled $^{15}N - {}^{1}H$ HSQC spectra observed after photo-activation of the peptide-binding MII state (Fig. 1). This asymmetry is caused by relaxation interference between the ^{15}N chemical shift anisotropy and the one-bond $^{15}N - {}^{1}H$ dipolar interaction [135].

Observation of TrDCs is possible only, if the dissociation rate constant (off-rate) of the bound complex exceeds a certain limit, i.e., the peptide must leave the binding site prior to complete transverse relaxation. This off-rate limit for peptide binding to MII has been estimated [128]. The micrometer-sized disk membrane moves very slowly in solution, for NMR purposes it is a solid, non-tumbling object. In the absence of internal motion the decay rate of transverse magnetization is determined by the largest static dipolar interactions which amount to roughly 20 kHz. However, rhodopsin undergoes fast axial rotation about the membrane normal with a rotational correlation time of $\sim 20\,\mu s$ [136]. This rotational diffusion is faster than the inverse of

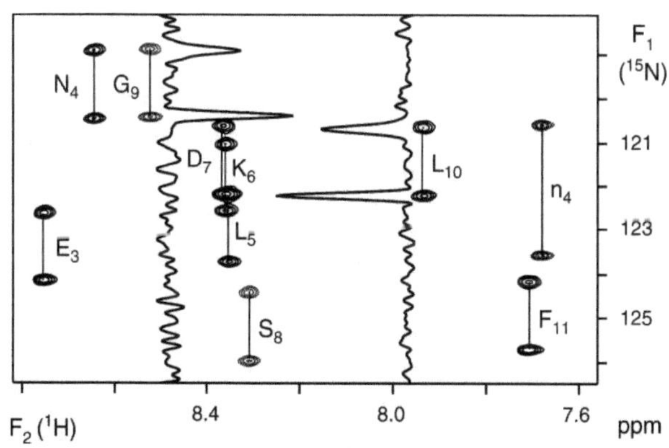

Fig. 1 Section of a ^{15}N–^{1}H HSQC of ^{15}N-labeled S2 peptide (2.6 mM) in the presence of rhodopsin-rich disk membranes recorded at 20 °C without proton decoupling in the F1 dimension. Acquisition was started immediately after photo-activation of rhodopsin (63 μM). Cross sections through the Gly9 and Leu10 resonances show strong doublet asymmetry due to relaxation interference indicating transient binding of the small peptide to a very slowly tumbling target. Gly9, Ser8, and Asn4 sidechain signals (labeled n4) are folded as a result of the small spectral width in F1. (Reprinted with permission from [146])

even the largest static dipolar interactions. As a consequence, all dipolar couplings relevant for the MII-bound peptide will be scaled down by a factor of $\left(3\cos^2\alpha - 1\right)/2$, where α is the angle between the local bilayer normal and the B_0 field of the NMR magnet. Both intermolecular dipolar couplings in the bound peptide $\left(D_{\mathrm{B}}^{\mathrm{PQ}}\right)$ and the decay rate of transverse relaxation vary with α, i.e., they depend on where on the surface of the aligned oblate disk membrane the peptide binds. The most rapid decay of transverse magnetization on a time scale of ca. 50 μs occurs for α close to zero. Peptide magnetization will survive much longer in regions where α is close to the magic angle. For bound times exceeding 50 μs only a decreasing fraction of the released peptides will contribute to the measured NMR spectrum.

Let's consider a subset of rhodopsin molecules specified by a uniform angle α. The TrDCs observed as a result of peptide binding to this subset of receptor molecules will be related to the corresponding bound-state dipolar couplings by a uniform scaling factor if the peptide off-rate is much larger than the dipolar couplings in the bound state [128]. This condition is fulfilled for the interaction between MII and the S2 peptide, which is characterized by a dissociation constant of about 1 mM [137]. Assuming a diffusion-controlled on-rate of $\sim 10^8\ \mathrm{M}^{-1}\,\mathrm{s}^{-1}$ one obtains an off-rate on the order of 10^5 Hz and a bound time of $\sim$ 10 μs. Successive peptide binding to rhodopsin molecules that are oriented at different angles α causes averaging of TrDCs that simply changes the magnitude of the uniform scaling factor.

3.3.3 Choosing the Ligand-to-Target Ratio

A simplified scheme of the TrDC experiment is shown in Fig. 2. Dipolar couplings of a ligand undergoing unrestricted isotropic tumbling average to zero. If the ligand is a small peptide it may not adapt a well-defined structure in solution. Instead, small peptides often have considerable flexibility in the unbound state (cf. Fig. 2a). The target-bound peptide adopts a unique conformation that gives rise to a single set of dipolar couplings $\left(D_{\mathrm{B}}^{\mathrm{PQ}}\right)$ provided all target molecules are uniformly aligned with respect to the external magnetic field. Very large dipolar couplings that are unsuitable for liquid-state NMR detection may arise in the case of an essentially immobile target molecule (Fig. 2b). However, if a large excess of weakly binding ligand competes for a small number of aligned target molecules then only a small fraction of ligand will be bound (Fig. 2c). The observed TrDC is a weighted average over the couplings that apply to the two states (Eq. 3) provided the off-rate is sufficiently fast. Observed TrDCs and bound-state RDCs are simply related by a scaling factor if the unbound ligand does not show any preferred alignment, i.e., $D_{\mathrm{F}}^{\mathrm{PQ}} = 0$.

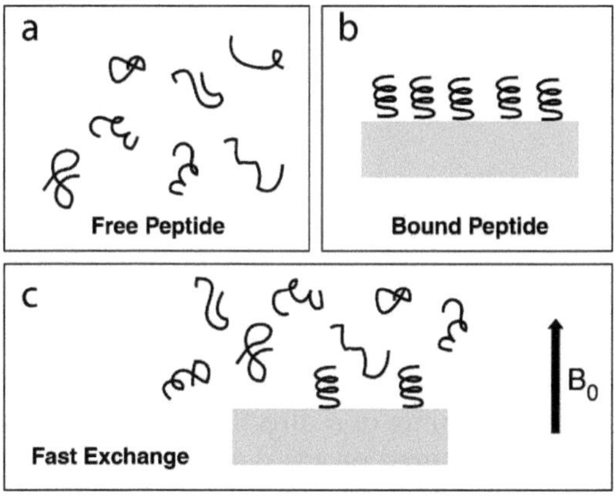

Fig. 2 Simplified scheme of the TrDC experiment. Dipolar couplings average to zero for small flexible peptides tumbling unrestricted in aqueous solution (**a**). Target binding stabilizes the active peptide conformation (**b**). A unique set of strong dipolar couplings arises if the immobile target is macroscopically aligned relative to the magnetic field. However, liquid-state NMR is not compatible with an immobile molecular complex. Fast exchange of the peptide with a minor population transiently bound to the aligned target and a large majority of peptide in the free state enables detection of TrDCs on the free form of the peptide (**c**). (Reproduced in modified form with permission from reference [146])

Several benefits arise from a large excess of ligand in the sample. It reduces the size of the observed TrDCs. Ideally, only the strongest one-bond couplings should show up in the spectrum amounting to just a couple of Hz. This prevents dipolar line broadening and avoids undue spectral complexity. Instead, the high-resolution character of the liquid-state NMR spectrum is preserved. A large excess of ligand also reduces the overall time that a given peptide spends in the bound state which reduces the risk of complete loss of magnetization. Finally, a high peptide concentration increases sensitivity. On the other hand, precise measurement of TrDCs requires that the magnitude of the largest TrDCs does not fall below a few Hz.

The assumption of completely isotropic tumbling of the unbound ligand is not always justified. Interactions of the ligand with the alignment medium or the membrane particles carrying the target may result in measurable RDCs even in the absence of specific target binding. Indeed, non-vanishing RDCs have been reported for small flexible peptides in bicelle solution and strained polyacrylamide gel [138]. TrDC analysis should start with Eq. 3 if measurable RDCs are observed in the absence of specific binding. For example, incubation of the S2 peptide with dark-adapted disk membranes, i.e., without the peptide-binding MII state present, induces small but measurable RDCs. These dark-state RDCs are most likely caused by unspecific binding of the peptide to the membrane and/or steric alignment. Analysis of Eq. 3 indicates that the bound-state RDCs are proportional to the difference of the RDCs measured after light-activation of MII and the RDCs obtained in the same sample prior to photo-activation if both the small size of the free-state RDCs and the almost 100-fold excess of S2 peptide over MII are taken into account.

3.3.4
RDCs of Metarhodopsin II-Bound Peptide

Dipolar couplings were derived from $^{15}N - {}^{1}H$ HSQC and $^{13}C - {}^{1}H$ CT-HSQC spectra recorded without proton decoupling in the indirect dimension (Fig. 1). Each observed signal splitting provides the sum of the scalar and, if any, residual dipolar coupling of a directly bonded spin pair. Subtraction of the signal splittings measured on the dark-adapted sample from the signal splittings obtained after light-activation provides a set of TrDCs that are directly proportional to the bound-state RDCs.

The peptide-binding MII state decays in an exponential fashion with half-life times of 15 and 54 min at 20 and 10 °C, respectively. A series of short HSQC experiments were recorded to follow the time dependence and to extrapolate all couplings to the time of photo-activation. The need to measure as many TrDC constraints as possible for structure refinement and the fast decay of the MII state made it mandatory to use uniformly ^{15}N- and ^{13}C-labeled peptide. Recombinant S2 peptide was efficiently produced as part of a fusion protein in *E. coli* and subsequently released from the fusion partner by enzymatic cleavage [139]. In total 38 TrDCs (9 backbone N – H, 9 Cα–Hα, and 20 sidechain C – H) were determined for the 11-residue S2 peptide.

Rapid rotational diffusion of rhodopsin about the membrane normal results in an axially symmetric alignment tensor for the MII-bound S2 peptide with the unique axis being parallel with the membrane normal. The magnitude A_a of this alignment tensor must be known in order to employ RDCs as restraints in structure refinement. An estimate of the magnitude of the dipolar coupling tensor $D_a^{NH} = D_0^{NH} A_a$ was obtained from the distribution of the observed dipolar one-bond $^{1}H - {}^{15}N$ and $^{13}C\alpha - {}^{1}H\alpha$ backbone couplings. The difference in bond length and gyromagnetic ratio between N – H and Cα–Hα spin pairs was accounted for by multiplying Cα–Hα couplings by a factor of – 0.48. A more precise value of D_a^{NH} was derived from a series of peptide structure simulations conducted with different D_a^{NH} values. The total energy of the calculated conformations assumes a minimum if the correct D_a^{NH} is used [128].

3.3.5
TrDC-Based Peptide Structure Refinement

Calculation of the MII-bound peptide structure followed a simulated annealing (SA) protocol implemented in XPLOR-NIH [128]. The measured 38 TrDCs alone do not sufficiently restrain the conformational space of the bound peptide. Preliminary peptide structures were based exclusively on 121 intramolecular $^1H - ^1H$ distances and 12 dihedral angles derived from TrNOE data. The NOESY cross-peak pattern is dominated by the bound peptide structure. Quantitative separation of cross-peaks that characterize the bound state and cross-peaks that are already observed for the free peptide is most easily achieved by recording NOESY spectra both in the presence and absence of binding and calculating the difference spectrum [140]. In the S2 peptide study these two spectra were acquired prior to and after photo-generation of MII, respectively, using the same sample.

The probability of indirect magnetization transfer in the bound state, i.e., spin diffusion, increases with the size of the bound complex. Spin diffusion can result in severely biased distance restraints and may lead to compromised structures. To reduce the effects of potential spin diffusion NOESY spectra were acquired with a relatively short mixing time of 48 ms, a r^{-4} distance dependence of NOE cross-peak intensity was assumed, rather generous tolerance limits of distance restraints were used, and weak cross-peaks between protons A and C were ignored if strong cross-peaks between A and B and between B and C were observed. Most importantly, the refined peptide structure must also satisfy the large set of dipolar coupling restraints, which are not contaminated by spin diffusion [128].

The parallel use of TrDCs and TrNOEs in the final stage of structure refinement aides in defining force constants and error limits for the two sets of restraints. It also provides a safeguard against structure perturbations caused by artifacts in either one of the two data sets. Defining the subtle details of the structure calculation protocol was an iterative process that resulted in a narrow bundle of S2 conformations with a non-hydrogen backbone atom rmsd of 0.17 Å. Use of TrDCs in addition to TrNOEs improves the local peptide geometry as reflected by an increased Ramachandran score with 100 rather than 88% of residues in the most favored region. Upon completion of the structure refinement protocol, the unique axis of the alignment tensor frame of the bound peptide, which reorients freely during SA, shall point along the axis of fast rotation of the bound complex, i.e., the membrane normal. This information on peptide alignment is encoded in the TrDCs but not in the TrNOE data.

A nearly ideal α-helix is formed by residues Ile1 through Ser8 of the bound S2 peptide. The helix is terminated by an α_L-type capping motif with Gly9 at the C′ position in the center of the reverse turn. The C-terminus forms a rather open structure and provides ample opportunity for hydrogen bonding and hydrophobic interaction with rhodopsin. This bound S2 conformation is very similar to the overall fold of the MII-bound $G_t\alpha$ (340–350) fragment determined by Kisselev et al. [141]. S2 differs from $G_t\alpha$ (340–350) by two conservative amino acid replacements. The long axis of the S2 helix is tilted by $(40 \pm 4)°$ with respect to the membrane normal. The rotation of the S2 peptide relative to the helix axis is also narrowly defined by the measured TrDCs [128].

3.3.6 Orientation of Transducin and Metarhodopsin II in the Bound Complex

Activation by a single photon causes structural changes in the G protein-coupled receptor rhodopsin that triggers binding of transducin, the first step in the visual signal transduction cascade. A crystal structure of the active MII transducin complex has remained out of reach until now. However, individual high-resolution X-ray structures of ground-state rhodopsin [130] and of transducin in the GDP-bound form [142], both crystallized in the absence of the interaction partner, have been published. Interestingly, the last eight residues of the α-subunit of transducin, which correspond to the C-terminal eight residues of the S2 peptide, are not structured in the crystal. The observed distinct conformation of the S2 peptide is apparently induced by binding to the MII state of rhodopsin and does very likely reflect the bi-

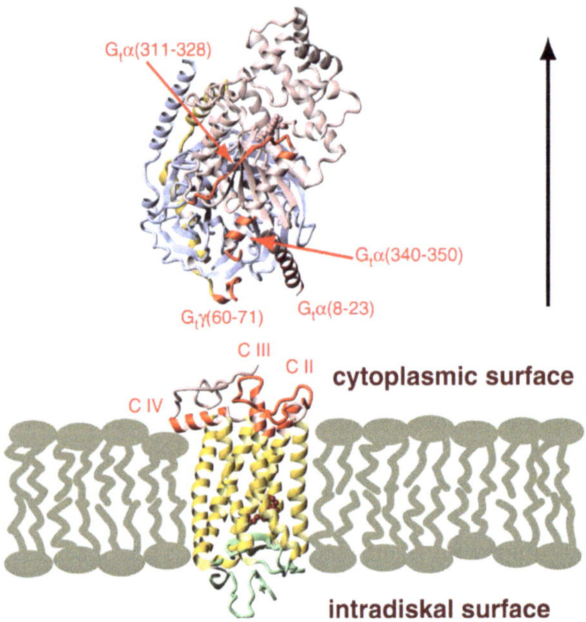

Fig. 3 Schematic depiction of the metarhodopsin II transducin interaction. Ribbon representations of the two proteins are based on the X-ray structures of ground-state rhodopsin [130] and GDP-bound transducin [142]. The unstructured C-terminus of the α-subunit of transducin was replaced by the NMR structure of the largely homologous S2 peptide [labeled with $G_t\alpha$ (340–350)]. S2 peptide orientation with respect to the membrane normal (black arrow) was derived from TrDCs. Color coded in *red* are potential interaction sites on both proteins. Three cytoplasmic loops of rhodopsin [147–149] and four distinct regions of transducin [129, 150] are likely binding sites according to mutational assays and peptide studies. (Reprinted from reference [128] with permission from Elsevier)

ologically active structure of the MII-bound C-terminus of the α-subunit of transducin.

In the crystal structure of transducin helix 5 of the α-subunit ends at residue 342. A simple model of MII-bound transducin is obtained by docking the S2 peptide NMR structure to the crystal structure of transducin based on an overlap of two residues that are α-helical in both structures [128, 141]. Assuming that no global changes occur in the transducin structure upon binding to MII, and that the elongated helix 5 is not kinked, the orientation of transducin with respect to the membrane normal is determined by the peptide orientation (Fig. 3). There remains just one rotational degree of freedom around the membrane normal for the orientation of transducin in the bound complex. The crystal structure of dark-adapted rhodopsin is included for illustration purposes in the figure, but this structure is expected to change significantly upon light-activation. The complimentary binding site for $G_t\alpha$ (340–350) was located near Ser240 in the third cytoplasmic loop of light-activated rhodopsin in a cross-linking study [143]. Future TrDC/TrNOE studies on peptide analogs representing other known binding sites on transducin (Fig. 3 and references in the figure caption) may provide clues to the relevance of the transducin crystal structure for the bound complex. For example, a recent TrNOE study revealed that a farnesylated peptide mimicking the C-terminus of the γ-subunit of transducin [$G_t\gamma$ (60–71)] forms an amphipathic helix upon binding to MII [144]. However, the published TrNOE data provide no information on the orientation of the bound $G_t\gamma$ (60–71) fragment relative to rhodopsin.

4 Conclusions

Residual dipolar couplings have been established as a unique and very rich tool in high-resolution NMR analysis of biomolecular structure and dynamics over the last decade. The TrDC method can be used for investigation of weakly bound complexes. Particularly attractive is the opportunity to study ligand-binding interactions of membrane proteins in a natural membrane environment using well-established high-resolution liquid-state NMR experiments. Application of the TrDC methodology to a diverse range of membrane proteins will require target reconstitution into membranes or membrane fragments that can be aligned in the NMR magnet. Partial alignment of small membrane fragments in strained polyacrylamide gels might provide a solution to this challenge [145]. Future applications may also benefit from heterologous expression of membrane proteins.

Acknowledgements This work was supported by the Deutsche Forschungsgemeinschaft (KO2143/3-1).

References

1. Clore GM, Gronenborn AM (1998) Proc Natl Acad Sci USA 95:5891
2. Tolman JR, Flanagan JM, Kennedy MA, Prestegard JH (1995) Proc Natl Acad Sci USA 92:9279
3. Tjandra N, Bax A (1997) Science 278:1111
4. Pervushin K, Riek R, Wider G, Wuthrich K (1997) Proc Natl Acad Sci USA 94:12366
5. Saupe A (1968) Angew Chem Int Ed Engl 7:97
6. Losonczi JA, Andrec M, Fischer MW, Prestegard JH (1999) J Magn Reson 138:334
7. Emsley JW, Lindon JC (1975) NMR spectroscopy using liquid crystal solvents. Pergamon Press, Oxford
8. Clore GM, Gronenborn AM, Bax A (1998) J Magn Reson 133:216
9. Bax A (2003) Protein Sci 12:1
10. Ramirez BE, Bax A (1998) J Am Chem Soc 120:9106
11. Clore GM, Starich MR, Gronenborn AM (1998) J Am Chem Soc 120:10571
12. Lipari G, Szabo A (1982) J Am Chem Soc 104:4546
13. Tjandra N, Omichinski JG, Gronenborn AM, Clore GM, Bax A (1997) Nat Struct Biol 4:732
14. Blackledge M (2005) Prog Nucl Magn Reson Spectrosc 46:23
15. Gayathri C, Bothner-By AA, Van Zijl PCM, Maclean C (1982) Chem Phys Lett 87: 192
16. Beger RD, Marathias VM, Volkman BF, Bolton PH (1998) J Magn Reson 135:256
17. Contreras MA, Ubach J, Millet O, Rizo J, Pons M (1999) J Am Chem Soc 121:8947
18. Bertini I, Felli IC, Luchinat C (2000) J Biomol NMR 18:347
19. Pintacuda G, Keniry MA, Huber T, Park AY, Dixon NE, Otting G (2004) J Am Chem Soc 126:2963
20. Ma C, Opella SJ (2000) J Magn Reson 146:381
21. Feeny J, Birdsall B, Bradbury AF, Biekofsky RR, Bayley PM (2001) J Biomol NMR 21:41
22. Wohnert J, Franz KJ, Nitz M, Imperiali B, Schwalbe H (2003) J Am Chem Soc 125:13338
23. Ikegami T, Verdier L, Sakhaii P, Grimme S, Pescatore B, Saxena K, Fiebig KM, Griesinger C (2004) J Biomol NMR 29:339
24. Tjandra N, Grzesiek S, Bax A (1996) J Am Chem Soc 118:6264
25. Worcester DL (1978) Proc Natl Acad Sci USA 75:5475
26. Kung HC, Wang KY, Goljer I, Bolton PH (1995) J Magn Reson B 109:323
27. Sanders CR, Schwonek JP (1992) Biochemistry 31:8898
28. Sternin E, Nizza D, Gawrisch K (2001) Langmuir 17:2610
29. Gaemers S, Bax A (2001) J Am Chem Soc 123:12343
30. Nieh MP, Glinka CJ, Krueger S, Prosser RS, Katsaras J (2002) Biophys J 82:2487
31. Ottiger M, Bax A (1998) J Biomol NMR 12:361
32. Bax A, Tjandra N (1997) J Biomol NMR 10:289
33. Rückert M, Otting G (2000) J Am Chem Soc 122:7793
34. Hansen MR, Rance M, Pardi A (1998) J Am Chem Soc 120:11210
35. Hansen MR, Mueller L, Pardi A (1998) Nat Struct Biol 5:1065
36. Sass HJ, Musco G, Stahl SJ, Wingfield PT, Grzesiek S (2000) J Biomol NMR 18:303
37. Tycko R, Blanco FJ, Ishii Y (2000) J Am Chem Soc 122:9340
38. Meier S, Haussinger D, Grzesiek S (2002) J Biomol NMR 24:351
39. Ulmer TS, Ramirez BE, Delaglio F, Bax A (2003) J Am Chem Soc 125:9179
40. Prestegard JH, Bougault CM, Kishore AI (2004) Chem Rev 104:3519

41. Bax A, Vuister GW, Grzesiek S, Delaglio F, Wang AC, Tschudin R, Zhu G (1994) Methods Enzymol 239:79
42. Billeter M, Neri D, Otting G, Qian YQ, Wuthrich K (1992) J Biomol NMR 2:257
43. Pham TN, Liptaj T, Bromek K, Uhrin D (2002) J Magn Reson 157:200
44. Tolman JR, Prestegard JH (1996) J Magn Reson B 112:245
45. Kontaxis G, Clore GM, Bax A (2000) J Magn Reson 143:184
46. Griesinger C, Sorensen OW, Ernst RR (1986) J Chem Phys 85:6837
47. Meissner A, Duus JO, Sorensen OW (1997) J Biomol NMR 10:89
48. Vander Kooi CW, Kupce E, Zuiderweg ER, Pellecchia M (1999) J Biomol NMR 15:335
49. Sorensen MD, Meissner A, Sorensen OW (1999) J Magn Reson 137:237
50. Meissner A, Duus JO, Sorensen OW (1997) J Magn Reson 128:92
51. Ottiger M, Delaglio F, Bax A (1998) J Magn Reson 131:373
52. Kontaxis G, Bax A (2001) J Biomol NMR 20:77
53. Ding K, Gronenborn AM (2003) J Magn Reson 163:208
54. Andersson P, Annila A, Otting G (1998) J Magn Reson 133:364
55. Cordier F, Dingley AJ, Grzesiek S (1999) J Biomol NMR 13:175
56. Permi P, Annila A (2000) J Biomol NMR 16:221
57. Vallurupalli P, Moor PB (2002) J Biomol NMR 24:63
58. Ding K, Gronenborn AM (2004) J Magn Reson 167:253
59. Yang D, Tolman JR, Goto NK, Kay LE (1998) J Biomol NMR 12:325
60. Ottiger M, Bax A (1998) J Am Chem Soc 120:12334
61. Ottiger M, Delaglio F, Marquardt JL, Tjandra N, Bax A (1998) J Magn Reson 134:365
62. McFeeters RL, Fowler CA, Gaponenko VV, Byrd RA (2005) J Biomol NMR 31:35
63. Wang YX, Marquardt JL, Wingfield P, Stahl SJ, Lee-Huang S, Torchia D, Bax A (1998) J Am Chem Soc 120:7385
64. Yang D, Venters RA, Mueller GA, Choy WY, Kay LE (1999) J Biomol NMR 14:333
65. Chou JJ, Delaglio F, Bax A (2000) J Biomol NMR 18:101
66. Tugarinov V, Kay LE (2003) J Mol Biol 327:1121
67. Jain NU, Noble S, Prestegard JH (2003) J Mol Biol 328:451
68. Luy B, Marino JP (2003) J Magn Reson 163:92
69. Tjandra N, Bax A (1997) J Magn Reson 124:512
70. Carlomagno T, Peti W, Griesinger C (2000) J Biomol NMR 17:99
71. Chou JJ, Bax A (2001) J Am Chem Soc 123:3844
72. Prestegard JH, al Hashimi HM, Tolman JR (2000) Q Rev Biophys 33:371
73. Bax A, Kontaxis G, Tjandra N (2001) Methods Enzymol 339:127
74. Griesinger C, Meiler J, Peti W (2003) Angular restraints from residual dipolar couplings for structure refinement. In: Rama Krishna N, Berliner LJ (eds) Protein NMR for the Millenium. Kluwer Academic/Plenum Publishers, New York, p 163
75. Bryce DL, Bax A (2004) J Biomol NMR 28:273
76. Clore GM, Gronenborn AM, Tjandra N (1998) J Magn Reson 131:159
77. Valafar H, Prestegard JH (2004) J Magn Reson 167:228
78. Zweckstetter M, Bax A (2000) J Am Chem Soc 122:3791
79. Zweckstetter M, Hummer G, Bax A (2004) Biophys J 86:3444
80. Schwieters CD, Kuszewski JJ, Tjandra N, Clore GM (2003) J Magn Reson 160:65
81. Banci L, Bertini I, Cremonini MA, Gori-Savellini G, Luchinat C, Wuthrich K, Guntert P (1998) J Biomol NMR 12:553
82. Brunger AT, Adams PD, Clore GM, DeLano WL, Gros P, Grosse-Kunstleve RW, Jiang JS, Kuszewski J, Nilges M, Pannu NS, Read RJ, Rice LM, Simonson T, Warren GL (1998) Acta Crystallogr D Biol Crystallogr 54:905
83. Tsui V, Zhu L, Huang TH, Wright PE, Case DA (2000) J Biomol NMR 16:9

84. Clore GM, Starich MR, Bewley CA, Cai M, Kuszewski J (1999) J Am Chem Soc 121:6513
85. Skrynnikov NR, Goto NK, Yang D, Choy WY, Tolman JR, Mueller GA, Kay LE (2000) J Mol Biol 295:1265
86. Goto NK, Skrynnikov NR, Dahlquist FW, Kay LE (2001) J Mol Biol 308:745
87. Evenas J, Tugarinov V, Skrynnikov NR, Goto NK, Muhandiram R, Kay LE (2001) J Mol Biol 309:961
88. Bewley CA, Clore GM (2000) J Am Chem Soc 122:6009
89. Clore GM, Bewley CA (2002) J Magn Reson 154:329
90. Mal TK, Skrynnikov NR, Yap KL, Kay LE, Ikura M (2002) Biochemistry 41:12899
91. Contessa GM, Orsale M, Melino S, Torre V, Paci M, Desideri A, Cicero DO (2005) J Biomol NMR 31:185
92. Revington M, Zhang Y, Yip GN, Kurochkin AV, Zuiderweg ER (2005) J Mol Biol 349:163
93. Jain NU, Wyckoff TJ, Raetz CR, Prestegard JH (2004) J Mol Biol 343:1379
94. Braddock DT, Cai M, Baber JL, Huang Y, Clore GM (2001) J Am Chem Soc 123:8634
95. Fischer MW, Losonczi JA, Weaver JL, Prestegard JH (1999) Biochemistry 38:9013
96. Bolon PJ, al Hashimi HM, Prestegard JH (1999) J Mol Biol 293:107
97. Hus JC, Marion D, Blackledge M (2001) J Am Chem Soc 123:1541
98. Tian F, Valafar H, Prestegard JH (2001) J Am Chem Soc 123:11791
99. Giesen AW, Homans SW, Brown JM (2003) J Biomol NMR 25:63
100. Valafar H, Mayer KL, Bougault CM, LeBlond PD, Jenney FE Jr, Brereton PS, Adams MW, Prestegard JH (2004) J Struct Funct Genomics 5:241
101. Prestegard JH, Mayer KL, Valafar H, Benison GC (2005) Methods Enzymol 394:175
102. Annila A, Aitio H, Thulin E, Drakenberg T (1999) J Biomol NMR 14:223
103. Delaglio F, Kontaxis G, Bax A (2000) J Am Chem Soc 122:2142
104. Andrec M, Du P, Levy RM (2001) J Am Chem Soc 123:1222
105. Kontaxis G, Delaglio F, Bax A (2005) Methods Enzymol 394:42
106. Meiler J, Peti W, Griesinger C (2000) J Biomol NMR 17:283
107. Rohl CA, Baker D (2002) J Am Chem Soc 124:2723
108. Qu Y, Guo JT, Olman V, Xu Y (2004) Nucleic Acids Res 32:551
109. Meiler J, Baker D (2005) J Magn Reson 173:310
110. Walsh JD, Kuszewski J, Wang YX (2005) J Magn Reson 177:155
111. Ni F (1994) Prog Nucl Magn Reson Spectrosc 26:517
112. Post CB (2003) Curr Opin Struct Biol 13:581
113. Mayer M, Meyer B (1999) Angew Chem Int Ed Engl 38:1784
114. Blommers MJJ, Stark W, Jones CE, Head D, Owen CE, Jahnke W (1999) J Am Chem Soc 121:1949
115. Carlomagno T, Felli IC, Czech M, Fischer R, Sprinzl M, Griesinger C (1999) J Am Chem Soc 121:1945
116. Carlomagno T (2005) Annu Rev Biophys Biomol Struct 34:245
117. Shimizu H, Donohue-Rolfe A, Homans SW (1999) J Am Chem Soc 121:5815
118. Thompson GS, Shimizu H, Homans SW, Donohue-Rolfe A (2000) Biochemistry 39:13153
119. al Hashimi HM, Bolon PJ, Prestegard JH (2000) J Magn Reson 142:153
120. Sanders CR, Landis GC (1994) J Am Chem Soc 116:6470
121. Rinaldi F, Lin M, Shapiro MJ, Petersheim M (1997) Biophys J 73:3337
122. Zandomeneghi G, Meier BH (2004) J Biomol NMR 30:303
123. Fischer D, Geyer A (2005) Magn Reson Chem 43:893
124. Koenig BW, Gawrisch K (2005) J Phys Chem B 109:7540

125. Chen Z, Reif B (2004) J Biomol NMR 29:525
126. Bloom M, Burnell EE, MacKay AL, Nichol CP, Valic MI, Weeks G (1978) Biochemistry 17:5750
127. Bocian DF, Chan SI (1978) Annu Rev Phys Chem 29:307
128. Koenig BW, Kontaxis G, Mitchell DC, Louis JM, Litman BJ, Bax A (2002) J Mol Biol 322:441
129. Hamm HE, Deretic D, Arendt A, Hargrave PA, Koenig B, Hofmann KP (1988) Science 241:832
130. Palczewski K, Kumasaka T, Hori T, Behnke CA, Motoshima H, Fox BA, Le TI, Teller DC, Okada T, Stenkamp RE, Yamamoto M, Miyano M (2000) Science 289:739
131. Dratz EA, Hargrave PA (1983) Trends Biochem Sci 8:128
132. Hong FT, Mauzerall D, Mauro A (1971) Proc Natl Acad Sci USA 68:1283
133. Neugebauer DC, Blaurock AE (1977) FEBS Lett 78:31
134. Koenig BW, Mitchell DC, König S, Grzesiek S, Litman BJ, Bax A (2000) J Biomol NMR 16:121
135. Goldman M (1984) J Magn Reson 60:437
136. Cone RA (1972) Nat New Biol 236:39
137. Aris L, Gilchrist A, Rens-Domiano S, Meyer C, Schatz PJ, Dratz EA, Hamm HE (2000) J Biol Chem 276:2333
138. Ohnishi S, Shortle D (2003) Proteins 50:546
139. Koenig BW, Rogowski M, Louis JM (2003) J Biomol NMR 26:193
140. Anglister J, Levy R, Scherf T (1989) Biochemistry 28:3360
141. Kisselev OG, Kao J, Ponder JW, Fann YC, Gautam N, Marshall GR (1998) Proc Natl Acad Sci USA 95:4270
142. Lambright DG, Sondek J, Bohm A, Skiba NP, Hamm HE, Sigler PB (1996) Nature 379:311
143. Cai K, Itoh Y, Khorana HG (2001) Proc Natl Acad Sci USA 98:4877
144. Kisselev OG, Downs MA (2003) Structure (Cambridge MA, US) 11:367
145. Chou JJ, Kaufman JD, Stahl SJ, Wingfield PT, Bax A (2002) J Am Chem Soc 124:2450
146. Koenig BW (2002) Chembiochem 3:975
147. Konig B, Arendt A, McDowell JH, Kahlert M, Hargrave PA, Hofmann KP (1989) Proc Natl Acad Sci USA 86:6878
148. Franke RR, Sakmar TP, Graham RM, Khorana HG (1992) J Biol Chem 267:14767
149. Ernst OP, Meyer CK, Marin EP, Henklein P, Fu WY, Sakmar TP, Hofmann KP (2000) J Biol Chem 275:1937
150. Kisselev OG, Ermolaeva MV, Gautam N (1994) J Biol Chem 269:21399

Top Curr Chem (2007) 272: 217–257
DOI 10.1007/128_2006_089

Published online: 29 November 2006

Glycosyltransferase Structure and Function

Brock Schuman · Javier A. Alfaro · Stephen V. Evans (✉)

Department of Biochemistry & Microbiology, University of Victoria, Victoria, BC Canada
svevans@uvic.ca

Abstract The biosynthesis of the oligosaccharides and polysaccharides observed in any organism requires the existence of a repertoire of glycosyltransferase enzymes that catalyze the sequential transfer of sugars from a specific activated donor to a specific acceptor molecule to form regio- and stereospecific glycosidic linkages. A viral genome may encode just one glycosyltransferase, while a mammalian genome encodes hundreds of these enzymes. It is notable that approximately 1% of open reading frames over all sequenced genomes have been found to be glycosyltransferases, which is a fraction comparable to that allotted to kinases. Glycosyltransferases are a highly diverse group of enzymes with little homology even among enzymes that share the same substrate specificity. Classification of glycosyltransferases according to sequence homology reveals at least 86 families; however, to date only 27 of these families have members with known structure. This is in sharp contrast with glycosylhydrolases, which to date have published structures for 70 of the so far described 102 classes. The paucity of structural data for glycosyltransferases has been attributed to their membrane-associated character and low expression levels, but even with the relatively limited number of available structures it is possible to see emerging trends that offer a glimpse of the principles of enzyme structure.

1 Introduction

Recognition of the full significance of complex carbohydrate biosynthesis has lagged in comparison to that of other biological macromolecules. While the intricacies of simple sugar metabolism forged the foundations of biochemistry, many of the details of complex carbohydrate anabolism and function, called "the last frontier of molecular and cell biology" [1], are only now being elucidated. There are ample prospects in the field for both basic and applied research.

Glycosyltransferases are a broad range of anabolic enzymes that synthesize simple and complex carbohydrates by sequentially adding monosaccharides from activated donors in a stereo- and regio-specific manner to target acceptors. Glycosyltransferases specific for acceptors of every macromolecular class have been described.

Carbohydrates are by far the most variable and complex of all biological polymers. First, simple monosaccharides like ribose and glucose have 2^{n-2} possible epimers (where n is the number of carbon atoms), though mammalian glycosyltransferases predominantly utilize only 9 monosaccharide donors. Second, these units can in turn be *O*-linked through different oxidized stereocenters to yield an ordinal combination of linear or branched configurations [2]. The alternative connections and the possibility of branching distinguish carbohydrates from linear amino acid and nucleic acid polymers. Third, carbohydrates may contain substitutions and/or reductions, such as *N*-acetylation. Given such complexity, and the similar reactivity of a sugar's multiple functional groups, chemical synthesis of complex carbohydrates is challenging and often a rate limiting step in carbohydrate research. This alone has prompted a great deal of examination into the enzymes involved in carbohydrate and glycan biosynthesis.

The acceptor substrate of a glycosyltransferase may be as simple as a second monosaccharide homologous to the donor, or as complex as a component of a heteropolysaccharide requiring hundreds of glycosylation steps by a number of enzymes for complete biosynthesis [3, 4]. The acceptor is not always another carbohydrate, and can be nucleic acid [5, 6], lipid [7, 8], or a topological protein epitope [9, 10]. More than half of all mammalian proteins are glycosylated [11], most often *N*-linked to asparagine amines as part of an Asn-X-Ser/Thr consensus sequence required but not sufficient for *N*-linked glycosylation [12], or *O*-linked to serine, threonine [13–15], tyrosine [16], hydroxylysine [17] or hydroxyproline [18]. Initial glycosylation usually occurs in the ER during or immediately following translation [19], and hydrophilic oligosaccharides can facilitate correct folding in part by sheltering hydrophobic patches of the nascent protein from aggregation in the ER [2].

Glycosyltransferases are found as soluble globular proteins [20], integral membrane bound proteins, or as lipid-anchored enzymes [21] in almost all eukaryotic cellular compartments including the inter-membrane space of mitochondria and chloroplasts [22]. Glycosylation is the most common form of post-translational modification, and the most variable from individual to individual, which has prompted the development of many elaborate expression vectors with overstuffed cassettes designed to mimic endogenous synthesis conditions [23–25].

Glycosylation of acceptor molecules is required for many different functions, including development [26], signal transduction [27], cell adhesion [28], to aid alleopathic warfare as both inhibitory molecules [29–31] and in providing resistance [32, 33], as well as afford molecular refuge from heat shock [34], nucleases [6] and desiccation [35].

Glycosyltransferases comprise a dominant fraction (approximately 1%) of expressed eukaryotic open reading frames [36], which is comparable to the fraction currently allocated to kinases. The exploration of glycogenomics promises to elucidate many of the perplexing anomalies of total phenomics and lead to novel approaches to scores of biotechnological issues. Just a few of the current and proposed applications of glycosyltransferases include bioremediation [37–39], isospecific antimicrobial toxins [31, 40, 41], herbicides [42, 43] and pesticides [44], reducing the effects of aging [45], powerful idiospecific adjuvants and immunogens [46], as well as a broad range of biomedical applications including organ transplants [47]. Glycosyltransferase function has been extensively reviewed (for example, see [48, 49]) and we concentrate here on the structure-function correlations that have emerged in the last few years.

Classification. The International Union of Biochemistry and Molecular Biology (IUBMB) enzyme nomenclature and classification is founded on the reactions catalyzed, not on the structures of the enzymes. While this is intuitive and useful for most enzymes with specific molecular substrates, it is not sufficient to classify glycosyltransferases into a reasonable number of families. As little as a single point mutation [50–52], a conformational shift induced by a regulatory factor [53], or even the utilization of alternative acceptors [54] can alter enzyme substrate specificity while leaving the stereochemical mechanism and even the underlying form of the fold unchanged. As such, the Carbohydrate Active enZyme databank (CAZy) has utilized general amino acid sequence homology to categorize glycosyltransferases into 86 families to date [55–57]. Family rosters are increasing in parallel with genomic revelation, and several enzymes have been discovered that have yet to be formally classified. Families contain members who appear to share common evolutionary origins and presumably tertiary structure as well. Out of 86 families, only 27 currently have one or more members with known structure.

Functionally, glycosyltransferases have always been segregated into "retaining" or "inverting" enzymes, according to whether the stereochemistry of the donor's anomeric bond is retained ($\alpha \rightarrow \alpha$) or inverted ($\alpha \rightarrow \beta$) during the transfer [58], Fig. 1. With the advent of several structure determinations in the last few years, it has become possible to go beyond this simple demarcation to reveal some common trends in mechanism, activity and substrate recognition.

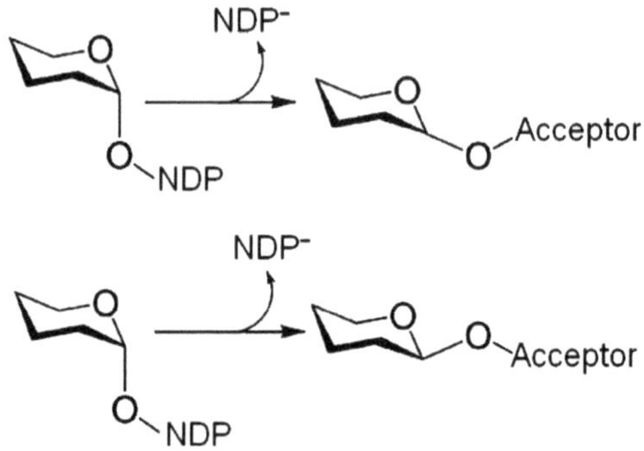

Fig. 1 Glycosyltransferases are divided into two fundamental classes, where they either **a** invert or **b** retain the configuration of the anomeric bond in the transfer of the donor sugar to an acceptor

Observed Glycosyltransferase Fold Types. Given their principal biological roles and locations in the cell, it is not surprising that most glycosyltransferases are membrane associated. Most integral glycosyltransferases share type II membrane topology, i.e. a cytosolic N-terminus separated from the C-terminal catalytic domain by a single transmembrane pass and a stem region, Fig. 2 [59].

The modest degree of sequence homology within and sometimes among the various families has made the prediction of tertiary structures difficult; however, structural determinations in recent years have revealed that the catalytic domains of most glycosyltransferases display one of two fold types designated GT-A or GT-B [56, 60]. Examples of retaining and inverting enzymes have been observed in both the GT-A and GT-B type folds, thus this structural characteristic is not determinative of stereospecificity.

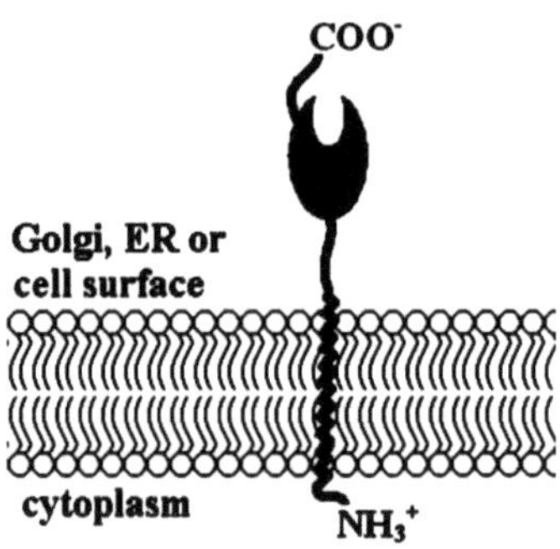

Fig. 2 Typical type II transmembrane topology observed for most glycosyltransferases, with a small intracellular N-terminal domain, a transmembrane region, a stem region, and a catalytic C-terminal domain that may be extracellular or within the lumen of membranous compartments

Examples of the catalytic domains of the GT-A and GT-B fold types are presented in Fig. 3, where GT-A and GT-B fold types consist of two closely associated domains at least one of which contains a Rossmann-fold responsible for donor nucleotide recognition. The Rossmann fold is a nucleotide binding domain [61], and is a ubiquitous structural motif among enzymes with either GT-A or GT-B fold type where it constitutes a dominant portion of the catalytic center in a cleft between the two domains. The Rossmann fold often contains much of the limited sequence homology that is observed across many glycosyltransferase families due to a finite repertoire of donor nucleotides utilized [55].

The GT-A type fold, Fig. 3a, is believed to be ancestral to enzymes with the GT-B type fold [56, 62]. The donor usually binds the N-terminal domain's Rossmann fold, as for example with SpsA [63] and the human ABO(H) blood group glycosyltransferases GTA and GTB (not to be confused with the fold types GT-A and GT-B [64]. Typically one or two of this domain's β-sheets extend into the C-terminal domain, rendering definite separation of the folds difficult. The second domain is usually responsible for acceptor recognition and has greater sequence and structural variability among the different families than the nucleotide binding domain. This is presumably due to the limited number of different donor nucleotide sugars, but a vast number of different acceptor molecules. The second domain occasionally has superficial resemblance to a Rossmann fold as SpsA does [63] but many motifs have been observed, including all α-helix. The two domains in the GT-A type fold cooperate to form the active-site cleft.

The GT-B type fold, Fig. 3b, has two somewhat homologous Rossmann or Rossmann-like folds in distinct domains which are usually separated by

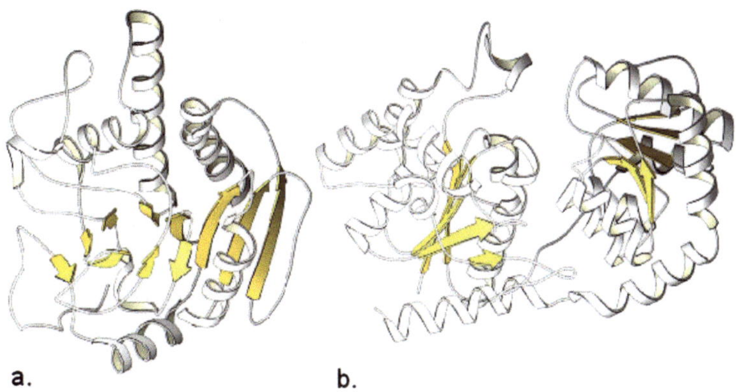

Fig. 3 SETOR [164] diagrams of the **a** GT-A-fold type showing human ABO blood group A enzyme with a single Rossmann fold on the *right*, and **b** GT-B fold type with two Rossmann folds showing vancomycin glycosylation enzyme GtfB. In both enzymes the active catalytic site is housed in a cleft between the two domains, albeit much more cavernous and apparent in the GT-B fold

a deep wide crevice while the enzyme is in an unliganded "open" conformation. Enzymes with the GT-B type fold have been suggested to have step-wise reaction mechanisms, where substrate binding in the "open" form can induce a conformational shift by a pair of main chain rotations to generate the "closed" form to align the nucleophile and substrates in the active site for catalysis [56, 62, 65].

There have been two further fold types predicted using iterative BLAST searches and other methods [57, 66, 67]. These so called GT-C and GT-D folds are thought to contain catalytic centers within transmembrane loops. As the catalytic domains themselves consist of integral-membrane proteins they are intrinsically difficult to crystallize, and there is currently no structurally characterized example of either the GT-C or GT-D fold types. A few enzymes, such as the Family 42 (or GT-42) sialyltransferases described below, do not display any of these fold types and are currently unclassified.

Enzyme Mechanism. The retaining and inverting enzymes require different mechanisms to explain the product stereochemistry, Fig. 4. The inverting reaction is mechanistically straightforward, and requires only nucleophilic attack on the non-hydroxylated face of the donor anomeric carbon by the enzymatically deprotonated acceptor, releasing the diphosphonucleotide (a favorable leaving group) and resulting in inversion of the anomeric center, Fig. 4a.

There is more debate concerning the mechanism of the retaining reaction, which has not been unambiguously resolved. The favored method for retention is the "double-displacement" mechanism [68] that requires two sequential S_N2 substitutions that invert then invert again the anomeric con-

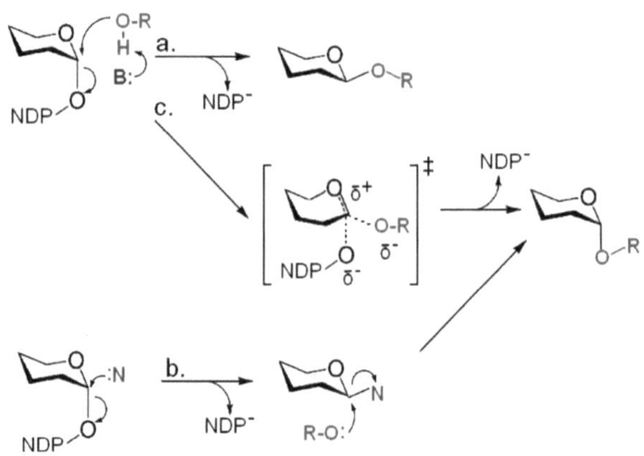

Fig. 4 Detailed proposed mechanisms for inverting and retaining glycosyltransferases. **a** There is general consensus that the "inverting" glycosyltransferases proceed through a single-step S_N2 nucleophilic attack following deprotonation of the acceptor by a general base. There is more debate concerning the "retaining" enzymes, where two mechanisms have been proposed. **c** The double displacement mechanism requires two sequential S_N2 reactions, **b** whereas the S_{Ni} mechanism involves an energetically unfavorable "internal-return" intermediate

figuration of the donor sugar yielding a net retention of stereochemistry. This mechanism requires an initial nucleophilic attack, presumably by the enzyme, to form a covalent intermediate which is subsequently directly attacked by the acceptor, Fig. 4b.

However, doubts concerning the double displacement mechanism have arisen based firstly on an inability to capture the inverted enzyme-sugar covalent intermediate. While homologous glycoside hydrolase intermediates have been trapped in crystallographic studies using fluoridated substrates [69, 70], a comparable glycosyltransferase intermediate has been elusive. Liquid chromatographic/mass spectrometric analyses of fragmented proteolytic digests have identified an unexpected potential nucleophile in the retaining enzyme LgtC, where a galactosyl moiety has been observed covalently bound to an aspartate residue 9 Å away from the active site [71]. Skeptics suggest that this is an artifact of the digest and electrospray conditions; however, it does force consideration of potential nucleophiles about the active site that may have previously been disregarded outright due to their distance.

Additional doubt concerning the double displacement mechanism has been expressed given the occasional lack of a strong nucleophile anywhere in the active site of the enzyme [72]. However, as will be discussed in this review, glycosyltransferases often exhibit high levels of molecular motion and disorder in portions of their polypeptide chains about their active sites, and it is possible that these disordered regions contain appropriate nucleophiles [65].

It has also been proposed that retaining enzymes may utilize the unusual S_N2-like mechanism dubbed S_{Ni}, Fig. 4c. It would require nucleophilic attack by the acceptor with concomitant release of NDP, with both bound to the anomeric carbon to form an energetically unfavorable "internal return" intermediate [58]. Such a mechanism is not without merit, though it is extremely rare in synthetic organic chemistry and there is no evidence that it is utilized in any other biological system.

Domain Movement, Disorder, and "Bioactive Conformations". Like most specific carbohydrate processing enzymes glycosyltransferases can be required to surround their substrates during recognition as the substrates can be chemically identical and differ only in stereochemistry (like UDP-Glc and UDP-Gal). Glycosyltransferases typically utilize a disordered region of polypeptide chain to allow substrates access to and products egress from the active site pocket. In crystallographic experiments this disorder is most often seen in the unliganded state. As well, the C-terminal tail (which has often been shown to be required for catalysis, and has been observed to participate in substrate recognition) is also frequently disordered [64, 73–76]. Some investigators have also suggested that the disordered loop functions to restrict the access of water to the active site, thereby preventing wasteful hydrolysis of the activated donor [76]. Hydrolysis of donor corresponds to using water as an acceptor, which would be expected to be chemically favored as water is less sterically hindered.

Stacking Interactions, DXD Motifs and Donor Coordination. While there are exceptions, glycosyltransferases primarily transfer monosaccharide donors activated by phosphorylation or phospho-nucleotidylation, most often utilizing UDP. Structural studies of glycosyltransferases in complex with the UDP-sugar donor or with UDP itself show that the donor's uracil is usually aligned with an aromatic residue, typically tyrosine, in the Rossmann nucleotide binding fold that forms the N-terminal domain in GT-A and GT-B fold type enzymes, Fig. 5. This tyrosine residue is adjacent to an array of hydrophilic residues as well as additional aromatic residues complementary to the enzyme's specific donor sugar [77–81]. Next to this aromatic residue there is usually found a "DXD motif" that consists of the amino acid triplet Asp-X-Asp [77]. The DXD motif is sometimes absent in retaining GT-B fold type enzymes, but was thought until quite recently [82] to be essential for catalysis in the GT-A fold-type enzymes and has been suggested to play a more direct role than simple donor coordination [78].

The DXD motif coordinates a metal center which, in turn, coordinates to the diphosphate moiety of the diphosphonucleotide donor. The metal coordination observed for the DXD motif has so far been conserved for all inverting enzymes, where typically only one O_δ of the second aspartate residue participates in Mn^{+2} binding Fig. 6b. Retaining enzymes usually display a metal-binding geometry with one contact between the metal and the first Asp and bidentate coordination to the second Asp (Fig. 6a), although an exception has been observed for Kre2p/Mnt1p, a yeast α-(1→2)-mannosyltransferase involved in mannoprotein biosynthesis in which the second aspartate is not involved in metal coordination [83].

There are glycosyltransferases that clearly do not utilize a metal center in the binding of donor sugar. These include the GT-C and GT-D folds mentioned, as well as GT family 42 (α-(2→3)-sialyltransferases) that undergoes catalysis in the absence of metal binding [57, 66, 67, 84].

Recognition of the donor sugar moiety itself varies from enzyme to enzyme, but is always in a hydrophilic pocket at the catalytic center. Though

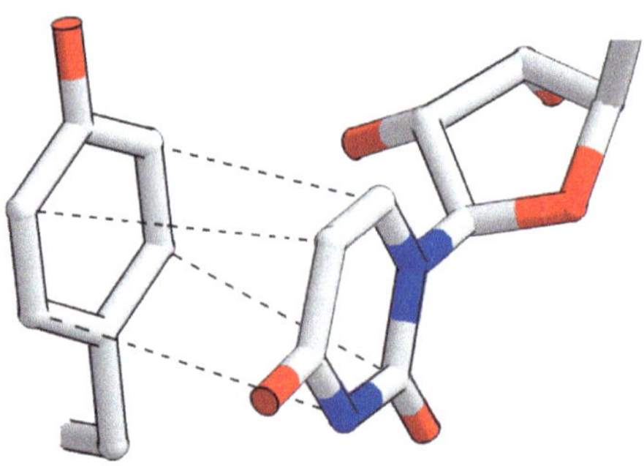

Fig. 5 Typical stacking interactions of tyrosine and the uracil moiety of UDP often utilized by glycosyltransferases to recognize the base moiety of the UDP-sugar donor

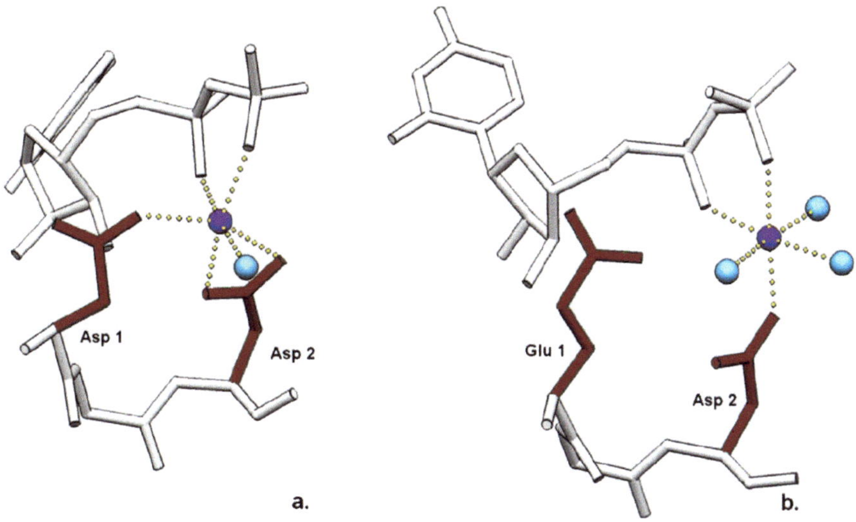

Fig. 6 Typical DXD coordination of the manganese ion by **a** both aspartate residues of the retaining (human ABO(H) blood group B glycosyltransferase GTB) enzymes as well as **b** just one aspartate residue inverting (plant *N*-acetylglucosaminyltransferase GnTI) enzymes

the UDP moiety is often incorporated in co-crystallizations, structures with the entire donor bound are less common. Available examples have demonstrated some common themes, such as hydrogen bonds to short aliphatic contiguous regions of the main chain, stacking interactions with an aromatic residue as discussed above, and van der Waals interactions with several isolated polar residues which are, of course, topologically specific to the stereochemistry of the donor. The UDP-donor substrate is often characterized as lying in a "tucked under" conformation, Fig. 7. This is true for both inverting

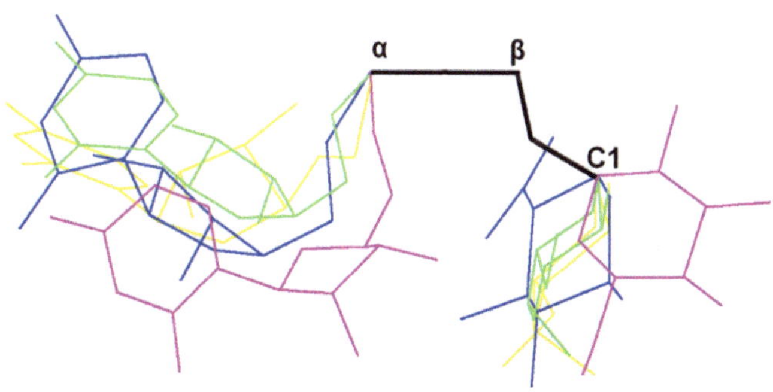

Fig. 7 The geometry of donor nucleotide-sugars bound to glycosyltransferases presenting the group for nucleophilic attack is usually described as "tucked under". Donor sugar C1 and phosphates are superimposed, and extra phosphate oxygen atoms have been omitted for clarity. *yellow* GT-7, *green* GT-6, *blue* GT-43, *violet* GT-63

Table 1 Glycosyltransferases discussed in this work

CAZy Family	Enzyme Example	Product stereochemistry	Fold type	Refs.
GT2	SpsA	Inverting	GT-A	[89]
GT7	Gal-T1	Inverting	GT-A	[93]
GT43	GlcAT-I	Inverting	GT-A	[104]
GT43	GlcAT-P	Inverting	GT-A	[104]
GT1	GtfA	Inverting	GT-B	[65]
GT63	BGT	Inverting	GT-B	[122]
GT6	GTA	Retaining	GT-A	[64]
GT6	GTB	Retaining	GT-A	[64]
GT64	mEXTL2	Retaining	GT-A	[135]
GT78	MGS	Retaining	GT-A	[140]
GT5	Glycogen synthase	Retaining	GT-B like	[148]
GT8	Glycogenin*	Retaining	GT-A	[144]
GT20	OtsA	Retaining	GT-B like	[162]
GT42	CSTII	Inverting	unclassified	[163]

* For readability, this enzyme is discussed with retaining GTB-like glycogen synthase

and retaining enzymes, and reflects the geometry utilized for a nucleophilic attack.

Acceptor Specificity and Known Structures. The biochemical, physiological and clinical relevance of these enzymes is seen in the myriad of acceptor molecules that are recognized by the glycosyltransferase repertoire. However, even with the diversity in acceptors and to some extent donors, and with the dearth of known structures, it is possible to see patterns emerging. The following sections describe a selection of enzymes (Table 1) chosen for their diversity in acceptor and in function and arranged according to product stereochemistry (inverting or retaining) and fold type (GT-A or GT-B). Although most of the structures of these enzymes were determined within the last 10 years, many of these molecules have scientific histories that date back 50 or even 100 years, and we have tried to place the structural studies into their historical context.

2 Examples of Inverting Enzymes with the GT-A Fold: GT-2, GT-7, GT-43

SpsA and the Origin of GTs. CAZy family-2 (GT-2) enzymes appear to be utilized by all life on earth, and chiefly polymerize diverse structural carbohydrates including chitin [85], hyaluronan [86] and cellulose, as well as the carbohydrate moieties of *O*-antigen, lipopolysaccharide and endospores [63].

GT-2 is the only class of glycosyltransferase currently recognized in the archaea species *Archaeoglobus fulgidus* and *Aeropyrum pernix* [87], and is thus arguably the most ancient derived family of glycosyltransferase. The species-specific extracellular products of these enzymes make family members potential targets for unique pesticides [44], herbicides [42, 43], antimicrobials [29–31, 40, 41] and for chemoenzymatic vaccine production [88].

SpsA participates in *Bacillus subtilis* spore coat polysaccharide biosynthesis, and currently is the only GT-2 with solved structure, Fig. 8. The donor and acceptor have yet to be indisputably identified despite extensive research on *B. subtilis* as the dominant Gram positive model organism [63, 89]. As such, SpsA cannot be classified by IUBMB standards; however, it was easily assigned to be an inverting enzyme in the GT-2 family via shared sequence homology. GDP-mannose and dTDP-rhamnose are known to be utilized by some GT-2 enzymes [90, 91], so it is possible SpsA may utilize an unusual nucleotide donor.

The dTDP and UDP substrates crystallized in complex with SpsA display the usual stacking interactions with an aromatic residue in the N-terminal nucleotide-binding domain, which has a traditional Rossmann fold. A disulfide bond confines the C-terminal tail, which is labile in many other glycosyltransferases [89].

Multi-iterative BLAST searches and site-directed mutagenesis experiments [91] have identified functional sequence homology between GT-2 enzymes and several glycosyltransferase families; most notably the N-terminal fold of retaining family GT-27. This has been suggested as evidence of either

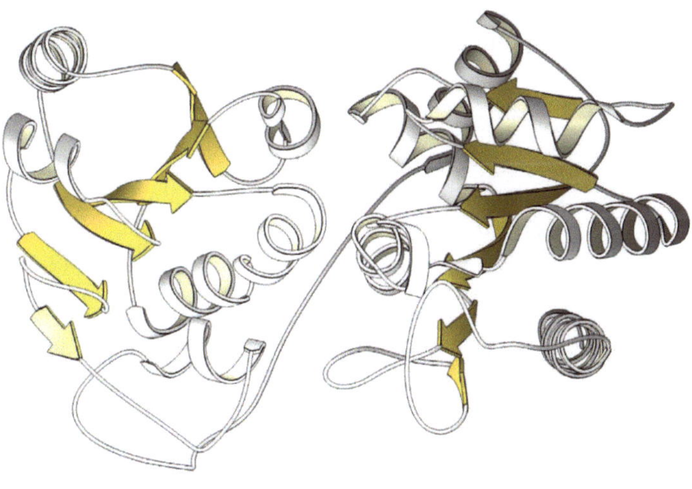

Fig. 8 SETOR diagram of SpsA secondary structure. SpsA belongs to the GT-A fold family where the unknown donor is bound in the cleft face of the N-terminal Rossmann fold (*right*); however, SpsA is unusual in that the second (acceptor-binding, *left*) domain also resembles a Rossmann fold

GT-27 as a prototypic retaining enzyme ancestrally derived from GT-2, or domain swapping between inverting and retaining enzymes [91]. The observation of structural homology between inverting and retaining enzymes is occurring with increasing frequency, and such comparisons are likely to give key insights into residues critical for retention as well as the underlying mechanisms.

Gal-T1 and Lactose Synthase. Family 7 glycosyltransferases (GT-7) are inverting hexosyltransferases, predominantly galactosyltransferases of glycoprotein and glycolipid core glycans [87]. The $\beta(1{\rightarrow}4)$-galactosyltransferase (Gal-T) subfamily synthesize conserved eukaryotic glycoside cores or attach intercellular oligosaccharides used for cell-surface adhesion, particularly during fertilization and development [26]. The most prevalent core structure of *O*-linked glycoconjugates is the disaccharide Gal-$\beta(1{\rightarrow}4)$-GlcNAc (LacNAc), formed by the galactosylation of GlcNAc terminated molecules by the enzyme Gal-T1 [92].

Gal-T1 is a type II integral trans-Golgi protein with the C-terminal GT-A fold type catalytic domain normally found in the lumen of the Golgi as a homodimer [93]. The catalytic domain's N-terminal fold is Rossmann-like; however only two of its β-sheets are separated by α-helices. Instead, core structure is maintained by four disulfide bonds between proximal cysteine residues, stabilizing turns between the sheets to define the fold, Fig. 9a [93]. The enzyme contains one unpaired cysteine that is surface-exposed and forms disulfide bonds with other molecules of Gal-T1 in solution forcing aggregation into inclusion bodies. While it is possible to renature crystallographic grade protein from these bodies, it has been demonstrated that mutation of this cysteine to threonine not only enhances in vitro folding by preventing aggregation but also increases enzymatic activity [94].

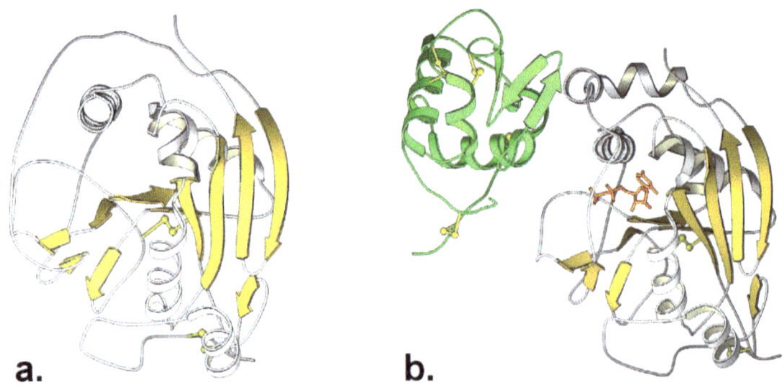

Fig. 9 SETOR diagram of Gal-T1 secondary structure **a** alone and **b** in complex with α-lactalbumin (*green*) in the lactose synthase ternary complex. The formation of a ternary complex between the two proteins occludes the binding of glycans to bring about a remarkable change in the conformation and the acceptor specificity of Gal-T1 to produce the disaccharide lactose for breast milk in mammalian breast tissue

Although the usual enzymatic function of Gal-T1 is to galactosylate glycans *en route* to the plasmalemma, a more interesting study of substrate specificity is found in milk production. In the glandular breast tissue of lactating mammals the preferred acceptor for this enzyme is altered from GlcNAc terminated oligosaccharides to glucose monosaccharides for lactose synthesis [95–97].

This change in Gal-T1 substrate specificity is due to an irreversible conformational shift imparted by the calcium dependant regulatory protein α-lactalbumin [95, 96, 98], a structural homologue of lysozyme [99]. α-lactalbumin can only bind Gal-T1 to form the "lactose synthase ternary complex" (Fig. 9b) after the initial binding of UDP-Gal [97], whereupon it becomes a fully functional enzyme in lactose synthesis.

α-lactalbumin does not sterically hinder or directly interact with the affected substrates to alter the specificity of Gal-T1 in the lactose synthase ternary complex to the production of lactose. Instead, it triggers an alteration in the secondary structure of Gal-T1, where residues 359–365 change conformation from loop to helix in the acceptor binding region to exclude the larger substrates involved in glycan extension. In the loop conformation, Arg-359, Phe-360 and Ile-363 form an *N*-acetyl binding pocket that is destroyed in the helix conformation [100]. Upon donor binding, this conformational shift exposes a hydrophobic patch which interacts with α-lactalbumin, which accounts for the donor binding prerequisite.

The ordered construction of the lactose synthase ternary complex from Gal-T1 and α-lactalbumin can only occur after Gal-T1 binds UDP-Gal [97]; however, once formed the complex will continue to use UDP-Gal in the production of lactose.

In an interesting aside, α-lactalbumin also serves as an excellent calcium delivery vehicle susceptible to proteolysis by the infant where milk lactose provides a selective advantage for *Lactobacillus bifidus* to expand as the primary biofilm of an infant's gastrointestinal tract in order to limit direct colonization by potential pathogens [101].

***β*-Glucuronyltransferases GlcAT-I and GlcAT-P, and the Synthesis of Heparan and HNK-1.** GT-43 enzymes consist of a group of type-II membrane-bound inverting β-glucuronyltransferases. Two examples from this family have been crystallized and been shown to have the GT-A-type fold. The first, glucuronyl transferase I (GlcAT-I) is involved in heparan/chondroitin sulfate biosynthesis and catalyzes the transfer of glucuronic acid (GlcUA) from UDP-GlcUA to the O3 oxygen of a galactose residue in a growing linkage region [74, 102, 103]. The second enzyme, GlcAT-P, is required for the synthesis of the carbohydrate epitope HNK-1 which is found on many neural cell adhesion molecules [104, 105], where it catalyzes the transfer of glucuronic acid from UDP-GlcUA to the O3 of the terminal galactose in Galβ(1$\rightarrow$4)-GlcNAc-R [104].

The structure of GlcAT-I revealed the basis for glucuronic acid transfer to the linkage region of various glycosaminoglycans, and spawned an intensive period of structural studies of heparan/chondroitin sulfate biosynthesis and glycosaminoglycan biology [102].

The structure of GlcAT-I displays only 7% sequence homology with the well-characterized glycosyltransferase SpsA; however, the two enzymes share a high degree of similarity in tertiary structure, including the position of the prospective nucleophiles in both enzymes. This was early evidence supporting the theory that many glycosyltransferases share structural homology despite low sequence homology [102].

GlcAT-I exists in vivo as a dimer each of which consists of two domains. The N-terminal domain contains a Rossmann fold that is mainly involved in donor recognition. The C-terminal domain consists predominantly of a mixed β-sheet. As with all glycosyltransferases with known structure the active site lies along the inter-domain cleft. The two domains of the enzyme are separated by a DXD motif comprised of Asp-194, Asp-195 and Asp-196. The dimer is oriented in such a way that both active sites are displayed on the same face of the enzyme, Fig. 10. The C-terminus itself has been shown to be critical for catalysis [102], even though in the crystal lattice it forms an α-helix that is not associated with either molecule of the dimer.

The donor GlcUA moiety is recognized by the conserved residues His-308 and Asp-252, which both form hydrogen bonds with the 2-OH of GlcUA, whereas the 3-OH is specifically recognized by Asp-194 from the DXD motif,

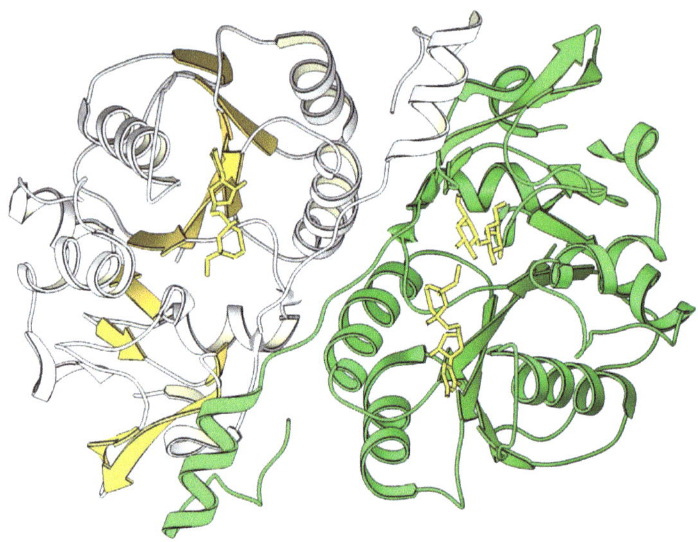

Fig. 10 SETOR diagram of GlcAT-I secondary structure, which belongs to the GT-A fold family, depicting the active site orientation and monomer interactions of the dimer

and the 4-OH forms hydrogen bonds with Arg-161 and Arg-156. Finally, the carboxylate group of GlcUA forms a hydrogen bond with the backbone amide of Arg-156.

GlcAT-I has been crystallized in the presence of UDP and the glycone of the acceptor fragment corresponding to the linker region Gal-$\beta(1\rightarrow3)$-Gal-$\beta(1\rightarrow4)$-Xyl-Ser, and in the presence of UDP-GlcUA, Fig. 11a [102]. Specificity for the enzyme toward the acceptor is provided primarily by the terminal galactose, where hydrogen bonds are formed between O3 and Glu-281, O4 and Asp-252, and O6 and Glu-227 and Arg-247. Glu-281, Asp-252 and Glu-227, which form the specific interactions with the GlcUA residue, are conserved among all Family-43 glucuronyltransferases [102]. The central galactose moiety of the acceptor makes only one interaction, which is between the O6 and Gln-318. No electron density at all is observed for the Xyl moiety, indicating that it is probably not important for acceptor recognition.

GlcAT-P and Neurological Function. The second family GT-43 enzyme with known structure is GlcAT-P, which is homologous to GlcAT-I. GlcAT-P is involved in the synthesis of the HNK-1 carbohydrate epitope, which is found on many neural cell adhesion molecules [106]. The HNK-1 carbohydrate is a sulfated glucuronic acid attached to the *N*-acetyllactosamine structure, HSO_3-3GlcA-Gal-$\beta(1\rightarrow4)$-GlcNAc, on some glycolipids, and glycoproteins [107]. Mutations in GlcAT-P have been assessed in the general population as potential risk factors for schizophrenia and related conditions [108]. Like GlcAT-I, GlcAT-P is a type-II membrane protein that exists naturally as a dimer and extends its active site into the lumen of the Golgi.

All the residues involved in donor and substrate binding are conserved, both in primary sequence and in structure, between GlcAT-P and GlcAT-

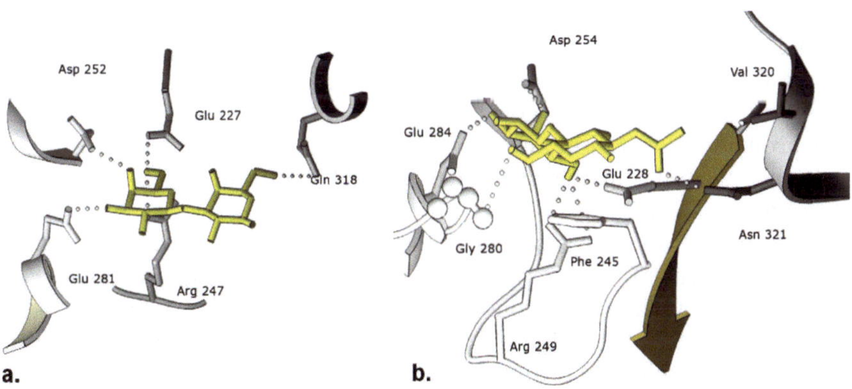

Fig. 11 **a** GlcAT-I acceptor fragment coordination by hydrogen bonds with Glu-227, Arg-247, Asp-252, Glu-281, and Gln-318. **b** Hydrogen bonding and stacking interactions observed in the acceptor binding site of GlcAT-P involves Glu-284, Asp-254, Glu-228, Asn-321, Phe-245, Arg-249, and Gly-280. Interestingly, the hydrogen bond provided by Asn321 comes from a long C-terminal tail belonging to the second enzyme in the dimer

I [104], and so it is not surprising that donor binding is almost identical to that of GlcAT-I (the rms deviation between N-terminal Rossmann-like folds of the two proteins is 0.4 Å).

GlcAT-P demonstrates the functional requirement for dimerization, as a long C-terminal loop extends from the one enzyme in the dimer to the acceptor binding site of the other where it interacts with the acceptor GlcNAc through Val-320 and Asn-321, Fig. 11b. [104].

3
Examples of Inverting Enzymes with the GT-B Fold: GT-1, GT-63

Gtf and Vancomycin-Class Antibiotics. CAZy family 1 members (GT-1) encompass enzymes covering a breadth of peptide and lipid acceptors. For example, some of these enzymes glycosylate lipophilic metabolites and xenobiotics to solubilize them for elimination [109]. Others are involved in sphingosine [110] and steroid biosynthesis. Most GT-1 members to date with solved structures are the Gtf-class glycosyltransferases GtfA [65], GtfB [111], and GtfD [112], each of which glycosylate vancomycin-class antibiotics.

Vancomycin is a trimacrocyclic oxidatively cross-linked heptapeptide antibiotic containing both D and L chiral centers (D-Leu-D-Cyt-L-Asn-D-Hpg-D-Hpg-L-Cyt′-L-DHpg, where Cyt = (2R, 3R) m-chloro-3-hydroxytyrosine; Cyt′ = (2S, 3R) m-chloro-3-hydroxytyrosine; DHpg = dihydroxyphenylglycine), Fig. 12, which is glycosylated with a deoxydisaccharide on the peptide residue hydroxyphenylglycine-4 (Hpg 4), a non-standard amino acid [75, 113], Fig. 13. The complexity of vancomycin makes chemical synthesis challenging and impractical for mass production. This has provoked great interest in its biosynthesis, which requires a series of anabolic enzymes thought to act in three discrete ordered steps. First, the heptapeptide core is assembled by non-ribosomal peptide synthetase. Second, its five aromatic side chains are oxidatively cross-linked providing a rigid aglycone substrate. In the third step this substrate undergoes a series of glycosylations [73, 75].

The finished product is secreted by *Actinomycetes* to provide an advantage over Gram positive competitors. Vancomycin prevents peptidoglycan glycosylation by binding to conserved D-alanyl-D-alanine residues in order to inhibit the transfer of the normal acceptors *N*-acetylmuramic acid and *N*-acetylglucosamine. As a clinical antibiotic it was saved as a last resort for methicillin-resistant infections, in part due to its nephrotoxicity [114]. Until resistant staphylococci were identified in 1987 vancomycin was the only totally effective unmodified antibiotic against Gram positives [115]. Resistant pathogens, particularly nosocomial infections, are increasingly endemic in the western world [116]. At a moderate energy cost to the bacteria [117], re-

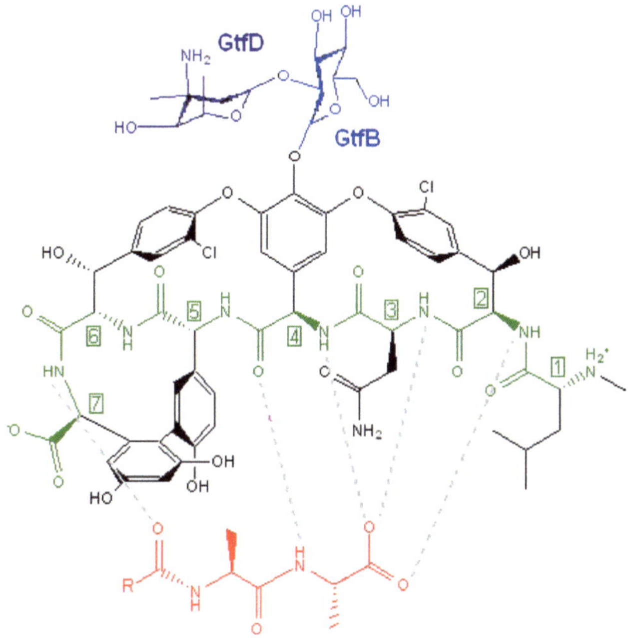

Fig. 12 Structure of the antibiotic vancomycin showing the proteinaceous backbone (*green*), 5 hydrogen bonds formed with peptidoglycan D-alanyl-D-alanine (*red*), and glycosylation sites of the glycosyltransferases GtfB and GtfD (*blue*)

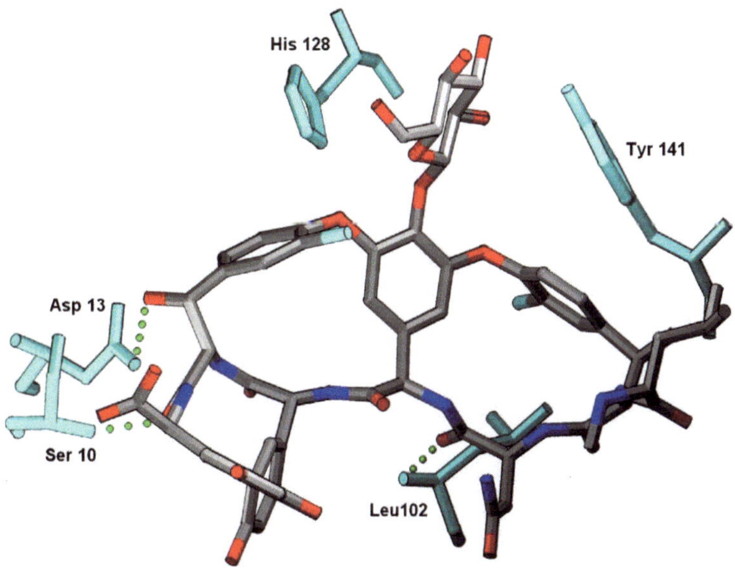

Fig. 13 The closed form of GtfA is generated by hydrogen bonds between the rigid DVV acceptor and residues Ser-10 and Asp-13, Leu-102, His-128 and Tyr-141 (*cyan*), causing the observed rotations about the Pro-204-Ala-373 hinge to generate the closed form of the enzyme

sistant peptidoglycan can be achieved by changing the terminus from D-Ala-D-Ala to D-Ala-D-lactate to radically lower vancomycin affinity and render the antibiotic ineffective.

Changing vancomycin's natural disaccharide has been shown to enhance potency against resistant strains; but this has so far required minimum inhibitory concentrations [118, 119] on par with a lethal dose. Furthermore, altering the sugar may not be the most direct way to combat resistance as it is the rigid aglycone moiety that appears to confer substrate efficacy [120].

The Gtf proteins with known structure contain typical GT-B Rossmann folds, Fig. 14a. The long C-terminal tail, known to be required for catalysis, crosses the cleft behind the domain interface and forms an α-helix that lies along the surface of the N-terminal domain. These structures display the expected disordered polypeptide loop about their catalytic sites in the unliganded, or "open", form.

This open-form architecture of glycosyltransferases was first observed while investigating GtfB, which transfers the initial sugar from UDP-Glc to 4-OH-Phe-Gly-4 of the vancomycin aglycone. Open-form structures with donor bound were not observed even when crystals were grown in the presence of vast molar excess, which suggested prerequisite binding of the acceptor [111].

The closed form, Fig. 14b, is generated upon binding of donor and acceptor, and results from the N and C terminal domains pivoting about two residues, Pro-204 and Ala-373. The closed form is thought to best correspond to the active conformation of the enzyme and has been proposed as a useful vehicle for addressing mechanistic features and substrate binding specificity [65].

Gtfs have revealed many molecular firsts, such as the first structure determination to show captured acceptor bound in the closed conformation, which is relevant to all GT-B fold type glycosyltransferases.

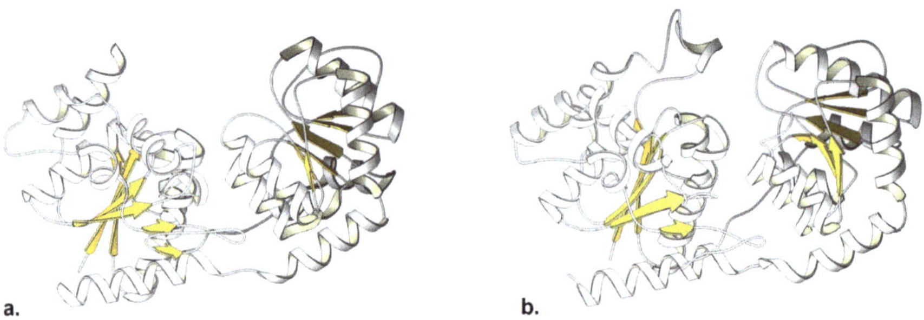

Fig. 14 SETOR diagrams of GtfA secondary structure in the **a** unliganded open conformation and **b** closed conformation induced by DVV (not shown) binding. The bisubstrate dependant closed conformation caps the deep wide crevice between the two domains of the GT-B fold by two main chain rotations and the ordering of a labile polypeptide loop

BGT, Unfolding DNA, and more Open and Closed Forms. CAZy family-63 (GT-63) so far consists of the single member β-glucosyltransferase (BGT), which is a DNA modifying enzyme encoded by T4 bacteriophage that transfers glucose from UDP-Glc to DNA. The importance of BGT to the virion becomes immediately evident when examining its lifecycle. Upon host invasion, T4 bacteriophage eliminates all host macromolecular synthesis by nuclease digestion of host DNA. Both host and T4 nuclease activities impose a protective DNA modification requirement on T4 bacteriophage itself, which is accomodated by inverting transfer of glucose from uridine-diphosphoglucose (UDP-Glc) to cytosine by BGT to form 5-hydroxymethylcytosine (5-HMC) in double stranded DNA [121].

BGT is a monomer of 351 amino acid residues and, like other members of the GT-B fold family, consists of two non-identical flexible Rossmann-like domains [122, 123]. Although it is not true for all glycosyltransferases, the reaction kinetics suggest that BGT can bind acceptor and donor in any order. The C-terminal domain binds UDP-Glc while the N-terminal domain binds the DNA acceptor substrate. Although the open form can bind DNA, BGT undertakes a large conformational change upon UDP-Glc binding, and adopts a closed conformation that displays increased affinity for subsequent DNA binding, Fig. 15 [123–125]. Domain closure appears to be necessary for glucose transfer [126]. Part of the conformational change in generating the closed form involves a UDP-glucose induced movement in a loop formed by residues 189 to 195 [124]. Acceptor and donor have been proposed to come together one of two ways: If DNA binds the N terminus first, then UDP-glucose brings the C-terminal domain, containing loop 189–195, to form the complete active site of BGT. If UDP-glucose binds first, the loop is moved into the catalytic position and the BGT-UDP-glucose complex becomes pre-formed for DNA binding and catalysis [124].

UDP-Glc has been crystallized in the active site of the enzyme. In the closed conformation, the uridine base of the UDP-Glc bound donor is buried

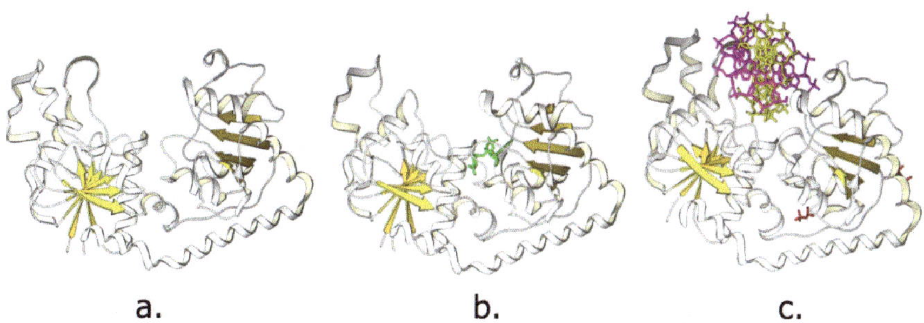

Fig. 15 SETOR diagram of BGT secondary structure **a** in open unliganded state, **b** closed form with UDP-Glc bound and increased DNA affinity, and **c** closed conformation with DNA bound

within the protein where it forms several hydrogen bonds along with a weak stacking interaction (in this case with Phe-213). In contrast, the ribose is somewhat solvent exposed. The donor glucose moiety itself is situated in a pocket defined by residues from both domains where it makes several interactions and is accessible to external solvent through a channel [122].

The contrast in size between the UDP-Glc donor and the enzyme itself begs the question as to how such a small molecule can induce the large conformational change from the open to the closed form. The answer appears to lie in the negatively-charged phosphate moieties of the donor that form electrostatic interactions with the side chains of two arginine residues (Arg-191 and Arg-195). These arginine residues are part of a network of three inter-domain salt bridges that exist in the closed form of the enzyme (Arg-191 to Asp-258; Arg-195 to Asp-258; Asp-100 to Arg-191). The binding of UDP-Glc to these residues induces the formation of these salt bridges and triggers the inter-domain conformational changes, Fig. 16 [122, 127].

Recognition by BGT of the DNA acceptor itself requires an extra-helical flipped 5-HMC base, which occurs by thermodynamic chance in natural double stranded DNA [124]. Once BGT locates a flipped cytosine, the base is stabilized in its flipped position by Asn-70 and Phe-71, which are part of loop 70–74 that penetrates into the DNA duplex via the major groove. Asn-70 fills the empty space left by the extra-helical base, where it forms hydrogen bonds to the guanine on the complementary strand and distorts the DNA backbone

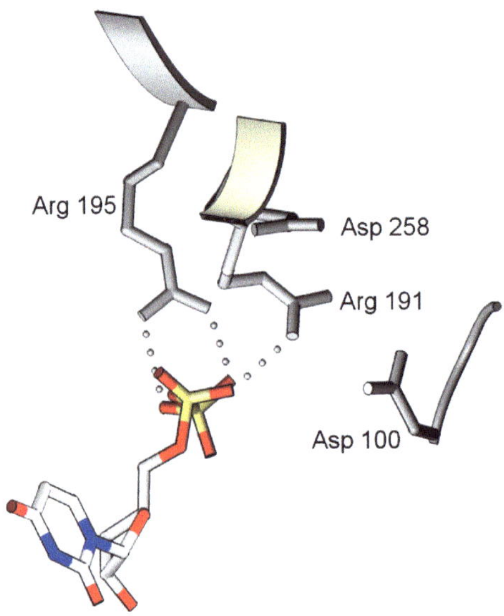

Fig. 16 The binding of UDP-Glc to UDP induces the formation of several charged residue interactions that result in a conformational change to generate the closed form of the enzyme

to stabilize the flipped base while Phe-71 completes the base stacking interactions in DNA.

A final point concerns the role of the divalent metal cation found in many glycosyltransferases. In most nucleoside diphosphate or triphosphate containing enzymes the metal ion serves to stabilize the negatively charged diphosphate group [126]. This is true for glycosyltransferases in the GT-A fold family. However, the roles of the divalent metal ion have been suggested to differ between glycosyltransferases in the GT-A fold family and those in the GT-B fold family [125], where enzymes with the GT-B fold have a reduced need for this function because either the helix dipole effect of the $\alpha/\beta/\alpha$ motif or positively charged side-chains serve the same purpose [125]. BGT itself contains three arginine residues that interact with the diphosphate moiety to reduce the net charge. Instead of binding the diphosphate, the metal cofactor has been suggested to facilitate the cleavage of the UDP-sugar linkage and stabilize the leaving group [126]. In most GT-B fold family members, divalent cations appear to be required only to achieve full catalytic activity [125]. Interestingly, superposition of crystal structures for the active site of BGT bound to UDP in the presence of Mn^{2+} and BGT bound to UDP-Glc indicate that the manganese and glucose binding sites conflict. Attempts to obtain a BGT structure with UDP-Glc and Mn^{2+} simultaneously bound failed to show electron density corresponding to the manganese ion, and it has been suggested that the metal ion may be involved in product release after glucose transfer [125].

4 Examples of Retaining Enzymes with the GT-A Fold: GT-6, GT-64, GT-78

GTA, GTB, and Blood Transfusions. Family 6 (GT-6) glycosyltransferases include two human enzymes GTA and GTB (named long before the GT-A and GT-B fold types were discovered) that are responsible for the generation of the human ABO(H) blood group A and B antigens. The human ABO(H) blood group antigens are abundant erythrocyte cell surface oligosaccharides conjugated as both glycoprotein and glycolipid [128]. GTA and GTB both utilize the same acceptor molecules, which all terminate in the H-antigen disaccharide α-L-Fuc-(1$\rightarrow$2)-β-D-Gal-*O*R. GTA forms the A antigen by the $\alpha(1\rightarrow3)$ transfer of *N*-acetylgalactosamine from UDP-GalNAc to the H antigen, while GTB forms the B antigen by the corresponding transfer of galactose from UDP-Gal. This means that the A and B antigens differ *only* in the replacement of an acetamido group by an hydroxyl group on the terminal galactose residue. The ABO(H) antigens comprise one of the most prevalent human alloantigens and the first to be discovered [129], and this small difference between the A and B antigens is sufficient to cause a lethal immune response in a mismatched blood transfusion. Interestingly, the O blood group itself corresponds to the

H antigen, and is usually the result of an inactive or truncated GTA or GTB enzyme [130].

GTA and GTB are the two most homologous naturally occurring glycosyltransferases reported that utilize distinct naturally occurring donors, differing by only 4 "critical" amino acid substitutions Arg/Gly-176, Gly/Ser-235, Leu/Met-266 and Gly/Ala-268 [131, 132]. Given that the enzymes share the same acceptor it was initially thought that the four critical residues would be involved exclusively in donor recognition.

With the two wild type enzymes differing by only four amino acids, the generation of all possible recombinant chimera is straightforward. It has become convenient when discussing the GTA/GTB chimera to designate the four critical amino acid residues of GTA as AAAA and the four critical amino acid residues of GTB as BBBB. Surprisingly, kinetic studies of all 14 chimera and the two wild type enzymes showed (well in advance of any structural determinations) that of the four critical amino acids, only the final two (Leu/Met-266 and Gly/Ala-268) significantly affect donor specificity [51]. That is, any chimera XXAA (where X can be either A or B) would selectively transfer A donor (UDP-GalNAc), and any chimera XXBB would selectively transfer the B donor (UDP-Gal). Similarly, chimera that mixed the final two critical amino acid residues had mixed A and B activity. That is, any chimera XXAB or XXBA would transfer either the A or the B donor [51].

Although the first two critical amino acids (Arg/Gly-176 and Gly/Ser-235) did not affect donor specificity they clearly impacted enzyme turnover. The chimera BAAA, BBAA, ABAA all have GTA activity exceeding that of wild type GTA enzyme. The chimera BAAA has an 11-fold increase in K_{cat}, which remains the highest increase in K_{cat} reported for a single amino acid change in a glycosyltransferase [51].

The rationale for enzyme specificity and turnover had to wait for the structural studies [64]. GTA and GTB have nearly identical structures with all four critical amino acids residing in the ~ 13 Å wide active site cleft, Fig. 17. While Arg/Gly-176 is found in the N-terminal UDP-donor binding Rossmann fold, the other three are found on the C-terminal fold. Although the structure of either enzyme has not been reported in complex with a complete donor they have been crystallized with UDP, which has allowed for the likely modeling of the position of the sugar moieties, Fig. 18. Only residues 266 and 268 of the four critical amino acids are positioned to directly interact with the donor, and so confer specificity [64].

Given their high homology, GTA and GTB are ideal probes of substrate specificity. For example, a single point mutation of the non-critical amino acid residue P234S in GTB surprisingly leads to a complete switch in donor specificity from B to A [52], so much so that the GTB/P234S mutant transfers the A donor faster, and the B donor slower than, wild type GTA. The rational for this switch of specificity is not yet clear.

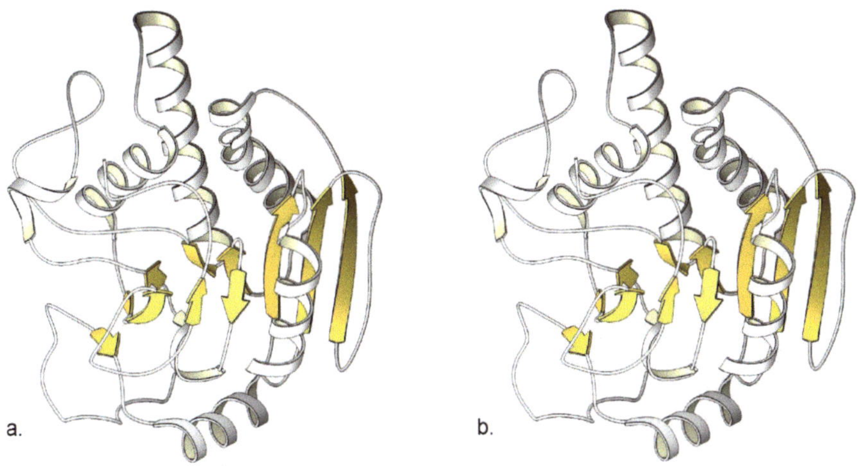

Fig. 17 SETOR diagrams of the GT-A folds of the human ABO(H) blood group A and B glycosyltransferases **a** GTA and **b** GTB. Differing by only four "critical" amino acid residues, GTA and GTB are the two most homologous naturally occurring glycosyltransferases that utilize different naturally occurring nucleotide donors. The GTA and GTB enzymes were named long before the GT-A and GT-B fold types were described

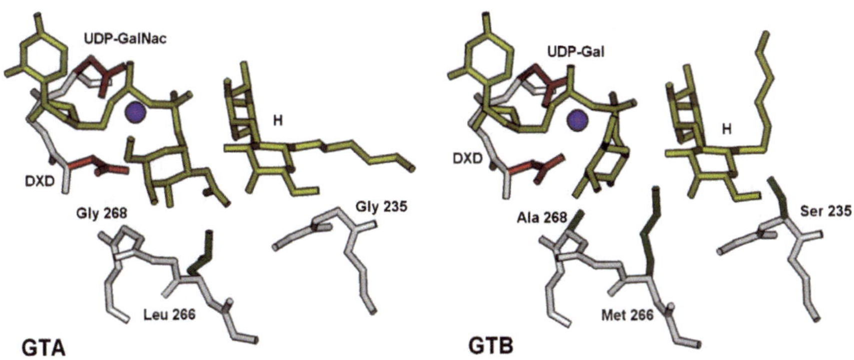

Fig. 18 Modeled donor coordination in the human ABO(H) blood group A and B enzymes **a** GTA and **b** GTB, displaying the relative locations of the three "critical" residues that are located in the active site relative to their respective donor sugars. The presence of Leu/Met-266 and Gly/Ala-268 are sufficient to change the specificity of the enzyme toward donor from UDP-GalNAc in GTA to UDP-Gal in GTB. The position of the acceptor (H) and DXD motifs are also shown

GTA and GTB have also provided some of the few glimpses of the three dimensional structures of glycosyltransferases with inhibitors bound [133]. One of the primary clinical applications of glycosyltransferases is to develop methods of inhibiting pathogen enzymes; however, this has been hampered by the relative paucity of structural data.

mEXTL2 and Heparan Sulfate. Family 64 (GT-64) includes a group of retaining glycosyltransferases involved in the synthesis of heparan sulfate. EXT1 and EXT2 elongate the heparan sulfate chain by alternating the addition of GlcUA and GlcNAc to their respective non-reducing ends. EXT1 and EXT2 have been linked to hereditary multiple exostosis, which is characterized by the appearance in growing children of several cartilaginous hard protrusions originating from the growth centers of bone marrow, and thus have direct clinical relevance [134]. Although these two enzymes have yet to be crystallized the structure for the EXT-like enzyme mouse EXTL2 (mEXTL2) has been reported [135]. mEXTL2 is one of the most homologous EXT-like enzymes to EXT1 and EXT2 which makes it an ideal candidate for the study of glucosaminyltransferases. mEXTL2 has been described as an α-GalNAc transferase from evidence showing the retention of configuration in the transfer of α-GalNAc to form GalNAc-α(1$\rightarrow$3)-Gal-β(1$\rightarrow$3)-Gal-β(1$\rightarrow$4)-Xyl-; however, the exact in vivo function of mEXTL2 has yet to be characterized [134]. In vitro, mEXTL2 displays α-GlcNAc and α-GalNAc transferase activity to the acceptor GlcUA-β(1$\rightarrow$3)-Gal-β-1-*O*-naphthalenemethanol. The mEXTL2 enzyme is suspected to be capable of transferring GlcNAc to a tetrasaccharide linker of the core protein in heparan sulfate biosynthesis [136]. It has been postulated that the transferase mEXTL2 GalNAc functions to terminate glycosaminoglycan biosynthesis. As a result mEXTL2 is an excellent model for the α(1$\rightarrow$4)-*N*-acetylglucosaminyltransferase reaction catalyzed by some members in this α-*N*-acetylhexosaminyltransferase heparan synthesis family [135].

mEXTL2 is a type II Golgi membrane protein that contains a C-terminal catalytic domain that has been crystallized in the apo form, in complex with UDP-GlcNAc, UDP-GalNAc and UDP and GlcUA-β(1$\rightarrow$3)-Gal-β-1-*O*-naphthalenemethanol [135]. mEXTL2 is a globular protein with two domains, a UDP-binding domain (Ala-63 to Val-150) and an acceptor binding domain (Thr-154 to Lys-327) that flank a DXD motif (Asp-151 to Asp-153). The UDP-binding domain consists of a typical Rossmann-like fold with four β-strands alternating with α-helices. The acceptor binding domain consists of two β-sheets oriented about 90° to each other and flanked by α-helices. Residues between these two sheets form a hydrophobic core, Fig. 19 [135].

UDP-GlcNAc and UDP-GalNAc bind almost identically to mEXTL2. The UDP moieties of these two donors form stacking interactions and hydrogen bonds consistent with those seen in other glycosyltransferases. Although the position of C1 is nearly identical for the pyranose moiety of the donor molecules, these C4 epimers bind in slightly different orientations. The 3-OH and 4-OH groups of the GalNAc moiety display the same hydrogen bond partners as the 4-OH and 6-OH groups of the GlcNAc moiety. The residues involved in donor recognition are highly conserved in this family and clustered around the two completely conserved cysteine residues Cys-244 and Cys-296 [135].

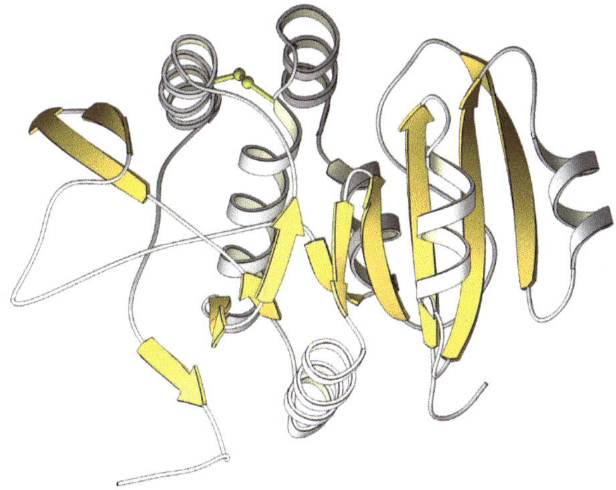

Fig. 19 A secondary structure model of mEXTL2 displaying a typical N-terminal nucleotide binding Rossmann fold on the *right*. The acceptor binding site (*left*) depicts two β-sheets oriented about 90° to each other forming a hydrophobic core

The polypeptide loop Glu-275 to Glu-288 that is observed to be disordered in the unliganded structure is ordered upon acceptor binding, which results in the formation of two new hydrogen bonds between UDP and the protein. As with many glycosyltransferases, interactions with the acceptor are limited to a few residues. Enzyme acceptor interactions involve the carboxylate of GlcUA and the Gal-O4 (to distinguish it from glucose). The orientation of the acceptor analogue is preserved through stacking interactions with Phe-290 and Trp-284. The naphthalenemethanol moiety in the acceptor was not clearly defined in the structure [135].

mEXTL2 is one of the retaining transferases that has been suggested to have an S_{Ni}-like reaction mechanism. The double displacement mechanism does not readily lend itself to the observed geometry in the active site as there does not appear to be an appropriately positioned nucleophile to accept the first transfer. The authors suggest that Arg-293 may act as the enzyme's nucleophile; however, they also point out that this is speculative given the lack of evidence that arginine is able to act as a strong nucleophile in any biological system [135]. The authors also put forward the possibility of a double displacement pathway involving a water molecule as the "enzyme" nucleophile with Asp-246 as an activating base; however, the requirement of a conformational change in Arg-293 in order to accommodate a water molecule in a position to make a nucleophilic attack was finally held to be unfavorable [135].

MGS and Surviving in Hot Water. Family-78 (GT-78) currently contains a single member, a retaining mannosyltransferase, mannosylglycerate synthase (MGS), which is a key stress-induced enzyme in the hyperthermophilic bacteria *Rhodothermus marinus*. Many prokaryotes accumulate low molecular weight compounds in order to maintain protein stability during osmotic and temperature shocks. Hypothermophiles tend to produce negatively charged soluble small molecules such as mannosylglycerate. Mannosylglycerate and its derivative mannosylglyceramide are exploited by *R. marinus* to protect against the osmotic stress and high temperatures found in alkaline submarine hot springs [137]. Mannosylglycerate and the other negatively charged osmolytes have been suggested to raise the Gibbs free energy of both the native and denatured state of the protein by their exclusion from its solvation shell. However, since the denatured protein has a larger solvation shell than the native protein, the Gibbs free energy of the denatured state is increased to a greater extent than that of the native protein [137, 138]. As a consequence, the folded form of the protein has a significantly lower relative Gibbs free energy than the unfolded form, giving the folded form an advantage under thermodynamic stress.

MGS is a GT-A fold enzyme that catalyzes the retaining transfer of mannose from GDP-mannose to D-glycerate during the synthesis of 2-O-α-D-mannosylglycerate, Fig. 20 [139]. Recognition of D-glycerate is imparted mainly by hydrogen bonds through all 4 oxygen atoms, where the carboxylate group forms hydrogen bonds to the main chain amide of Ile-138 and Thr-139 and the side chain hydroxyl of Thr-139. The C2-OH is held in position by Arg-131, while the C3-OH forms hydrogen bonds to the peptide amide of Met-137 and Ala-136 [140].

Mannosylglycerate synthase is another example of a retaining glycosyltransferase where there appears to be a lack of an appropriate enzyme nucleophile, and has been suggested to utilize an S_{Ni} mechanism. However, the

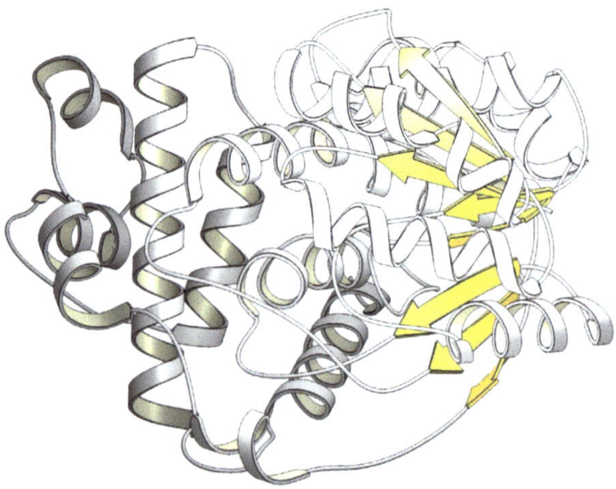

Fig. 20 Mannosylglycerate synthase (involved in the biosynthesis of mannosylglycerate) has an acceptor binding domain consisting mostly of α-helices

close relationship between the observed structures of retaining and inverting enzymes has raised the possibility of engineering an enzyme of one sort to undertake a transfer of the other. It has been speculated that only a small change in the position of the acceptor in MGS could place the substrates in position for an inverting instead of retaining transfer. This would require the introduction of a basic residue to deprotonate the acceptor, and the authors made suggestions of how this could be accomplished [140].

This point is especially interesting in light of the history of family-78, as MGS was initially placed in the inverting family GT-2 based on 30% sequence homology. A new family, GT-78, was created when it was shown that MGS was a retaining enzyme. In view of this relatively high sequence homology, it has been argued that only a few evolutionary steps should be required to transform MGS into an inverting glycosyltransferase [140].

5 Examples of Retaining Enzymes with the GT-B Fold: GT-5, GT-20

Glycogen Synthase and other Glycosyltransferases of Glycogen Metabolism. Family 5 (GT-5) enzymes polymerize starches, cellulose, glycogen, and other glucans. Glucans are all-glucose glycans that can serve major structural roles such as cellulose in plant cell walls [141, 142]; however, their most prevalent function is fixed carbon biomass accumulation and energy storage [143]. The main storage glucan of animals is glycogen, which is poly $\alpha(1\rightarrow4)$ glucose with $\alpha(1\rightarrow6)$ branches every 8–12 residues.

Three enzymes, two of them glycosyltransferases with known structure, are sufficient for glycogen anabolism from UDP-Glc monomers. An initial octo-$\alpha(1\rightarrow4)$ glucose primer necessary for a new glycogen molecule is prepared by the enzyme glycogenin, which is a retaining GT-A enzyme belonging to family GT-8 [144], Fig. 21a. Glycogenin initially catalyzes an unusual self glycosylation to an internal tyrosine residue, then makes sequential $\alpha(1\rightarrow4)$-Glc additions. A second enzyme, glycogen synthase, a member of retaining family GT-5 (with GT-B fold type), Fig. 21b, then catalyzes additional linear polymerization of this primer by $\alpha(1\rightarrow4)$ retaining transfer of glucose from UDP-Glc to the non-reducing end of the elongating polysaccharide [145]. Finally, glycogen branches are formed by transfer of an $\alpha(1 \rightarrow 4)$ gluco-oligomer (at least 11 residues) from the reducing end of poly-$\alpha(1 \rightarrow 4)$-glucose to an internal glucose by a transglycosylase commonly known as "branching enzyme" [145]. Though it does technically transfer sugar moieties from the activated linear precursor, this branching enzyme has been placed in the glycoside *hydrolase* family-13 (GH-13) along with amylases due to the hydrolytic mechanism of transfer [146].

Inherited metabolic disorders in many of the enzymes involved in glycogen synthesis and degradation are known as glycogenosis, or glycogen stor-

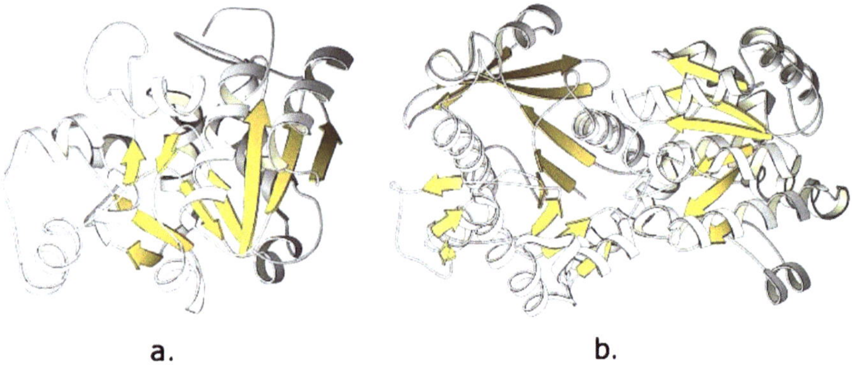

Fig. 21 SETOR diagrams of two glycosyltransferases involved in glycogen anabolism; **a** Glycogenin is a GT-A fold type enzyme responsible for the construction of the poly-Clc glycogen primer. **b** Glycogen Synthase has a GT-B like fold type and is the enzyme responsible for the bulk of glycogen polymerization by sequential addition of glucose units

age diseases (GSDs) [147]. Although glycogen storage diseases are quite rare (~ 1 in 20 000–25 000), their affects can be dire. For example, mutation in GYS2, the gene encoding a liver isozyme of GT5's glycogen synthase is a catastrophic disease (GSD type 0) which, unlike other forms of GSD, does not involve the storage of excessive or abnormal glycogen and is characterized by decreased glycogen stores, primarily in the liver, resulting in growth retardance, neurological damage and developmental delay from decreased liver glycogen anabolism [147].

An *Agrobacterium tumefaciens* glycogen synthase with solved structure has a fold similar to the GT-B fold family, with two homologous Rossmann folds [148]. Only the open form of this enzyme has been reported, both unliganded and in complex with ADP. ADP binds the DXD triplet on the C-domain face of the open crevice [148].

The closed form may be induced by crystallization of glycogen synthase in complex with the octameric $\alpha(1 \rightarrow 4)$ glucan primer. Unfortunately, this has not yet been reported. Nor have the corresponding structures of the mutated human GSD type 0 enzymes, of which there are at least three distinct mutations identified [149].

The Trehalose Synthase OtsA. Family 20 glycosyltransferases have an unclassified compact GT-B-like fold. They catalyze the formation of trehalose, a non-reducing disaccharide Glc-$\alpha(1,1)$-α-Glc found in abundance, sometimes as the dominant storage glucan, in cryptobiotic organisms able to withstand severe environmental stresses such as desiccation, pressure [150], osmotic stress [151], extreme heat [152] or cold [153], reduction [154] or oxidation [155, 156].

Trehalose has a tremendous water binding capacity, giving it some unique stabilizing physico-chemical properties. Trehalose is not easily hydrolyzed by

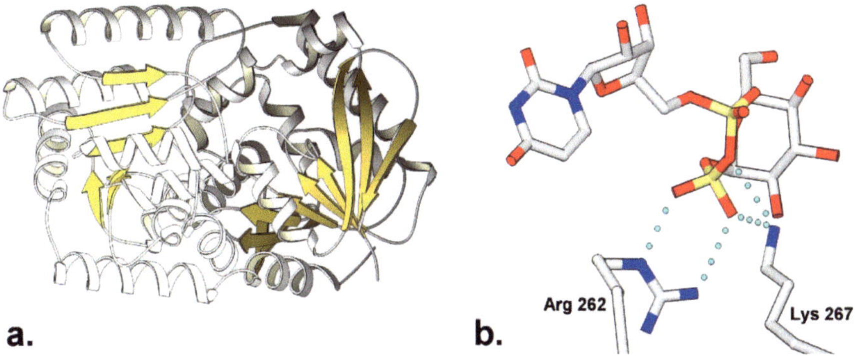

Fig. 22 **a** SETOR diagram of the unclassified compact GT-B like fold type of OtsA, a Trehalose synthase. **b** OtsA lacks a DXD motif, donor diphosphate is instead coordinated by the basic side-chains of Arg-262 and Lys-267

acid, begins to melt at 97 °C, resolidifies as an anhydrous disaccharide at 130 °C, then melts again at 203 °C [157, 158]. These factors allow a hydrogen bonding shell of trehalose to maintain membrane and protein integrity through an array of physiological stresses [159].

While some cryptobiotic bacteria support several pathways for trehalose synthesis, most utilize a two step process such as that of *E. coli* in which trehalose-6-phosphate synthase (OtsA, Fig. 22a) transfers UDP-Glc to α-Glc6P, which is subsequently dephosphorylated by OtsB [160].

OtsA lacks a DXD motif and divalent cation density, and instead the UDP donor diphosphate moiety forms a salt bridge with the basic side-chains of Arg-262 and Lys-267, Fig. 22b.

Though the Glc6P acceptor binding site lies mostly in the N-terminal domain, its phosphate group is sequestered in a positively charged pocket made by the side chains of Arg-9 and Arg-300 [161, 162]. OtsA selects the α-anomer of the acceptor to form its $\alpha(1 \rightarrow 1)\alpha$ product. The β-anomer is hydrophobically excluded by Ile155 and Trp85 and furthermore, only α-Glc6p O1 is correctly positioned to attack the donor [161, 162].

6
Neither GT-A nor GT-B—Example of Unclassified Folds, GT-42

CstII, Brain Function and Food Poisoning. Family-42 consists of a family of inverting 2,3-sialyltransferases that all use a CMP-β-*N*-acetylneuraminate as a sugar donor. Sialyltransferases are involved in the transfer of sialic acid (*N*-acetylneuraminic acid or NeuAc) to a range of glycoproteins, glycolipids and polysaccharides located on the cell surfaces of a variety of organisms. One important group of sialic acid containing glycosphingolipids is gangliosides. Gangliosides are membrane phospholipids with a complex sialic acid-containing oligosaccharide headgroup, and are a dominant glycan species in neural tissues. In microorganisms, gangliosides may behave as toxins or receptors and may modulate cell growth, differentiation, and cell-cell or cell-matrix interactions [84].

Structural data is available for one of these sialyltransferases, Cstll from *Campylobacter jejuni* strain OH4384. *C. jejuni* has been identified as the most frequent source of intestinal infections in the world and is usually caused by eating tainted food. Common symptoms include diarrhea and abdominal pain [163]. More significantly, *C. jejuni* has been closely linked to the degenerative autoimmune disease Guillian–Barre syndrome (GBS), which is an acute disease of the peripheral nervous system where the nerves in the arms and the legs become inflamed and, over time, stop working [163]. *C. jejuni* is an intestinal pathogen that is believed to express variable cell surface carbohydrate mimics of gangliosides that are associated to bacterial virulence. It is these ganglioside mimics that are suspected to trigger the auto immune response involved in GBS [84, 163].

The solved structure for CstII consists of 259 residues arranged in two unequally sized domains, Fig. 23. The first domain is a Rossmann like fold that is associated with nucleotide sugar binding. The active site of the enzyme is located at the cleft between these two domains which does not involve residues from any of the other participants of the CstII tetramer. The second domain is simple in structure containing two α-helices and is involved in acceptor recognition and sealing the active site on substrate binding.

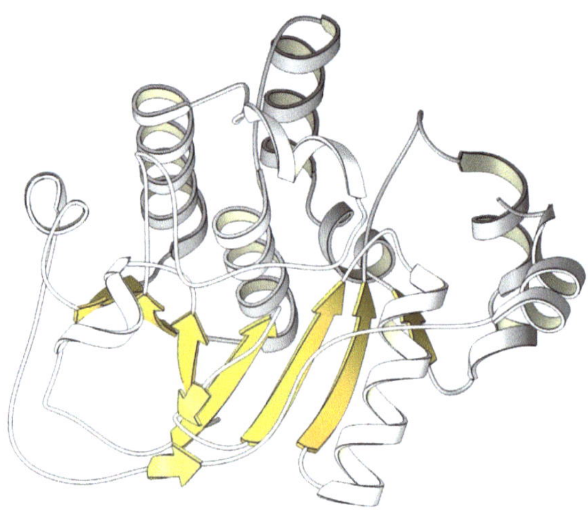

Fig. 23 Cstll secondary structure involves two unequally sized domains. The nucleotide binding domain is a typical Rossmann fold. The second small domain contains two α-helices that form a lid over the active site upon substrate binding. Interestingly, this enzyme lacks a DXD motif and therefore also lacks a metal cation at the catalytic center

Interestingly, CstII lacks a DXD motif, and does not have a metal cation at the catalytic center [84]. The donor CMP molecule binds in a cleft in the nucleotide binding domain at the protein membrane interface which facilitates the transfer of the sugar onto the terminus of the acceptor. The cytidine ring of CMP is held in position by hydrogen bonds, but with no DXD motif the CMP phosphate itself is held by hydrogen bond interactions with the hydroxyl groups of Tyr-156 and Tyr-162, ND2 of Asn-31 and by Asn-9 through a bridging water molecule, Fig. 24 [84].

The structure of CstII crystallized with donor analogue CMP-3FNeuAc maintains the interactions observed with CMP and induces a conformational change in residues 175–187 creating a "lid" that buries the donor away from the solvent and creates the acceptor binding site. Specificity for the NeuAc moiety is endowed mainly by the recognition of the two carboxylate groups. Specificity for one of these carbonyl groups arises from the complementarity to the positively charged region on an α-helix dipole [84].

Current evidence points towards an S_N2 mechanism for CstII. Some CstII variants are bifunctional enzymes that can catalyze either $\alpha(2\rightarrow3)$-sialyltransferase activity or $\alpha(2\rightarrow8)$-sialyltransferase activity. The two reactions have different acceptors. In the first, a nucleophilic attack on C2 of CMP-NeuAC by the 3-hydroxyl of galactose results in the formation of NeuAC-$\alpha(2\rightarrow3)$-Gal-$\beta(1\rightarrow3)$-GalNac. In the second, a nucleophilic attack on another CMP-NeuAC residue by the 8-hydroxyl of the *N*-acetylneuraminic acid residue of this sugar results in the elongation of this chain. The candidate enzyme nucleophile in both cases is His-188 that is located 4.8 Å from the anomeric carbon of the donor sugar.

The lack of a DXD motif in CstII provides a unique problem in catalysis. The DXD motif is involved mainly with cation binding. This metal cation is

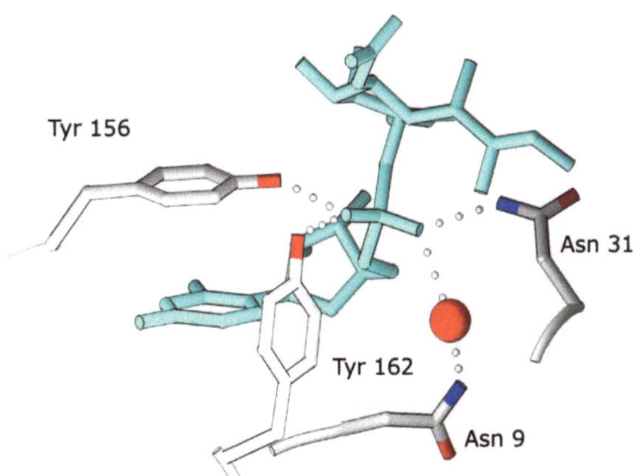

Fig. 24 CstII lacks a DXD motif and displays a unique method of binding the donor phosphate through hydrogen bonding interactions arising from Tyr-156, Tyr-162, Asn-31 and Asn-9 through a bridging water molecule

believed to act as a Lewis acid that stabilizes the leaving group in the reaction. The authors suggest that CstII resolves this problem through hydrogen bonding interactions provided by Tyr-156 and Tyr-162 that increase the leaving group character of CMP [84].

7
Conclusions

Glycosyltransferases utilize a limited repertoire of donors and a vast array of acceptors to produce a nearly countless variety of complex carbohydrate structures. There are now more identified genes for glycosyltransferases than any other type of enzyme save, possibly, kinases. Although the first glycosyltransferases were chemically characterized a generation ago it is only within the last decade that significant progress has been made in elucidating their three-dimensional structures. To date, all glycosyltransferases with known structure contain mobile polypeptide loops or protein domains that allow the enzyme to isolate the active site in order to exclude incorrect substrates as much as identify the correct ones. Although almost all glycosyltransferases fall within one of two fold families, there are exceptions. With the growing need to exploit these enzymes as targets for new therapeutics or as robust chemosynthetic tools it is imperative to carry structural studies to as many unique glycosyltransferase families as possible.

References

1. Sharon N (2001) The conquest of the last frontier of molecular and cell biology. Biochimie 83(7):555
2. Fukuda M (1994) In: Fukuda M, Hindsgaul O (eds) Molecular Glycobiology: Frontiers in Molecular Biology. 3–18. Oxford University Press, Oxford
3. Schwartz NB (1982) In: Varma RS (ed) Glycosaminoglycans and Proteoglycans in Physiological and Pathological Processes of Body Systems. Karger, Basel, New York, p 41–54
4. Geetha-Habib M, Campbell SC, Schwartz NB (1984) Subcellular localization of the synthesis and glycosylation of chondroitin sulfate proteoglycan core protein. J Biol Chem 259(11):7300–7310
5. Frederiksen S, Malling H, Klenow H (1965) Isolation of 3′-Deoxyadenosine (Cordycepin) from the liquid medium of Cordyceps militaris (L. Ex. Fr.). Biochim Biophys Acta 95:189–193
6. Kolchinsky AM, Mirzabekov AD, Gilbert W, Li L (1976) Preferential protection of the minor groove of non-operator DNA by lac repressor against methylation by dimethyl sulphate. Nucleic Acids Res 3(1):11–18
7. Osborn MJ, D'Ari L (1964) Enzymatic incorporation of N-acetylglucosamine into cell wall lipopolysaccharide in a mutant strain of Salmonella typhimurium. Biochem Biophys Res Commun 16(6):568–575

8. Cooper D, Manley RS (1975) Cellulose synthesis by Acetobacter xylinum. I. Low molecular weight compounds present in the region of synthesis. Biochim Biophys Acta 381(1):78–96
9. McGuire EJ, Jourdian GW, Carlson DM, Roseman S (1965) Incorporation of D-galactose into glycoproteins. J Biol Chem 240(10):4112–4115
10. Moczar E, Moczar M (1970) A micro-method for the determination of hydroxylysine and its glycosylated derivatives. J Chromatogr 51(2):277–281
11. Apweiler R, Hermjakob H, Sharon N (1999) On the frequency of protein glycosylation, as deduced from analysis of the SWISS-PROT database. Biochim Biophys Acta 1473:4–8
12. Hunt LT, Dayhoff MO (1970) The occurrence in proteins of the tripeptides Asn-X-Ser and Asn-X-Thr and of bound carbohydrate. Biochem Biophys Res Commun 39(4):757–765
13. Sadler JE (1984) In: Ginsburg V, Robbins PW (eds) Biosynthesis of glycoproteins: formation of O-linked oligosaccarides. Biology of carbohydrates (2):199. Wiley, NY
14. Hart GW, Haltiwanger RS, Holt GD, Kelly WG (1989) Glycosylation in the nucleus and cytoplasm. Annu Rev Biochem 58:841–874
15. Lehle L, Schwarz RT (1976) Formation of dolichol monophosphate 2-deoxy-D-glucose and its interference with the glycosylation of mannoproteins in yeast. Eur J Biochem 67(1):239–245
16. Schaffer C, Dietrich K, Unger B, Scheberl A, Rainey FA, Kahlig H, Messner P (2000) A novel type of carbohydrate-protein linkage region in the tyrosine-bound S-layer glycan of Thermoanaerobacterium thermosaccharolyticum D120-70. Eur J Biochem 267(17):5482–5492
17. Spiro RG (1973) Glycoproteins. Adv Protein Chem 27:349–467
18. Allen AK, Desai NN, Neuberger A, Creeth JM (1978) Properties of potato lectin and the nature of its glycoprotein linkages. Biochem J 171:665–674
19. Shore G, Maclachlan GA (1975) The site of cellulose synthesis. Hormone treatment alters the intracellular location of alkali-insoluble β-1,4-glucan (cellulose) synthetase activities. J Cell Biol 64(3):557–571
20. Dahm K, Breuer H (1966) Partial purification of a soluble UDP-glucuronyltransferase from human intestine. Biochim Biophys Acta 113(2):404–406
21. Ikeda M, Wachi M, Jung HK, Ishino F, Matsuhashi M (1991) The Escherichia coli mraY gene encoding UDP-N-acetylmuramoyl-pentapeptide: undecaprenyl-phosphate phospho-N-acetylmuramoyl-pentapeptide transferase. J Bacteriol 173(3): 1021–1026
22. Zgurskaya HI, Nikaido H (1999) AcrA is a highly asymmetric protein capable of spanning the periplasm. J Mol Biol 285(1):409–420
23. Prieto PA, Mukerji P, Kelder B, Erney R, Gonzalez D, Yun JS, Smith DF, Moremen KW, Nardelli C, Pierce M, Li Y, Chen X, Wagner TE, Cummings RD, Kopchick JJ (1995) Remodeling of mouse milk glycoconjugates by transgenic expression of a human glycosyltransferase. J Biol Chem 270(49):29515–29519
24. Palacpac NQ, Yoshida S, Sakai H, Kimura Y, Fujiyama K, Yoshida T, Seki T (1999) Stable expression of human β1,4-galactosyltransferase in plant cells modifies N-linked glycosylation patterns. Proc Natl Acad Sci USA 96(8):4692–4697
25. Dumon C, Priem B, Martin SL, Heyraud A, Bosso C, Samain E (2001) In vivo fucosylation of lacto-*N*-neotetraose and lacto-*N*-neohexaose by heterologous expression of Helicobacter pylori α-1,3 fucosyltransferase in engineered Escherichia coli. Glycoconj J 18(6):465–474
26. Shur BD, Evans S, Lu Q (1998) Cell surface galactosyltransferase: current issues. Glycoconj J 15(6):537–548

27. Takeuchi M, Inoue N, Strickland TW, Kubota M,Wada M, Shimizu R, Hoshi S, Kozutsumi H, Takasaki S, Kobata A (1989) Relationship between sugar chain structre and biological activity of recombinant human erythropoietin produced in chinese hamster overy cells. Proc Natl Acad Sci USA 86:7819–7822
28. Ohman R, Barker R, Hill R, Roseman S (1973) Glycosyltransferases - cellular studies. Birth Defects Orig Artic Ser 9(2):198–201
29. Bu'lock JD, Gregory H (1959) Pathways of sugar metabolism in relation to the biosynthesis of polyacetylenic antibiotics. Experientia 15:420–421
30. Reynolds PE, Somner EA (1990) Comparison of the target sites and mechanisms of action of glycopeptide and lipoglycodepsipeptide antibiotics. Drugs Exp Clin Res 16(8):385–389
31. Walsh CT, Losey HC, Freel Meyers CL (2003) Antibiotic glycosyltransferases. Biochem Soc Trans 31(Pt 3):487–492
32. Blumberg PM, Strominger JL (1974) Interaction of penicillin with the bacterial cell: penicillin-binding proteins and penicillin-sensitive enzymes. Bacteriol Rev 38(3):291–335
33. Gristina AG, Hobgood CD, Webb LX, Myrvik QN (1987) Adhesive colonization of biomaterials and antibiotic resistance. Biomaterials 8(6):423–426
34. Scholz S, Sonnenbichler J, Schafer W, Hensel R (1992) Di-myo-inositol-1,1′-phosphate: a new inositol phosphate isolated from Pyrococcus woesei. FEBS Lett 306(2–3):239–242
35. Clegg JS, Seitz P, Seitz W, Hazlewood CF (1982) Cellular responses to extreme water loss: the water-replacement hypothesis. Cryobiology 19(3):306–316
36. Shimma Y, Jigami Y (2004) Expression of human glycosyltransferase genes in yeast as a tool for enzymatic synthesis of sugar chain. Glycoconj J 21(1–2):75–78
37. Sandermann H Jr, Schmitt R, Eckey H, Bauknecht T (1991) Plant biochemistry of xenobiotics: isolation and properties of soybean O- and N-glucosyl and O- and N-malonyltransferases for chlorinated phenols and anilines. Arch Biochem Biophys 287(2):341–350
38. Chang HK, Zylstra GJ (1999) Role of quinolinate phosphoribosyl transferase in degradation of phthalate by Burkholderia cepacia DBO1. J Bacteriol 181(10):3069–3075
39. Brazier-Hicks M, Edwards R (2005) Functional importance of the family 1 glucosyltransferase UGT72B1 in the metabolism of xenobiotics in Arabidopsis thaliana. Plant J 42(4):556–566
40. Solenberg PJ, Matsushima P, Stack DR, Wilkie SC, Thompson RC, Baltz RH (1997) Production of hybrid glycopeptide antibiotics in vitro and in Streptomyces toyocaensis. Chem Biol 4(3):195–202
41. Somoza JR, Skillman AG Jr, Munagala NR, Oshiro CM, Knegtel RM, Mpoke S, Fletterick RJ, Kuntz ID, Wang CC (1998) Rational design of novel antimicrobials: blocking purine salvage in a parasitic protozoan. Biochemistry 37(16):5344–5348
42. Gamble W (1975) Mechanism of action of hypolipidemic and herbicidal aryloxy acids. J Theor Biol 54(2):181–190
43. Edwards R, Del Buono D, Fordham M, Skipsey M, Brazier M, Dixon DP, Cummings I (2005) Differential induction of glutathione transferases and glucosyltransferases in wheat, maize and Arabidopsis thaliana by herbicide safeners. Z Naturforsch [C] 60(3–4):307–316
44. Luckow VA, Lee SC, Barry GF, Olins PO (1993) Efficient generation of infectious recombinant baculoviruses by site-specific transposon-mediated insertion of foreign genes into a baculovirus genome propagated in Escherichia coli. J Virol 67(8):4566–4579

45. Burkle A, Diefenbach J, Brabeck C, Beneke S (2005) Ageing and PARP. Pharmacol Res 52(1):93–99
46. Lloyd KO (2000) The chemistry and immunochemistry of blood group A, B, H, and Lewis antigens: past, present and future. Glycoconj J 17(7–9):531–541
47. Bucher P, Morel P, Buhler LH (2005) Xenotransplantation: an update on recent progress and future perspectives. Transpl Int 18(8):894–901
48. Davies GJ (2001) Sweet secrets of synthesis. Nat Struct Biol 8(2):98–100
49. Rupprath C, Schumacher T, Elling L (2005) Nucleotide deoxysugars: essential tools for the glycosylation engineering of novel bioactive compounds. Curr Med Chem 12(14):1637–1675
50. Seto NO, Palcic MM, Compston CA, Li H, Bundle DR, Narang SA (1997) Sequential interchange of four amino acids from blood group B to blood group A glycosyltransferase boosts catalytic activity and progressively modifies substrate recognition in human recombinant enzymes. J Biol Chem 272(22):14133–14138
51. Seto NO, Compston CA, Evans SV, Bundle DR, Narang SA, Palcic MM (1999) Donor substrate specificity of recombinant human blood group A, B and hybrid A/B glycosyltransferases expressed in Escherichia coli. Eur J Biochem 259(3): 770–775
52. Marcus SL, Polakowski R, Seto NO, Leinala E, Borisova S, Blancher A, Roubinet F, Evans SV, Palcic MM (2003) A single point mutation reverses the donor specificity of human blood group B-synthesizing galactosyltransferase. J Biol Chem 278(14):12403–12405
53. Khatra BS, Herries DG, Brew K (1974) Some kinetic properties of human-milk galactosyl transferase. Eur J Biochem 44(2):537–560
54. Yoshida M, Itano N, Yamada Y, Kimata K (2000) In vitro synthesis of hyaluronan by a single protein derived from mouse HAS1 gene and characterization of amino acid residues essential for the activity. J Biol Chem 275(1):497–506
55. Campbell JA, Davies GJ, Bulone V, Henrissat B (1997) A classification of nucleotide-diphospho-sugar glycosyltransferases based on amino acid sequence similarities. Biochem J 326(Pt 3):929–939
56. Coutinho PM, Deleury E, Davies GJ, Henrissat B (2003) An evolving hierarchical family classification for glycosyltransferases. J Mol Biol 328(2):307–317
57. Rosen ML, Edman M, Sjostrom M, Wieslander A (2004) Recognition of fold and sugar linkage for glycosyltransferases by multivariate sequence analysis. J Biol Chem 279(37):38683–38692
58. Sinnott ML (1990) Catalytic mechanisms of enzymatic glycosyl transfer. Chem Rev 90:1171–1202
59. Paulson JC, Colley KJ (1989) Glycosyltransferases. Structure, localization, and control of cell type-specific glycosylation. Biol Chem 264(30):17615–17618
60. Bourne Y, Henrissat B (2001) Glycoside hydrolases and glycosyltransferases: families and functional modules. Curr Opin Struct Biol 11(5):593–600
61. Dodson E, Harding MM, Hodgkin DC, Rossmann MG (1966) The crystal structure of insulin. 3. Evidence for a 2-fold axis in rhombohedral zinc insulin. J Mol Biol 16(1):227–241
62. Unligil UM, Rini JM (2000) Glycosyltransferase structure and mechanism. Curr Opin Struct Biol 10(5):510–517
63. Charnock SJ, Davies GJ (1999) Structure of the nucleotide-diphospho-sugar transferase, SpsA from Bacillus subtilis, in native and nucleotide-complexed forms. Biochemistry 38(20):6380–6385

64. Patenaude SI, Seto NO, Borisova SN, Szpacenko A, Marcus SL, Palcic MM, Evans SV (2002) The structural basis for specificity in human ABO(H) blood group biosynthesis. Nat Struct Biol 9(9):685–690
65. Mulichak AM, Losey HC, Lu W, Wawrzak Z, Walsh CT, Garavito RM (2003) Structure of the TDP-epi-vancosaminyltransferase GtfA from the chloroeremomycin biosynthetic pathway. Proc Natl Acad Sci USA 100(16):9238–9243
66. Liu J, Mushegian A (2003) Three monophyletic superfamilies account for the majority of the known glycosyltransferases. Protein Sci 12(7):1418–1431
67. Kikuchi N, Kwon YD, Gotoh M, Narimatsu H (2003) Comparison of glycosyltransferase families using the profile hidden Markov model. Biochem Biophys Res Commun 310(2):574–579
68. Chelsky D, Parsons SM (1975) Stereochemical course of the adenosine triphosphate phosphoribosyltransferase reaction in histidine biosynthesis. J Biol Chem 250(14):5669–5673
69. Howard S, He S, Withers SG (1998) Identification of the active site nucleophile in jack bean α-mannosidase using 5-fluoro-β-L-gulosyl fluoride. J Biol Chem 273(4):2067–2072
70. Numao S, Kuntz DA, Withers SG, Rose DR (2003) Insights into the mechanism of Drosophila melanogaster Golgi α-mannosidase II through the structural analysis of covalent reaction intermediates. J Biol Chem 278(48):48074–48083
71. Lairson LL, Chiu CP, Ly HD, He S, Wakarchuk WW, Strynadka NC, Withers SG (2004) Intermediate trapping on a mutant retaining α-galactosyltransferase identifies an unexpected aspartate residue. J Biol Chem 279(27):28339–28344
72. Persson K, Ly HD, Dieckelmann M, Wakarchuk WW, Withers SG, Strynadka NC (2001) Crystal structure of the retaining galactosyltransferase LgtC from Neisseria meningitidis in complex with donor and acceptor sugar analogs. Nat Struct Biol 8(2):166–175
73. Li TL, Choroba OW, Charles EH, Sandercock AM, Williams DH, Spencer JB (2001) Characterisation of a hydroxymandelate oxidase involved in the biosynthesis of two unusual amino acids occurring in the vancomycin group of antibiotics. Chem Commun (Cambridge) 18:1752–1753
74. Pedersen LC, Darden TA, Negishi M (2002) Crystal structure of β1,3-glucuronyltransferase I in complex with active donor substrate UDP-GlcUA. J Biol Chem 277(24):21869–21873
75. Hubbard BK, Walsh CT (2003) Vancomycin assembly: nature's way. Angew Chem Int Ed 2(7):730–765
76. Yazer MH, Palcic MM (2005) The importance of disordered loops in ABO glycosyltransferases. Transfus Med Rev 19(3):210–216
77. Breton C, Bettler E, Joziasse DH, Geremia RA, Imberty A (1998) Sequence-function relationships of prokaryotic and eukaryotic galactosyltransferases. J Biochem (Tokyo) 123(6):1000–1009
78. Busch C, Hofmann F, Selzer J, Munro S, Jeckel D, Aktories K (1998) A common motif of eukaryotic glycosyltransferases is essential for the enzyme activity of large clostridial cytotoxins. J Biol Chem 273(31):19566–19572
79. Wiggins CA, Munro S (1998) Activity of the yeast MNN1 α-1,3-mannosyltransferase requires a motif conserved in many other families of glycosyltransferases. Proc Natl Acad Sci USA 95(14):7945–7950
80. Shibayama K, Ohsuka S, Sato K, Yokoyama K, Horii T, Ohta M (1999) Four critical aspartic acid residues potentially involved in the catalytic mechanism of Escherichia coli K-12 WaaR. FEMS Microbiol Lett 174(1):105–109

81. Hodson N, Griffiths G, Cook N, Pourhossein M, Gottfridson E, Lind T, Lidholt K, Roberts IS (2000) Identification that KfiA, a protein essential for the biosynthesis of the *Escherichia coli* K5 capsular polysaccharide, is an α-UDP-GlcNAc glycosyltransferase. The formation of a membrane-associated K5 biosynthetic complex requires KfiA, KfiB, and KfiC. J Biol Chem 275(35):27311–27315
82. Pak JE, Arnoux P, Zhou S, Sivarajah P, Satkunarajah M, Xing X, Rini JM (2006) X-ray crystal structure of leukocyte type core 2 beta 1,6-N-acetylglucosaminyltransferase. Evidence for a convergence of metal ion-independent glycosyltransferase mechanism. J Biol Chem 281(36):26693–26701
83. Lobsanov YD, Romero PA, Sleno B, Yu B, Yip P, Herscovics A, Howell PL (2004) Structure of Kre2p/Mnt1p: a yeast α1,2-mannosyltransferase involved in mannoprotein biosynthesis. J Biol Chem 279(17):17921–17931
84. Chiu CP, Watts AG, Lairson LL, Gilbert M, Lim D, Wakarchuk WW, Withers SG, Strynadka NC (2004) Structural analysis of the sialyltransferase CstII from Campylobacter jejuni in complex with a substrate analog. Nat Struct Mol Biol 11(2):163–170
85. Kamst E, Bakkers J, Quaedvlieg NE, Pilling J, Kijne JW, Lugtenberg BJ, Spaink HP (1999) Chitin oligosaccharide synthesis by rhizobia and zebrafish embryos starts by glycosyl transfer to O4 of the reducing-terminal residue. Biochemistry 38(13):4045–4052
86. Itano N, Kimata K (1996) Expression cloning and molecular characterization of HAS protein, a eukaryotic hyaluronan synthase. J Biol Chem 271(17):9875–9878
87. Coutinho PM, Henrissat B (1999) Carbohydrate-active enzymes: an integrated approach. In: Gilbert HJ, Davies GJ, Svensson B, Henrissat B (eds) Recent Advances in Carbohydrate Engineering. 3–12. Royal Society of Chemistry, Cambridge
88. Gillespie SH (1989) Aspects of pneumococcal infection including bacterial virulence, host response and vaccination. J Med Microbiol 28(4):237–248
89. Tarbouriech N, Charnock SJ, Davies GJ (2001) Three-dimensional structures of the Mn and Mg dTDP complexes of the family GT-2 glycosyltransferase SpsA: a comparison with related NDP-sugar glycosyltransferases. J Mol Biol 314(4):655–661
90. Focia PJ, Craig SP 3rd, Eakin AE (1998) Approaching the transition state in the crystal structure of a phosphoribosyltransferase. Biochemistry 37(49):17120–17127
91. Hagen FK, Hazes B, Raffo R, deSa D, Tabak LA (1999) Structure-function analysis of the UDP-N-acetyl-D-galactosamine:polypeptide N-acetylgalactosaminyltransferase. Essential residues lie in a predicted active site cleft resembling a lactose repressor fold. J Biol Chem 274(10):6797–6803
92. Ramakrishnan B, Balaji PV, Qasba PK (2002) Crystal structure of β1,4-galactosyltransferase complex with UDP-Gal reveals an oligosaccharide acceptor binding site. J Mol Biol 318(2):491–502
93. Gastinel LN, Cambillau C, Bourne Y (1999) Crystal structures of the bovine β4-galactosyltransferase catalytic domain and its complex with uridine diphosphogalactose. EMBO J 18(13):3546–3557
94. Boeggeman EE, Ramakrishnan B, Qasba PK (2003) N-terminal stem region of bovine and human β1,4-galactosyltransferase I increases the in vitro folding efficiency of their catalytic domain from inclusion bodies. Protein Expr Purif 30(2):219–229
95. Powell JT, Brew K (1976) Metal ion activation of galactosyltransferase. J Biol Chem 251(12):3645–3652
96. O'Keeffe ET, Hill RL, Bell JE (1980) Active site of bovine galactosyltransferase: kinetic and fluorescence studies. Biochemistry 19(22):4954–4962

97. Ramakrishnan B, Shah PS, Qasba PK (2001) α-Lactalbumin (LA) stimulates milk β-1,4-galactosyltransferase I (β4Gal-T1) to transfer glucose from UDP-glucose to N-acetylglucosamine. Crystal structure of β4Gal-T1 x LA complex with UDP-Glc. J Biol Chem 276(40):37665–37671
98. Brodbeck U, Denton WL, Tanahashi N, Ebner KE (1967) The isolation and identification of the B protein of lactose synthetase as α-lactalbumin. J Biol Chem 242(7):1391–1397
99. Qasba PK, Kumar S (1997) Molecular divergence of lysozymes and α-lactalbumin. Crit Rev Biochem Mol Biol 32(4):255–306
100. Ramakrishnan B, Qasba PK (2001) Crystal structure of lactose synthase reveals a large conformational change in its catalytic component, the β1,4-galactosyltransferase-I. J Mol Biol 310(1):205–218
101. Cinquin C, Le Blay G, Fliss I, Lacroix C (2004) Immobilization of infant fecal microbiota and utilization in an in vitro colonic fermentation model. Microb Ecol 48(1):128–138
102. Pedersen LC, Tsuchida K, Kitagawa H, Sugahara K, Darden TA, Negishi M (2000) Heparan/chondroitin sulfate biosynthesis. Structure and mechanism of human glucuronyltransferase I. J Biol Chem 275(44):34580–34585
103. Helting T, Rodén L (1969) Biosynthesis of Chondroitin Sulfate: Glucuronosyl Transfer In The Formation Of The Carbohydrate-Protein Linkage Region. J Biol Chem 244(10):2799–2805
104. Kakuda S, Shiba T, Ishiguro M, Tagawa H, Oka S, Kajihara Y, Kawasaki T, Wakatsuki S, Kato R (2004) Structural basis for acceptor substrate recognition of a human glucuronyltransferase, GlcAT-P, an enzyme critical in the biosynthesis of the carbohydrate epitope HNK-1. J Biol Chem 279(21):22693–22703
105. Tone Y, Kitagawa H, Imiya K, Oka S, Kawasaki T, Sugahara K (1999) Characterization of recombinant human glucuronyltransferase I involved in the biosynthesis of the glycosaminoglycan-protein linkage region of proteoglycans. FEBS Lett 459(3):415–420
106. Oka S, Terayama K, Kawashima C, Kawasaki T (1992) A novel glucuronyltransferase in nervous system presumably associated with the biosynthesis of HNK-1 carbohydrate epitope on glycoproteins. J Biol Chem 267(32):22711–22714
107. Chou DK, Ilyas AA, Evans JE, Costello C, Quarles RH, Jungalwala FB (1986) Structure of sulfated glucuronyl glycolipids in the nervous system reacting with HNK-1 antibody and some IgM paraproteins in neuropathy. J Biol Chem 261(25):11717–11725
108. Jeffries AR, Mungall AJ, Dawson E, Halls K, Langford CF, Murray RM, Dunham I, Powell JF (2003) β-1,3-Glucuronyltransferase-1 gene implicated as a candidate for a schizophrenia-like psychosis through molecular analysis of a balanced translocation. Mol Psychiatry 8(7):654–663
109. Tukey RH, Strassburg CP (2000) Human UDP-glucuronosyltransferases: metabolism, expression, and disease. Annu Rev Pharmacol Toxicol 40:581–616
110. Wysocki SJ, Segal W (1972) Influence of thyroid hormones on enzyme activities of myelinating rat central-nervous tissues. Eur J Biochem 28(2):183–189
111. Mulichak AM, Losey HC, Walsh CT, Garavito RM (2001) Structure of the UDP-glucosyltransferase GtfB that modifies the heptapeptide aglycone in the biosynthesis of vancomycin group antibiotics. Structure (Cambridge) 9(7):547–557
112. Mulichak AM, Lu W, Losey HC, Walsh CT, Garavito RM (2004) Crystal structure of vancosaminyltransferase GtfD from the vancomycin biosynthetic pathway: interactions with acceptor and nucleotide ligands. Biochemistry 43(18):5170–5180

113. Freund E, Vitali F, Linden A, Robinson JA (2000) Solid-phase synthesis using (Allyloxy)carbonyl(Alloc) chemistry of a putative heptapeptide intermediate in vancomycin biosynthesis containing m-chloro-3-hydroxytyrosine. Helv Chim Acta 83:2572–2579
114. Appel GB, Neu HC (1977) The nephrotoxicity of antimicrobial agents. N Engl J Med 296:722–728
115. Schwalbe RS, Stapleton JT, Gilligan PH (1987) Emergence of vancomycin resistance in coagulase-negative staphylococci. N Engl J Med 316(15):927–931
116. Martone WJ (1998) Spread of vancomycin-resistant enterococci: why did it happen in the United States? Infect Control Hosp Epidemiol 19(8):539–545
117. Ramadhan AA, Hegedus E (2005) Survivability of vancomycin resistant enterococci and fitness cost of vancomycin resistance acquisition. J Clin Pathol 58(7):744–746
118. Jones RN, Barrett MS, Erwin ME (1997) In vitro activity and spectrum of LY333328, a novel glycopeptide derivative. Antimicrob Agents Chemother 41(2):488–493
119. Rodriguez MJ, Snyder NJ, Zweifel MJ, Wilkie SC, Stack DR, Cooper RD, Nicas TI, Mullen DL, Butler TF, Thompson RC (1998) Novel glycopeptide antibiotics: N-alkylated derivatives active against vancomycin-resistant enterococci. J Antibiot (Tokyo) 51(6):560–569
120. Brown JP, Feeney J, Burgen AS (1975) A nuclear magnetic resonance study of the interaction between vanomycin and acetyl-D-alanyl-D-alanin in aqueous solution. Mol Pharmacol 11(2):119–125
121. Josse J, Kornberg A (1962) Glycosylation of deoxyribonucleic acid, III, alpha- and beta-Glycosyl transferases from T4-infected *Escherichia coli*. J Biol Chem 237:1968–1976
122. Morera S, Imberty A, Aschke-Sonnenborn U, Ruger W, Freemont PS (1999) T4 phage β-glucosyltransferase: substrate binding and proposed catalytic mechanism. J Mol Biol 292(3):717–730
123. Lariviere L, Morera S (2002) A base-flipping mechanism for the T4 phage β-glucosyltransferase and identification of a transition-state analog. J Mol Biol 324(3): 483–490
124. Lariviere L, Morera S (2004) Structural evidence of a passive base-flipping mechanism for beta-glucosyltransferase. J Biol Chem 279(33):34715–34720
125. Lariviere L, Gueguen-Chaignon V, Morera S (2003) Crystal structures of the T4 phage β-glucosyltransferase and the D100A mutant in complex with UDP-glucose: glucose binding and identification of the catalytic base for a direct displacement mechanism. J Mol Biol 330(5):1077–1086
126. Morera S, Lariviere L, Kurzeck J, Aschke-Sonnenborn U, Freemont PS, Janin J, Ruger W (2001) High resolution crystal structures of T4 phage β-glucosyltransferase: induced fit and effect of substrate and metal binding. J Mol Biol 311(3):569–577
127. Vrielink A, Ruger W, Driessen HP, Freemont PS (1994) Crystal structure of the DNA modifying enzyme β-glucosyltransferase in the presence and absence of the substrate uridine diphosphoglucose. EMBO J 13(15):3413–3422
128. Watkins WM, Morgan WT (1957) Specific inhibition studies relating to the Lewis blood-group system. Nature 180:1038–1040
129. Landsteiner K (1901) Ueber Agglutinationserscheinungen normalen menschlichen Blutes. Wien Klin Wochenschr 14:1132–1134
130. Lee HJ, Barry CH, Borisova SN, Seto NO, Zheng RB, Blancher A, Evans SV, Palcic MM (2005) Structural basis for the inactivity of human blood group O2 glycosyltransferase. J Biol Chem 280(1):525–529

131. Clausen H, White T, Takio K, Titani K, Stroud M, Holmes E, Karkov J, Thim L, Hakomori S (1990) Isolation to homogeneity and partial characterization of a histo-blood group A defined Fuc α1-2Gal α1-3-N-acetylgalactosaminyltransferase from human lung tissue. J Biol Chem 265(2):1139–1145
132. Yamamoto F, Clausen H, White T, Marken J, Hakomori S (1990) Molecular genetic basis of the histo-blood group ABO system. Nature 345:229–233
133. Nguyen HP, Seto NO, Cai Y, Leinala EK, Borisova SN, Palcic MM, Evans SV (2003) The influence of an intramolecular hydrogen bond in differential recognition of inhibitory acceptor analogs by human ABO(H) blood group A and B glycosyltransferases. J Biol Chem 278(49):49191–49195
134. Zak BM, Crawford BE, Esko JD (2002) Hereditary multiple exostoses and heparan sulfate polymerization. Biochim Biophys Acta 1573(3):346–355
135. Pedersen LC, Dong J, Taniguchi F, Kitagawa H, Krahn JM, Pedersen LG, Sugahara K, Negishi M (2003) Crystal structure of an α1,4-*N*-acetylhexosaminyltransferase (EXTL2), a member of the exostosin gene family involved in heparan sulfate biosynthesis. J Biol Chem 278(16):14420–14428
136. Kitagawa H, Shimakawa H, Sugahara K (1999) The tumor suppressor EXT-like gene EXTL2 encodes an α1,4-*N*-acetylhexosaminyltransferase that transfers *N*-acetylgalactosamine and *N*-acetylglucosamine to the common glycosaminoglycan-protein linkage region. J Biol Chem 274(20):13933–13937
137. Faria TQ, Lima JC, Bastos M, Macanita AL, Santos H (2004) Protein stabilization by osmolytes from hyperthermophiles: effect of mannosylglycerate on the thermal unfolding of recombinant nuclease a from Staphylococcus aureus studied by picosecond time-resolved fluorescence and calorimetry. J Biol Chem 279(47):48680–48691
138. Davis-Searles PR, Saunders AJ, Erie DA, Winzor DJ, Pielak GJ (2001) Interpreting the effects of small uncharged solutes on protein-folding equilibria. Annu Rev Biophys Biomol Struct 30:271–306
139. Martins LO, Empadinhas N, Marugg JD, Miguel C, Ferreira C, da Costa MS, Santos H (1999) Biosynthesis of mannosylglycerate in the thermophilic bacterium *Rhodothermus marinus*. Biochemical and genetic characterization of a mannosylglycerate synthase. J Biol Chem 274(50):35407–35414
140. Flint J, Taylor E, Yang M, Bolam DN, Tailford LE, Martinez-Fleites C, Dodson EJ, Davis BG, Gilbert HJ, Davies GJ (2005) Structural dissection and high-throughput screening of mannosylglycerate synthase. Nat Struct Mol Biol 12(7):608–614
141. Frydman RB, Cardini CE (1964) Biosynthesis of phytoglycogen from adenosine diphosphate D-glucose in sweet corn. Biochem Biophys Res Commun 14:353–357
142. Shukla RN, Sanwal GG (1971) Studies on UDP-glucose: D-fructose 2-glucosyltransferase from tapioca tuber. Arch Biochem Biophys 142(1):303–309
143. Goldemberg SH (1962) Specificity of uridine diphosphate glucose-glycogen glucosyltransferase. Biochim Biophys Acta 56:357–359
144. Cao Y, Steinrauf LK, Roach PJ (1995) Mechanism of glycogenin self-glucosylation. Arch Biochem Biophys 319(1):293–298
145. Barker SA, Bourne E, Peat S (1949) The enzymic synthesis and degradation of starch. Part IV The purification and storage of the Q-enzyme of the potato. J Chem Soc 1:1705–1711
146. Henrissat B (1991) A classification of glycosyl hydrolases based on amino-acid sequence similarities. Biochem J 280:309–316
147. Cori GT (1952) Glycogen structure and enzyme deficiencies in glycogen storage disease. Harvey Lect 48:145–171

148. Buschiazzo A, Ugalde JE, Guerin ME, Shepard W, Ugalde RA, Alzari PM (2004) Crystal structure of glycogen synthase: homologous enzymes catalyze glycogen synthesis and degradation. EMBO J 23(16):3196–3205
149. Bachrach BE, Weinstein DA, Orho-Melander M, Burgess A, Wolfsdorf JI (2002) Glycogen synthase deficiency (glycogen storage disease type 0) presenting with hyperglycemia and glucosuria: report of three new mutations. J Pediatr 140(6):781–783
150. Iwahashi H, Shimizu H, Odani M, Komatsu Y (2003) Piezophysiology of genome wide gene expression levels in the yeast Saccharomyces cerevisiae. Extremophiles 7(4):291–298
151. Porchia AC, Curatti L, Salerno GL (1999) Sucrose metabolism in cyanobacteria: sucrose synthase from Anabaena sp. strain PCC 7119 is remarkably different from the plant enzymes with respect to substrate affinity and amino-terminal sequence. Planta 210(1):34–40
152. Singer MA, Lindquist S (1998) Thermotolerance in Saccharomyces cerevisiae: the yin and yang of trehalose. Trends Biotechnol 16:460–468
153. Sahara T, Goda T, Ohgiya S (2002) Comprehensive expression analysis of time-dependent genetic responses in yeast cells to low temperature. J Biol Chem 277(51): 50015–50021
154. Chen Q, Ma E, Behar KL, Xu T, Haddad GG (2002) Role of trehalose phosphate synthase in anoxia tolerance and development in Drosophila melanogaster. J Biol Chem 277(5):3274–3279
155. Benaroudj N, Lee DH, Goldberg AL (2001) Trehalose accumulation during cellular stress protects cells and cellular proteins from damage by oxygen radicals. J Biol Chem 276:24261–24267
156. Pereira Ede J, Panek AD, Eleutherio EC (2003) Protection against oxidation during dehydration of yeast. Cell Stress Chaperones 8(2):120–124
157. Birch GG (1963) Trehalose. Adv Carb Chem 18:201–225
158. Elbein AD (1974) The metabolism of α,α-trehalose. Adv Carbohydr Chem Biochem 30:227–256
159. Kaushik JK, Bhat R (2003) Why is trehalose an exceptional protein stabilizer? An analysis of the thermal stability of proteins in the presence of the compatible osmolyte trehalose. J Biol Chem 278(29):26458–26465
160. De Smet KA, Weston A, Brown IN, Young DB, Robertson BD (2000) Three pathways for trehalose biosynthesis in mycobacteria. Microbiology 146(Pt 1):199–208
161. Gibson RP, Turkenburg JP, Charnock SJ, Lloyd R, Davies GJ (2002) Insights into trehalose synthesis provided by the structure of the retaining glucosyltransferase OtsA. Chem Biol 9(12):1337–1346
162. Gibson RP, Tarling CA, Roberts S, Withers SG, Davies GJ (2004) The donor subsite of trehalose-6-phosphate synthase: binary complexes with UDP-glucose and UDP-2-deoxy-2-fluoro-glucose at 2 Å resolution. J Biol Chem 279(3):1950–1955
163. Moore JE, Corcoran D, Dooley JS, Fanning S, Lucey B, Matsuda M, McDowell DA, Megraud F, Millar BC, O'Mahony R, O'Riordan L, O'Rourke M, Rao JR, Rooney PJ, Sails A, Whyte P (2005) Campylobacter Vet Res 36(3):351–382
164. Evans SV (1993) SETOR: hardware-lighted three-dimensional solid model representations of macromolecules. J Mol Graph 11(2):134–138

Top Curr Chem (2007) 272: 259–296
DOI 10.1007/128_2006_087

Published online: 19 October 2006

Exploiting Ligand and Receptor Adaptability in Rational Drug Design Using Dynamics and Structure-Based Strategies

Steven R. LaPlante

Department of Chemistry, Boehringer Ingelheim (Canada) Ltd.,
Research and Development, 2100 Cunard St., Laval, Quebec, H7S 2G5, Canada
slaplante@lav.boehringer-ingelheim.com

Abstract Interactions between a biological macromolecule and a small molecule ligand involve intricate recognition events which are strongly influenced by both structure and dynamics factors. The ligand and receptor must mutually fit and adapt to each other to form a strong complex, and detailed knowledge of these factors would certainly aid drug design efforts. This work describes our experience in the characterization and exploitation of these properties within the context of two medicinal chemistry programs that targeted the essential protease enzymes of the human cytomegalovirus (HCMV) and the hepatitis C virus (HCV). Central to the rational use of such data was the qualitative elucidation of the binding parameters that defined modes/roles of each substituent of the ligands. This required the development of dynamics- and structure-based strategies that logically considered data from structure–activity relationships, NMR spectroscopy, computational chemistry, and X-ray crystallography. An important emphasis was made to monitor the relationship between inhibitor activity and ligand flexibility using ^{13}C NMR T_1 relaxation data, within the context of promoting the bioactive conformation as a drug design tool. Besides these methods which focused on data from the ligand perspective, interesting ligand-induced adaptive features of the receptors were also observed, and their impacts on drug design were explored.

Keywords ^{13}C T_1 relaxation · Bioactive conformation · Drug design · Dynamics · HCMV · HCV

Abbreviations

NMR	nuclear magnetic resonance
^{13}C T_1	^{13}C NMR spin lattice relaxation
IC_{50}	inhibition constant at 50% concentration
HCV	hepatitis C virus
HCMV	human cytomegalovirus
C-terminal	C-terminus of a peptide chain
N-terminal	N-terminus of a peptide chain
NOESY	NMR nuclear Overhauser effect spectroscopy
ROESY	NMR rotating-frame nuclear Overhauser effect spectroscopy
AIDS	auto-immune deficiency syndrome
HSV	herpes simplex virus
DLB	NMR differential line broadening
SAR	structure-activity relationship
NS	non-structural
HSQC	NMR heteronuclear single-quantum correlation

1 Introduction and Overview

The binding of a ligand to a macromolecule involves intricate recognition events that are strongly influenced by forces which include shape complementarity, van der Waal contacts, electrostatic interactions, and solvation effects. Most structure-based, drug design efforts focus on these attributes. Also critical to these recognition events, and often ignored, are structural and flexibility adaptations of the ligand and receptor to attain the bioactive complex. Therefore, a more encompassing view of biomolecular recognition must include an extension of the classical "lock-and-key" model to one that incorporates the effects of dynamics and conformational changes.

Although rational drug design efforts would certainly be accelerated from knowledge of these adaptive processes, only limited applications have been reported. This is partly due to the fact that few experimental methods can provide this type of atomic-level information by practical and timely means, and as a consequence, its importance and potential impact in drug design have not yet been fully appreciated.

Here, two projects are reviewed that illustrate our findings in antiviral drug design in which the dynamics and structural principles that dictated ligand binding were explored and exploited. In addition, novel strategies and tools were developed to monitor the adaptive, biomolecular recognition processes, and it is expected that they will have general utility in other medicinal chemistry efforts.

The first account describes the identification and characterization of antiviral inhibitors that targeted the serine protease domain of the human cytomegalovirus (HCMV protease). Although it is written from a chronological point-of-view, it focuses primarily on the multidisciplinary strategies applied to monitor the impact of bioactive conformations on inhibitor activity. The account first provides an introduction to HCMV protease as an antiviral target. It then explains our strategy of identifying the first lead inhibitor whereby a C-terminal warhead was incorporated onto N-terminal product peptides. The warhead induced a reversible, covalent mode of binding that mimicked the transition-state of substrate cleavage.

Having identified the core scaffold of the ligand which was involved in direct binding to the protease, potent inhibitors with less peptidic character were discovered. NMR methods were developed to monitor the bioactive conformation of these inhibitors, and the dependence of potency on the free-state flexibility was discovered. The more potent compounds exhibited similar free and bound conformations, which was likely due to optimization of the bioactive conformation and, as a result, to a reduction in the entropic cost of binding.

Another important finding in this first account was that the active-site underwent important conformational adaptations upon binding the peptidomimetic ligands and substrates, which ultimately served to characterize HCMV protease as an induced-fit enzyme. Owing to this, the resultant entropic cost required to induce the "activated" state adversely limited attempts to further improve inhibitor potencies. Based on this, medicinal chemistry efforts shifted to an alternate series of compounds that were found to bind to the "unactivated" form of the protease, which presumably avoided paying expensive entropic costs. Throughout this campaign, novel strategies were developed to monitor the bioactive conformation, and changes in flexibility of the ligands and receptor. Given the general nature of these strategies, they were then applied with success to a highly related campaign that targeted the HCV protease.

The second account evaluates the bioactive conformation of peptide-based inhibitors of HCV protease, and it reviews the application of the ^{13}C T_1 method as a tool for monitoring changes to this bioactive conformation for drug design purposes. It begins with an introduction to the rationale for considering HCV protease as an antiviral target. Then a brief overview is provided on the design of the potent BILN 2061 family of inhibitors starting from an N-terminal product peptide. Rather than provide a chronological review, the subsequent sections focus on systematically evaluating the importance of the bioactive conformation with regard to potency. First it is shown that a potent precursor of BILN 2061 adopts similar free and bound conformations using transferred NOESY and ROESY NMR data. Then spin-lattice relaxation NMR experiments (^{13}C T_1) are introduced as a tool for monitoring atomic-level details of the flexibility of the inhibitors. This is followed by examples of ^{13}C T_1 collected on inhibitors that were pivotal along the pathway in designing BILN 2061. The data presented clearly show a striking correlation between ^{13}C T_1 times and inhibitor potency, which demonstrates the importance of optimizing the free state conformation via rigidification strategies, when and where appropriate.

2 HCMV Protease

2.1 HCMV Protease as an Antiviral Target

The human cytomegalovirus (HCMV) is a highly prevalent pathogen that poses a serious risk to immunocompromised individuals, notably AIDS patients, organ transplant recipients and neonates who acquire the infection congenitally [1, 2]. Typical of members of the herpesvirus family, HCMV encodes a unique protease involved in capsid assembly whose activity is essential to the production of infectious virions [3–6]. The enzyme is responsible, late in the viral cycle, for the processing of the assembly protein whose function is analogous to that of the "scaffolding" protein of bacteriophages [7]. In the case of HCMV's close homologue HSV-1, failure to process the assembly protein results in the accumulation of aberrant, noninfectious capsids [5].

The full-length HCMV protease precursor (Fig. 1) contains 708 amino acids encoded by the UL80 gene, which is co-terminal with the UL80.5 open reading frame [3]. The assembly protein precursor is encoded by UL80.5 and is in frame with the carboxyl-terminal amino acids of the protease precursor. It was found that the enzyme can process its own C-terminus at a site identi-

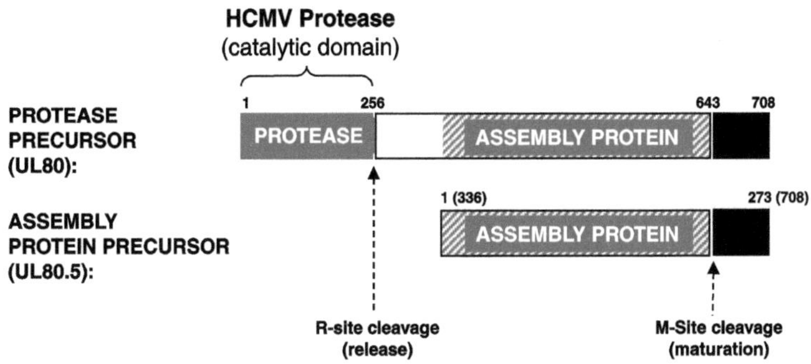

Fig. 1 Diagram depicts the UL80 and UL80.5 open reading frame products. The HMCV protease cleavage sites (R-site and M-site) are indicated on the HCMV protease and assembly protein precursors

cal to that of its assembly protein substrate, the maturation site (M-site), see Fig. 1. The protease also undergoes self-processing at the release site (R-site) near its amino terminus. This cleavage liberates the 256 amino acid catalytic domain, which will be referred to here as HCMV protease. Although this enzyme has been characterized as a member of the serine protease family, X-ray crystallographic analyses [8–11] have shown that it possesses a unique protein fold and an unusual catalytic triad in which the third member is a histidine rather than an aspartate. Also setting it apart from classical serine proteases is its existence as a dimer, which is believed to be the sole active species [12, 13]. Finally, our spectroscopic studies [14] demonstrated that the binding of substrate-based competitive inhibitors results in a conformational change in the enzyme and that catalysis by HCMV protease can be best described in terms of an "induced-fit" model [15, 16].

2.2 Mapping the Critical Binding Parts of the Substrate Peptides and Determining the First Structure of a Ligand Bound to HCMV Protease

Given that substrate hydrolysis by human cytomegalovirus (HCMV) protease was essential for viral capsid assembly, we embarked on a knowledge-building exercise to acquire a clear understanding of the structural elements of the substrates that were required for recognition and hydrolysis. Our enzymology studies demonstrated that peptides which corresponded to 17 amino acids of the R- and M-sites (and referred to as **R-site** and **M-site** peptides) were sufficient to induce hydrolysis by HCMV protease (Fig. 2) [16]. Enzymology studies also found that amino acid substitutions of the P′ residues (using standard nomenclature [17]) had a much less pronounced effect on oligopeptide substrate hydrolysis rates than those of P-side residues (data not shown) [18]. It was also discovered that a peptide spanning the P4–

	...P9 P8 P7 P6 P5 P4 P3 P2 P1 P1′ P2′ P3′ P4′...P8′	k_{cat}/K_M ($M^{-1}s^{-1}$)
R-site	Thr-Glu-Arg-Glu-Ser-Tyr-Val-Lys-Ala - Ser-Val-Ser-Pro...	42
R-mutant	...Asn...	282
M-site	Arg-Ala-Gln-Ala-Gly-Val-Val-Asn-Ala - Ser-Cys-Arg-Leu...	657
M-mutant	...Lys...	54
		IC_{50} (μM)
R-product	Thr-Glu-Arg-Glu-Ser-Tyr-Val-Lys-Ala	~3000
M-product	Arg-Ala-Gln-Ala-Gly-Val-Val-Asn-Ala	>1000
1	Gly-Val-Val-Asn-Ala-CF_3	1.8

Fig. 2 Amino acid sequences of peptides/inhibitors are provided along with enzymological (k_{cat}/K_M) and inhibitory activity (IC_{50}) data [16]. The nomenclature used to denote amino acid positions are given above (i.e. P9 to P4′) [17]. The peptide labeled as **R-site**, **R-mutant**, **M-site**, and **M-mutant** are 17 amino acids in length, but only the sequence to P4′ is provided for the sake of limited space. The full sequence can be found elsewhere [16]

P1′ M-site core alone was capable of competitively inhibiting catalysis with binding affinity only 5-fold less than that of the P4–P4′ substrate [19]. This supported the presumption that the structural elements of the substrate which were N-terminal to the scissile bond were crucial for complexation to the enzyme.

In addition to enzymology studies, differential line-broadening (DLB) NMR was employed to obtain a concise picture of which residues played direct binding roles in substrate and product recognition by the enzyme [16]. An important advantage of the DLB method [20] was that it provided atomic-resolution data that was easily acquired and interpreted. An example is shown in Fig. 3A using the N-terminal product peptide (**R-product**) of the R-site. Although it was not a potent inhibitor ($IC_{50} \sim 3000\,\mu M$), it nonetheless bound to the protease as revealed by the selective resonance perturbations observed when the hydrogen resonances of the peptide where compared in the absence versus presence of HCMV protease (Fig. 3A). The changes in the peak shape and intensity were a result of fast-exchange averaging between the free and bound states (Fig. 3B), which differ in chemical shift and line-shape. The broadened resonance of the methyl group of Ala 1 (compare "a" with "b" in Fig. 3A) indicated that this group became pocket exposed upon binding to HCMV (see the illustration in Fig. 3B). In contrast, the ^1H resonance of the methyl group of Thr 9 changed little, as expected for a group that was predominantly solvent exposed in the free and bound states. Using this method, the ensemble of DLB patterns were monitored for the **R-product** and **M-product** peptides, and it was discovered that the P4 to P1 residues

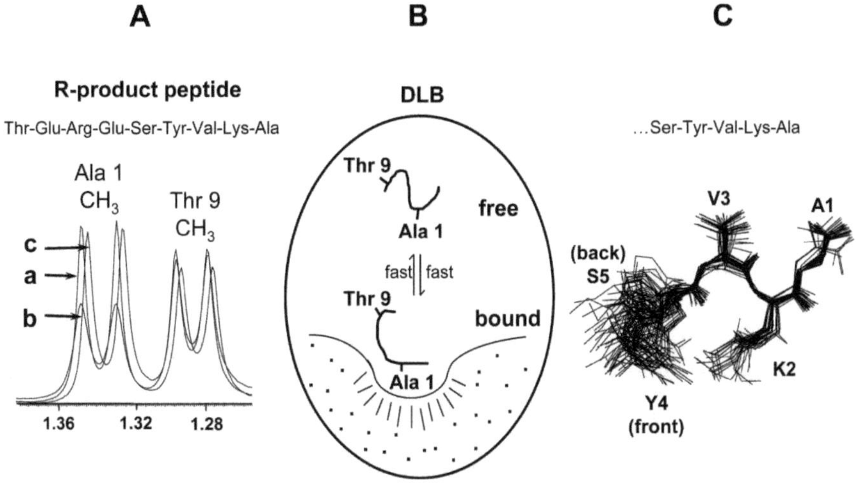

Fig. 3 DLB mapping that distinguished between residues that were solvent exposed versus those that directly contact the receptor. Shown is the methyl region of the ^{1}H NMR spectrum of **R-product** showing the P1 and P9 CH_3 side-chain doublets [16]. **A** "a" is of free **R-product**; "b" is of **R-product** after the addition of HCMV protease at a ratio of 7 : 1; and "c" is of **R-product** after the addition of HCMV protease ratio and inhibitor **6** at a 7 : 1 : 2 ratio. For the sake of clarity, the P1 methyl signals have been skewed slightly. **B** A diagram demonstrates that the DLB method requires fast-exchange binding between the free and bound states (on the NMR time-scale). **C** The 3D structure of the P5–P1 sequence of **R-product** when bound to HCMV protease. The 32 overlapped structures determined were derived using transferred NOESY restraints and simulated annealing calculations

directly contacted the protease whereas the P9 to P5 residues were solvent exposed [16].

It was later established that complexation was indeed at the active site of HCMV protease using DLB competition experiments. Upon the addition of a potent inhibitor of the enzyme, compound **6** (vide infra, Fig. 5), the P1 methyl signal was restored to its original uncomplexed form ("c" in Fig. 3A) owing to the ability of compound **6** to plug the active-site by forming a stable hemiketal, covalent adduct with the catalytic serine (vide infra) [16].

Moreover, the 3D structures of the **R-product** and **M-product** peptides when bound to HCMV protease were also determined [16] using another powerful NMR method, the "transferred NOESY" experiment [20]. The crosspeaks observed from this experiment reflected a finger-print of short, intra-molecular hydrogen distances of the peptides, and distance-restrained simulated-annealing calculations were performed to determine the bound structures. As an example, Fig. 3C shows that a single-family of structures of R-product bound to the protease in well-defined, extended conformations from P4 through P1, with no defined structure for the P9 to P5 segments. In summary, although these N-terminal product peptides were very weak inhibitors (Fig. 2), they were nevertheless found to be bound in similar, well-defined conformations.

Also of interest was the identification of the source of the different processing rates observed between the **R-site** and **M-site** peptides (Fig. 2), where the **M-site** was hydrolyzed significantly faster than the **R-site** substrate (657 versus $42k_{cat}/K_M$) [16]. To address this question, enzymology studies which utilized mutated **R-site** and **M-site** peptide substrates (i.e. P5 and P4 residues were separately exchanged) revealed that these positions had essentially no influence on the specificity constants (k_{cat}/K_M). In sharp contrast, substitution of the P2 residue of the **R-site** peptide to that of the **M-site** (Lys to Asn, and called **R-mutant** in Fig. 2) changed its specificity constant to resemble that of the **M-site** peptide substrate, and vice versa for the M-mutant. Thus, the P2 side-chain played a principal role in the observed variation in cleavage rates, suggesting that this side-chain somehow influenced the catalytic triad reactivity. This observation had practical drug design utility in our HCMV and HCV campaigns [36] (vide infra).

2.3
Inhibitors Designed by Attachment of C-Terminal Warheads on Peptides

Serine proteases are perhaps the most well-studied class of enzymes [21–23]. Despite significant differences in global protein architecture, they possess strikingly similar catalytic machinery (triad), a fact which is often cited as the paradigm of convergent structural evolution at the enzyme level. Owing to the functional and structural homology between serine proteases, a number of classic inhibitor-design approaches have been applied to various members of the enzyme family.

Our efforts focused on one such approach, namely substrate-based activated carbonyl inhibitors [24]. As generally applied, the strategy involved the synthesis of N-terminal cleavage products in which the acid of the C-terminus was replaced with an electrophilic ketone [25]. Upon attack by the active-site serine, covalent hemiketal adducts were formed that mimic the transition state of the tetrahedral intermediate formed during the catalytic reaction. This was confirmed by ^{13}C NMR studies (vide infra). The expected boost in potency was observed when this approach was applied, in that compound 1 was found to inhibit HCMV protease with an IC_{50} of 1.8 μM (Fig. 2) [14, 25], as compared to an IC_{50} of > 1000 μM found for the corresponding **M-product** peptide (Fig. 2) that possessed a C-terminal carboxylate. Having attained a reasonable potency, structure-activity relationships (SAR) then became possible. Our early findings were insightful and consistent with the conclusions from both NMR and enzymology studies mentioned above. For example, N-terminal truncation of P5 resulted in an inhibitor with similar potency (e.g. compare the activity of compound **1** versus **2** in Fig. 4), and sharp losses in potency were observed upon further truncation of the P4

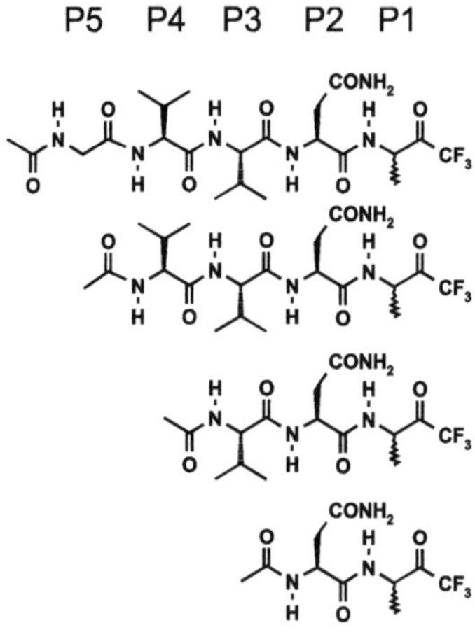

	IC$_{50}$ (μM)
1	1.8
2	3.0
3	80
4	>300

Fig. 4 Inhibitors **1–4** are shown along with inhibition constants. The amino-acid positions are designated on top as P5–P1

and P3 residues (e.g. compare the activities of compound **2** with **3** and **4** in Fig. 4) [25]. Thus, the P4 to P1 peptidyl segment played a critical role in ligand binding to the active-site of HCMV protease.

Subsequent medicinal chemist's efforts found limited success at optimizing compound **1** for drug design purposes. The best peptidyl portion discovered was *N*-*tert*-butylacetyl-L-*tert*-butylglycyl-L-N^{δ},N^{δ}-dimethylasparagyl-L-alanine, and the most effective activated carbonyl groups were those shown on compounds **5–7** in Fig. 5 [15, 25]. Despite the identification of compound **6** as the most potent inhibitor (IC$_{50}$ of 0.1 μM), a large variety of compounds were synthesized and tested without significant improvements in activity.

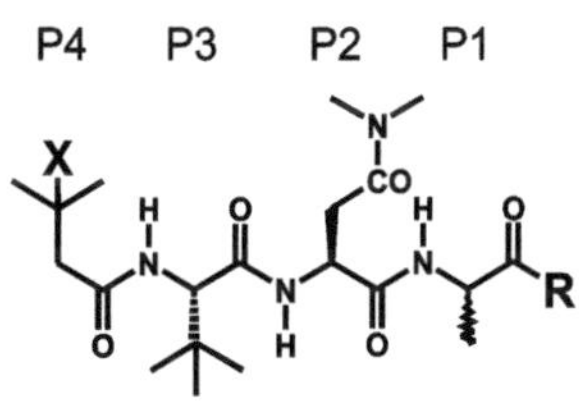

5	R = CF_3	X = CH_3	IC$_{50}$ = 1.1 μM
6	R = CF_2CF_3	X = CH_3	IC$_{50}$ = 0.1 μM
7	R = $CONHCH_2Ph$	X = CH_3	IC$_{50}$ = 0.2 μM
8	R = CH_3	X = CH_3	IC$_{50}$ = 2100 μM
9	R = CH_3	X = H	IC$_{50}$ = 5000 μM

Fig. 5 Inhibitors **5–9** are shown along with inhibition constants

Given this, there appeared to be an unidentified, physical limit to the potential in discovering more potent inhibitors of HCMV protease. It was thought that this may be related somehow to the binding mode, which warranted further investigation.

2.4
Elucidation of the Binding Mode of the Optimized Peptidyl Segment

Concurrent to the medicinal chemistry work which chemically probed the substituents that dictated binding and activity, structural research efforts were also focused on determining the binding mode of compounds such as **5**–**7** when bound to HCMV protease. While the transferred NOESY technique was clearly an expedient route to obtaining the bioactive 3D structure when

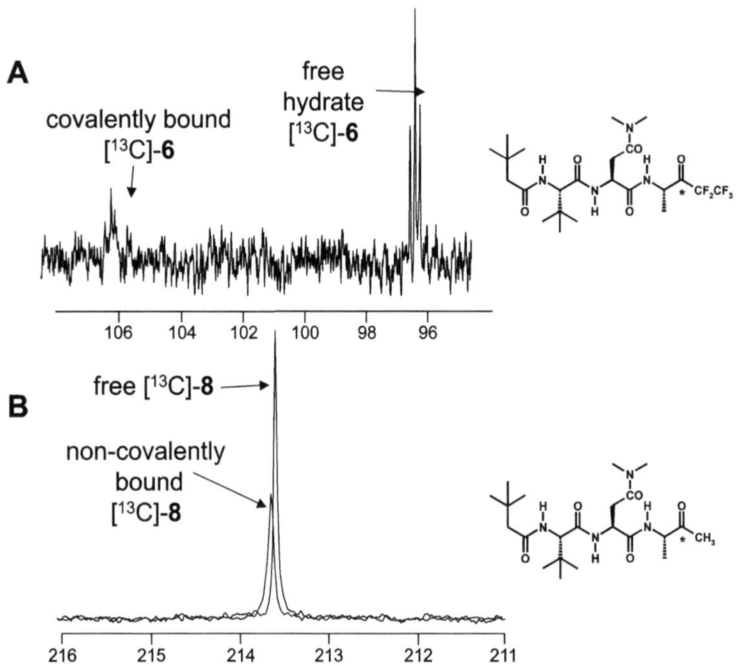

Fig. 6 Sections of the ^{13}C NMR spectra of **A** 1.6 mM [^{13}C]-**6** in the presence of 0.72 mM HCMV protease. Compound [^{13}C]-**6** was synthetically labeled with ^{13}C at the P1 carbonyl carbon. A * on the structure indicates the labeled position. The signals at 105.7 and 95.8 ppm correspond to the bound (ionized hemiketal) and free inhibitor, respectively. **B** 3.6 mM methyl ketone of free [^{13}C]-**8**. Compound [^{13}C]-**8** was synthetically labeled with ^{13}C at the P1 carbonyl carbon. A * on the structure indicates the labeled position. The spectrum of the free inhibitor is shown and denoted. The spectrum of excess inhibitor is a fast-exchange mix between the free and bound states, but the shift to the left clearly shows that a bound shift around 106 ppm is not possible

bound to a protein, there was an important constraint in applying the technique on these compounds. It could only be utilized if there was sufficiently fast and reversible exchange of an inhibitor between the free and bound states to allow for cross relaxation. Although activated carbonyl compounds such as **5–7** are reversible inhibitors, the exchange rates between their free and enzyme-bound states were very slow due to their mechanism of inactivating the enzyme (via covalent attachment), which conferred slow k_{off} rates (typically $7 \times 10^{-5}\ s^{-1}$ for this class of inhibitors). As a consequence, the transferred NOESY method could not be employed to determine the bound conformation [26].

To circumvent this kinetic constraint, we designed ligands that as closely as possible resembled the peptidyl portion of our inhibitors while not exhibiting its slow exchange properties. Thus, peptidyl methyl ketones **8** and **9** were designed [26] to fulfill these criteria, and this was confirmed by ^{13}C NMR experiments [14]. These experiments showed that fluoro ketone **5** was hydrated when free in aqueous solution (based on the ^{13}C chemical shift shown in Fig. 6A) and formed a covalent, hemiketal adduct in the bound state (based on the ^{13}C shift in Fig. 6A). On the other hand, methyl ketone **8** was not hydrated in the free state and did not form a covalent complex when bound to HCMV (based on the ^{13}C shifts in Fig. 6B).

As a result, replacing the C-terminal CF_3 with a CH_3 rendered the compounds unreactive toward attack by the active-site serine, and it eliminated

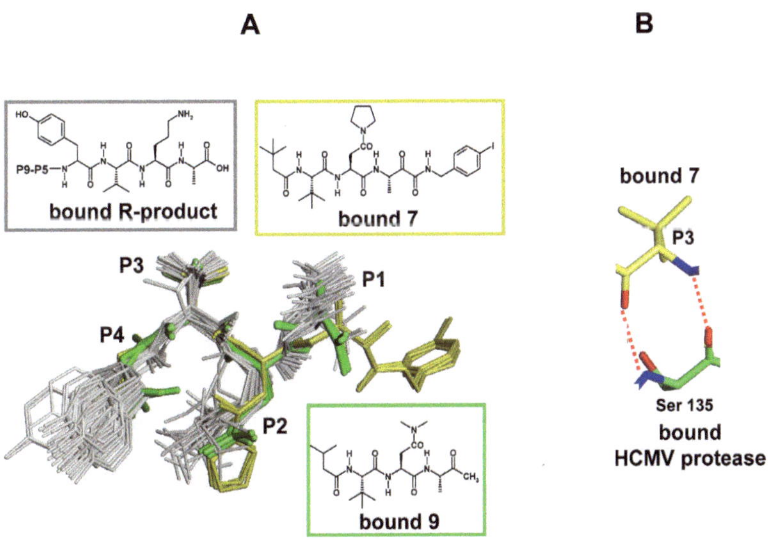

Fig. 7 **A** Overlap of NMR-derived bound conformations of compound **9** (*colored green*; 29 conformations), **R-product** peptide (*gray*; 32 structures), and the X-ray crystallographic conformations of bound compound **7** (*yellow*; 4 conformations). **B** Zoomed view of the P3–S3 interaction of compound **7** bound to HCMV protease

the covalent binding feature and provided a powerful structural probe of the bioactive conformation. Transferred NOESY data were then acquired on compound **9** (vide infra, Fig. 9A) and the derived distance restraints were applied to determine the family of bound structures shown in green in Fig. 7A [26]. Of particular interest was that these structures also resembled the bound structures of the **R-product** peptide shown in gray [16]. Furthermore, they also bore a striking resemblance to the much more potent and covalently-bound compound **7** (yellow, Fig. 7A) which was determined by X-ray crystallography [27]. They all bound in the extended conformation with a zigzagged backbone, and the P1 and P3 side-chains lie close to one another, as do the P2 and P4 side-chains. The commonality of this structural feature for all three compounds suggested that this bioactive conformation played an important role for binding and activity. The dramatic losses in potency observed upon N-terminal truncation of P4 and P3 was also consistent with this notion.

2.5
Ligand Adaptations Upon Binding

The importance of designing compounds that preferentially adopted this bioactive conformation in the free state was best exemplified by SAR and mechanistic studies performed at the P3 position. Dramatic improvements in potency were attained (Fig. 8) when a glycine at P3 (compound **10**, IC_{50} >

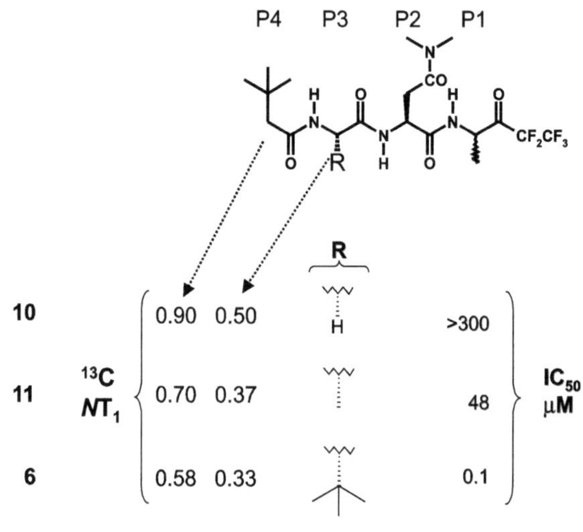

Fig. 8 Inhibitors **6**, **10**, and **11** are shown along with inhibition constants. ^{13}C T1 relaxation data are given for the P3 Ca carbon. For the methylene carbon which has two covalently attached hydrogens, *N*T1 values are provided where *N* is the number of attached to help in interpreting the relative flexibility between different carbon types

300 μM) was replaced with an alanine (compound **11**, IC_{50} 48 μM) or a tert-butyl group (compound **6**, IC_{50} 0.1 μM) [25]. These increases in potency could not be explained by direct contacts with the protease pocket, given that there was little S3 pocket to speak of as judged by an X-ray structure of a complex involving **7** which was solved at a later date [27]. Rather, the improvements in potency were a result of the incorporation of the bulkier side-chains which helped to rigidify the compounds to resemble the bioactive conformation.

The relative rigidity (or flexibility) of these compounds was monitored by an NMR technique called ^{13}C spin-lattice relaxation measurements (^{13}C T_1) which are sensitive to motions on the pico- to nanosecond time scale [15]. The interpretation of T_1 data is straightforward, shorter times are indicative of relatively slower segmental motion or flexibility of C – H vectors. Thus, the relatively shorter T_1 times observed for the $C\alpha$ of the *tert*-butyl analogue **6** (0.35 seconds) versus that of alanine (0.37 seconds, compound **11**) and that of glycine (0.50 seconds, compound **10**) was consistent with increased conformational restriction. Having found this, it became apparent that the conformational restriction induced by the bulkier P3 side-chain resulted in a minimization of the overall entropic cost of binding. The role played by the bulk was to force the critical P3 backbone into the bioactive, extended conformation in the free state, which aided in the facile formation of two hydrogen bonds that were required upon complexation with HCMV protease (Fig. 7B) [15, 27]. The T_1 data also shed light onto the fact that the bulkier P3 side-chain had important, long-range dynamic effects. The N-terminal

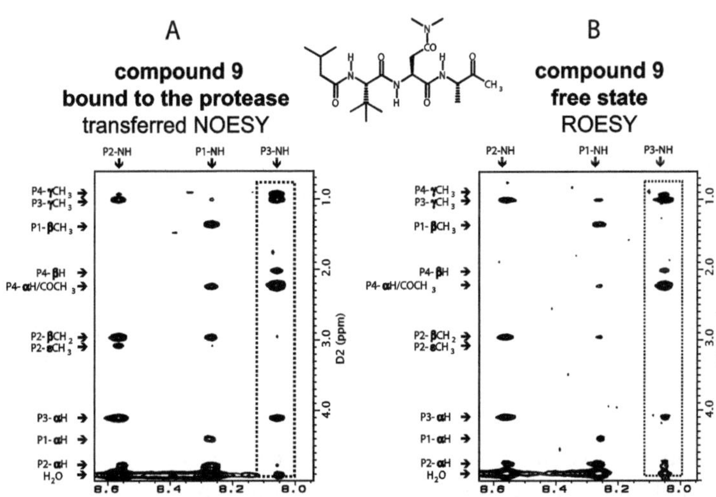

Fig. 9 Comparison of 2D NMR spectra of bound versus free compound **9**. **A** NH-aliphatic region of the transferred NOESY spectrum of compound **9** in the presence of HCMV protease at a ratio of 7 : 1. **B** ROESY spectrum of free compound **9**

methylene of **6** exhibited significantly lower T_1 times as compared to **10** (0.58 versus 0.90 seconds). The next section involving our HCV protease campaign will show excellent examples of how T_1 data was used to monitor and exploit similar long-range, dynamics effects.

These findings were corroborated by other complementary NMR methods, which determined that the optimized peptidyl segment adopted similar conformations in the free and bound states. For example, it was clearly evident from the data in Fig. 9 that the peptidyl part of compound **9** displayed the same crosspeak patterns in the ROESY and transferred NOESY spectra [15]. It also demonstrated that the optimized peptidyl segment underwent little conformational change upon binding HCMV protease, and as a result, likely paid little in entropic costs to bind HCMV protease.

2.6 HCMV Protease Active-Site Adaptations Upon Ligand Binding

Concurrent to our work aimed at designing more rigidified inhibitors, evidence was also mounting that suggested that HCMV protease underwent changes upon the binding of peptidyl ligands. Some unusually sharp resonances in the ^{1}H NMR spectrum of apo HCMV protease were broadened upon complexation. Furthermore, fluorescence emission studies demonstrated that HCMV protease underwent a conformational change upon complexation of inhibitors as demonstrated by the observed "blue shifts" shown in Fig. 10 [14]. These shifts were observed regardless of whether covalent complexes were formed (as observed for compounds **5**–**7**, Fig. 10A) or noncovalent complexes were formed (as observed for compound **8**, Fig. 10B). Thus, the induced structural alteration was elicited by the formation of contacts between the peptidyl portion of the inhibitors and the binding regions in the protease active site.

Interestingly, these shifts were also observed during complexation and catalysis of a peptidyl substrate, which helped in characterizing HCMV protease as an induced-fit enzyme [15]. A more detailed view of this conformational transition was later provided by a comparison of the X-ray structure of the apo form of HCMV protease (in gray, Fig. 11B) [8] with that of a complex with compound **7** (in green, Fig. 11A) [27]. The following differences were notable. There appeared to be a movement and/or ordering of the β5-strand, the L9-loop, the L2-loop, Arg-165, and Arg-166. Interestingly, the inhibitor-induced changes in the L9-loop resulted in the burying and immobilization of Trp-42, which in the free enzyme was, more or less, solvent exposed and clearly mobile. This change in microenvironment was fully consistent with the reported shifts in fluorescence emission upon peptidyl ligand binding (e.g. Fig. 1) [14, 15].

It was conferred from these X-ray structures (Fig. 11), that HCMV protease underwent substantial reorganization and, notably, rigidification upon

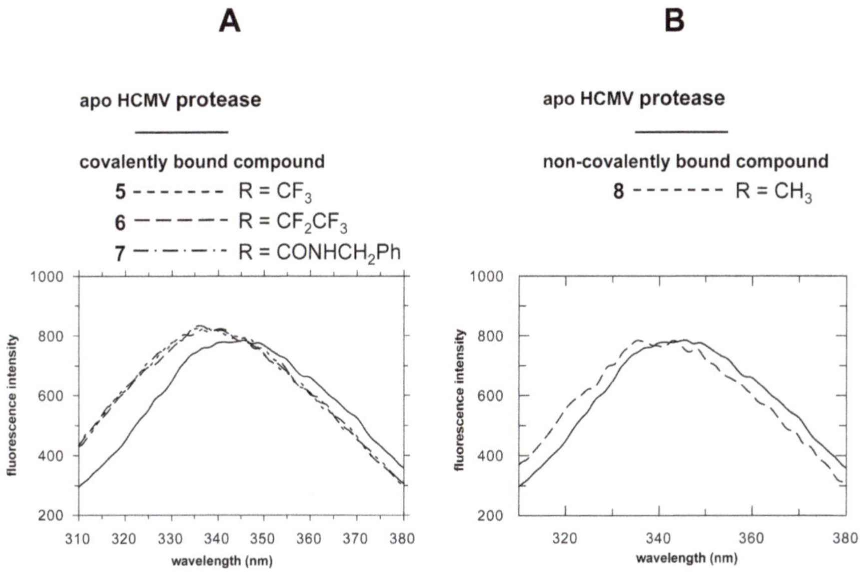

Fig. 10 **A** Changes in the fluorescence emission spectrum of HCMV protease in the presence of saturating amounts of peptidyl-activated carbonyl inhibitors. The C-termini of compounds are abbreviated, and R represents the peptidyl portion. The spectra shown were acquired after excitation at 280 nm. **B** Changes in the fluorescence emission spectrum of HCMV protease in the presence of saturating amounts of the non-activated carbonyl inhibitor **8**

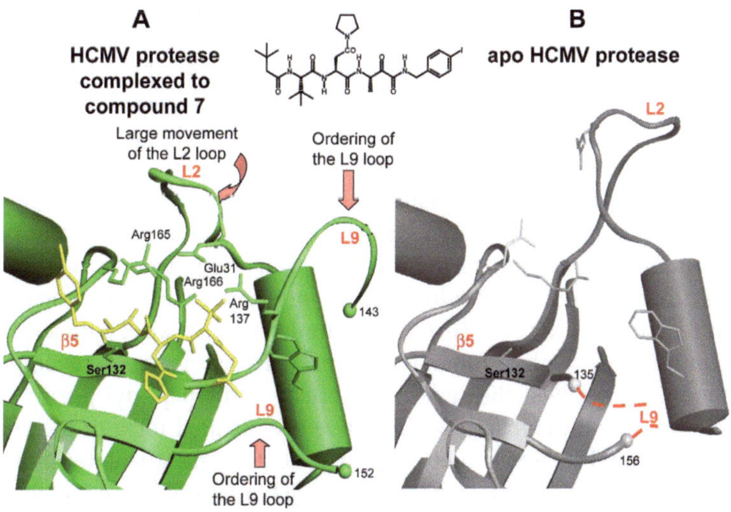

Fig. 11 Peptidyl ligand-induced conformational changes of HCMV protease. Views of the active site region of **A** the inhibited enzyme (*in green*) covalently bound with ketoamide 7 shown *in yellow*, and **B** the apoenzyme (*in gray*). Spheres indicate the points at which there was no electron density further along the L9-loop

peptidyl ligand binding. However, these are only "snapshots" of two discrete states in a crystal lattice. The exact manner by which these changes occur in solution are likely via an ensemble of rapidly interconverting structures, and the complexation with a rigid peptidyl ligand results in a restriction of the flexibility of the protein active-site. In this respect, the process bears elements of protein conformer selection [28]. In other words, the peptide guides the reorganization of the protein around it after association.

2.7 Implications of HCMV Protease as an Induced-Fit Enzyme on Inhibitor Design

Further efforts failed at improving the potency of the peptidyl activated carbonyl inhibitors described herein. This was particularly disappointing since much more potent dipeptidyl inhibitors had been reported for a related serine protease, chymotrypsin, using similar C-terminal warhead technology [29, 30]. Thus, a review and re-analysis of the binding mode was necessary in order to identify other reasonable approaches for drug design. Beginning from the inhibitor point-of-view, it was found that the rigidified and extended peptidyl segment was important for properly anchoring the inhibitor to the bottom of the active-site via hydrogen bonds with the $\beta 5$ strand. Also, the NMR studies demonstrated that these more optimized inhibitors possessed a fair degree of structure in solution that corresponded to their conformation when bound to HCMV protease, which effectively reduced the loss of conformational entropy upon complexation. However, we observed that the 1D proton NMR resonances of covalently enzyme-bound **5** remained relatively sharp, which suggested that, even as hemiketal adducts, the present peptidyl inhibitors retained a high degree of mobility when complexed to HCMV protease, particularly those involving the inhibitor side chains [15]. Therefore, the side-chain interactions were probably quite weak.

On the other hand, the formation of a covalent adduct was found to be a key factor that contributed to a strong interaction and to inhibition, owing to its structural analogy with the catalytic transition-state. This was clearly demonstrated by the $>$ 1000-fold increase in potency when comparing compounds **8** with **5** in which the C-terminal methyl was replaced with a warhead.

In summary [15], inhibition of HCMV protease by the peptidyl compounds described here involved the synergistic effect of the peptidyl portion to induce the transition to a catalytically active or activated form of the enzyme, and an electrophilic keto moiety that reacted to yield a stabilized covalent complex that mimicked the transition state. This rationale accounted for the relatively weak binding of the peptidyl chain and was consistent with the view that, in induced-fit catalysis, binding energy is expended to compensate for the energy required to convert the enzyme to a thermodynamically less favorable state.

Given this limitation, an alternate drug design strategy was therefore necessary. It was then thought that molecules which complex to, but do not induce the conformational change associated with an induced-fit enzyme, would be better suited as inhibitors. Such molecules would be expected to exhibit binding constants reflecting, more directly, the structural interactions between ligand and protein.

In fact, our medicinal chemistry efforts later focused on a class of inhibitors, the β-lactams [31–34] (e.g. compound **12** in Fig. 12), which inhibited HCMV protease by binding to the active site without inducing the structural reorganization as observed by fluorescence emission [15]. The blue-shift normally observed upon the binding of peptidyl inhibitors or substrate (as reported above) was not produced upon saturation of the enzyme with β-lactam inhibitors such as compound **12**, the spectrum remaining essentially identical with that of the free enzyme (data not shown here).

Furthermore, this class of compounds bound competitively with the peptidyl inhibitor **9** for the active-site of HCMV protease. Competition DLB studies showed that the DLB observed for compound **9** in the presence of the protease (b in Fig. 12) was lost upon addition of compound **12** (c in Fig. 12) [15]. Also, $^{1}H\{^{15}N\}$-edited NMR studies demonstrated that these molecules bound covalently (based on the observation of two bound amide resonances in Fig. 13B), and were very slowly cleaved as competitive substrates of HCMV protease (based on the observation of two free amide resonances in Fig. 13B) [35]. Other intermediate bound states were also observed and described in detail elsewhere [35], which provided other possible optimization avenues. Therefore, β-lactam inhibitors served as a new class of inhibitors of HCMV protease which had a greater potential given that its binding mode lacked a corollary entropic cost of induced-fit binding.

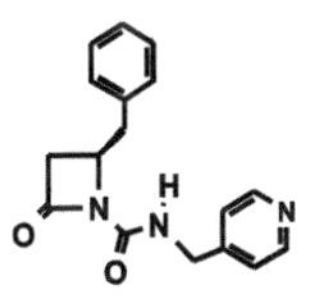

12
IC_{50} = 12 µM

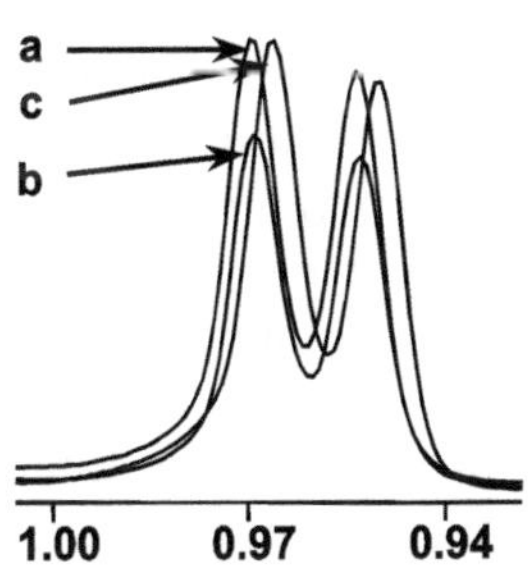

Fig. 12 β-lactam inhibitor **12** is shown along with inhibition constant. Also, ^{1}H NMR DLB experiments are provided that monitored competition binding to HCMV protease. Shown as "a" is the P4-γCH3 doublet of free **9**, and "b" is the same but after adding HCMV protease. The broadening observed in "b" is subsequently lost after the addition of compound **12** which competitively binds to the active-site of HCMV protease. For the sake of clarity, the methyl signals have been skewed slightly

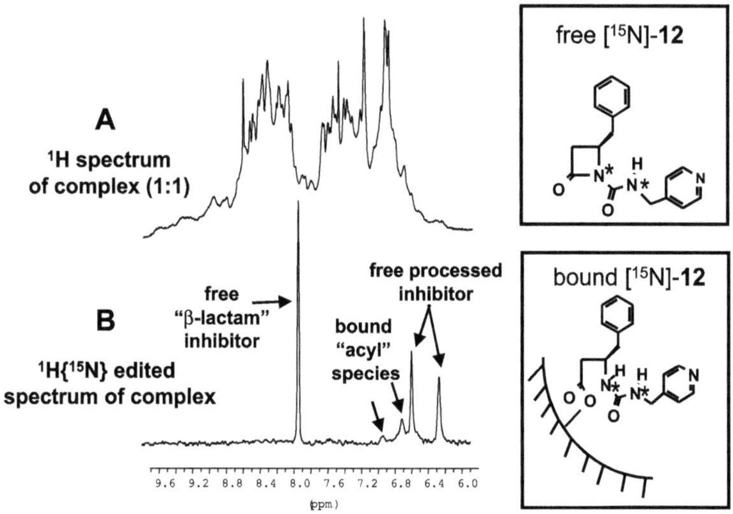

Fig. 13 NMR spectra of the interaction of the ^{15}N-labelled, β-lactam **12** with HCMV protease. **A** Downfield subregion of the ^{1}H NMR spectrum of HCMV protein in the presence of ^{15}N-**12**. ^{15}N-**12** was synthetically labeled with ^{15}N at the positions denoted by *stars* on the structure denoted as "free ^{15}N-**12**". **B** ^{15}N-filtered ^{1}H NMR spectrum acquired on the same sample as that used in **A**. Only the resonance of a ^{15}N nucleus that has a directly attached hydrogen is observed. The various species of compounds are indicated, and the covalently bound and ring-opened species of ^{15}N-**12** is illustrated

2.8 Conclusions

From the perspective of drug design, our medicinal efforts showed an interesting example of how two different classes of competitive inhibitors of comparable potency exercised their activity in two very different ways. In the case of the activated peptidyl ketones, inhibition was mechanism-based and relied on optimal functioning of the catalytic machinery. They depended on complexation of their peptidic chains to bring the enzyme to an activated state, even though this interaction did not contribute to a great degree to the overall binding.

In the case of the β-lactam inhibitors, the intrinsic binding energy appeared to be a much more important factor in providing potency. The absence of induced-fit activation was a particularly attractive feature given that a high entropic cost of binding would be averted. However, these alternate compounds were no longer pursued as HCMV antiviral compounds due to the introduction and fortunate success of HIV antiviral drugs. This was because health complications from HCMV infection were predominantly found in individuals whose immune systems were compromised by HIV infection.

Nonetheless, these compounds could be pursued in the future depending on developing needs. Also, the cumulative experience acquired during this campaign, such as the development of novel methods and strategies, proved to be beneficial early on in a concurrent program that targeted HCV protease [36] and will be described below.

3 HCV Protease—Monitoring Ligand Flexibility by ^{13}C NMR T_1 Relaxation, and the Importance of the Bioactive Conformation

This second account focuses on systematically evaluating the importance the bioactive conformation, and the impact that changes in flexibility have on the binding affinity of related analogues. As was introduced during the HCMV campaign (vide supra), ^{13}C T_1 NMR relaxation data are used as an important tool to monitor changes in molecular flexibility. This discussion begins with an introduction to the rationale for considering HCV protease as an antiviral target. It is then followed by a brief overview of the design of the potent BILN 2061 family of inhibitors starting from an N-terminal product peptide. The next section discusses the rationale for using ^{13}C T_1 data in a drug discovery environment, and the subsequent sections systematically correlate trends in inhibitor activity versus changes in flexibility as determined by ^{13}C T_1 data. Interesting trends are observed, such as unforeseeable long-range dynamics effects.

3.1 HCV as an Antiviral Target

Up to 200 million people around the world are infected with the hepatitis C virus. The majority of individuals with persistent HCV infection will develop chronic hepatitis C, which is a progressive liver disease that can lead to cirrhosis and hepatocellular carcinoma [37–40]. There currently is no antiviral treatment that specifically targets the HCV proteins that are essential for its replication, and the current recommended treatment involves prolonged administration of pegylated-interferon-α and ribavirin. This treatment has limited efficacy against some HCV genotypes and is associated with frequent and severe side effects [41–43].

HCV is an enveloped RNA virus belonging to the Flaviviridae family and Hepacivirus genus. As is typical for this family, its positive-sense RNA genome (9.5 kb) encodes a single precursor polyprotein which undergoes proteolytic maturation by enzymes that include host signalases, and the viral NS2/3 protease and NS3 protease. NS3 protease (which is also referred to here as HCV protease) is responsible for cleaving four of its non-structural (NS) proteins. We, and others, considered early on that HCV protease was an attractive target for the discovery of antiviral compounds. This assumption was later supported by studies that showed it was essential for viral replication in HCV-infected chimpanzees [44]. Moreover, clinical studies involving our HCV protease inhibitor BILN 2061 demonstrated the first proof-of-concept for the discovery of a compound having specific HCV antiviral activity in man [45, 46].

3.2 Overview of the Design of the Potent BILN 2061 Family of Inhibitors Starting from a Weak Peptidic Substrate

Concurrent to our early work during this campaign, literature reports classified HCV protease as a serine protease with a chymotrypsin-like, three-dimensional fold [47, 48]. The catalytic machinery consists of serine 139 (Ser 139), which is responsible for processing the HCV polyprotein, and its catalytic cohorts (His 57 and Asp 81). On the basis of the substrate cleavage sequences of HCV protease, we employed a peptide substrate having the sequence DDIVPC-SMSYTW [49] for in vitro enzymology studies/assays and for the discovery of our initial lead peptidic inhibitor.

An overview of the rational design of BILN 2061 starting from the lead N-terminal product peptide DDIVPC (Fig. 14) has been described elsewhere [36], and a brief highlight of some critical achievements are given below. First, there were the pivotal discoveries that this peptide exhibited product inhibition [36, 50, 51], and that its C-terminal acid contributed to its unusual potency and selectivity for HCV protease [36, 52]. In light of this, and given the failure to identify appropriate leads from screening a large collection of compounds, we chose to pursue a peptidomimetic approach to inhibitor design. This effort proved to be particularly challenging given the shallow and relatively featureless active-site of this unusual protease as viewed from the *apo* crystal structure [47].

The employed strategies [50, 52–68] included an early "knowledge-building" phase in which structural and dynamics data were acquired to,

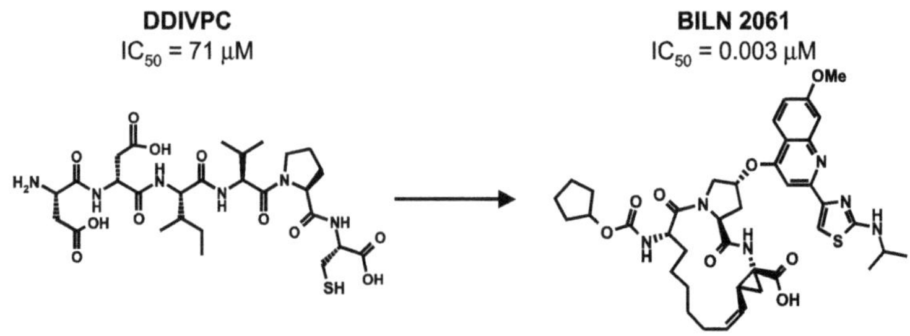

Fig. 14 Structures of the lead peptide DDIVPC and the clinical compound BILN 2061

(1) understand the bioactive conformation of lead peptides when bound to HCV protease, (2) to identify the important substituents that directly contact the protease pocket, and (3) to determine the differences in conformational flexibility between the free and bound states of ligands. With the rational use of this information, medicinal chemists identified potent hexapeptide compounds with improved P1, P2, and P5 substituents. Efforts to reduce the size and peptidic character resulted in N-terminal truncation to tetra- and tripeptidic compounds that had novel P1 and P2 substituents. The macrocyclic scaffold, which is a key characteristic of BILN 2061, was then designed to chemically rigidify the free-state conformation to further resemble the bound-like state, which resulted in a reduction in entropic costs of binding. Having extensive information regarding the binding mode of compounds, medicinal chemists exploited this knowledge in their campaign that eventually led to the BILN 2061 family of compounds.

3.3
A Close Precursor to BILN 2061 Adopts the Same Bioactive Conformation when Free and when Bound to HCV Protease

In the following sections, we will focus on monitoring features of the bioactive conformation to clarify and highlight its importance and its correlation with inhibitor potency. The analysis begins with compound **13** (Fig. 15) which is very similar to BILN 2061. An X-ray crystallographic structure of compound **13** within the active-site of HCV protease (Fig. 15) shows that it adopts an extended backbone conformation, forming several main-chain hydrogen bonds with the spine of HCV's β3 sheet. The P1 and part of the macrocycle lie in the only visible pocket, which appears quite shallow. The large aromatic P2 substituent lies over the surface, partially covering the protease catalytic triad. It is apparent that the proline ring at P2 is mostly solvent exposed and must play a central role as a scaffold that properly displays the three appendages to the protease surface.

This bioactive conformation can then be compared to the predominant conformation of **13** in the free state. To do this, hydrogens were modeled onto the bound structure shown in Fig. 15, and inter-hydrogen distances were measured and compared to those of the free state as determined from ROESY NMR data. The comparison can be found in Table 1, and it is clear that compound **13** adopts the same predominant free-state conformation (ROESY data) as that identified in the bound state by X-ray crystallography. Thus, this compound must pay minimal entropic costs upon binding, which likely contributes to its impressive potency.

13
$K_i = 0.011\ \mu M$

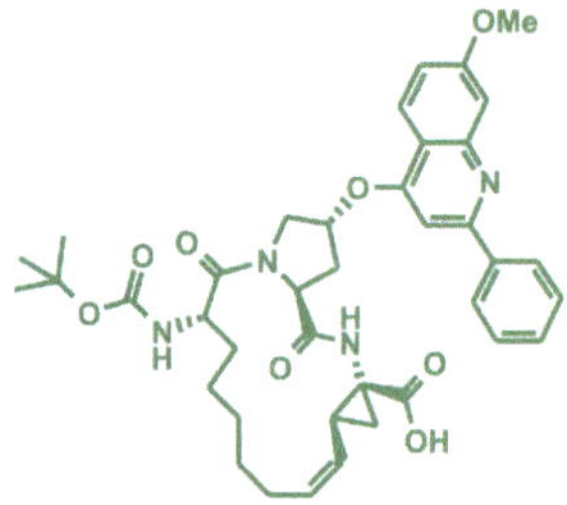

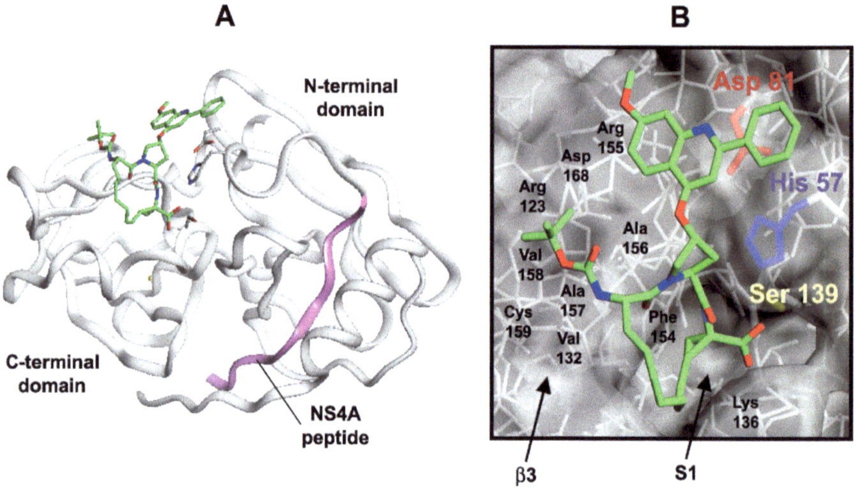

Fig. 15 Compound **13** is shown. **A** X-ray crystal structure of compound **13** (*green*) bound to HCV protease (*gray*). **B** An active-site view of compound **10** (*green*) bound to the HCV protease (*transparent surface*). Selected residues of HCV protease are labeled

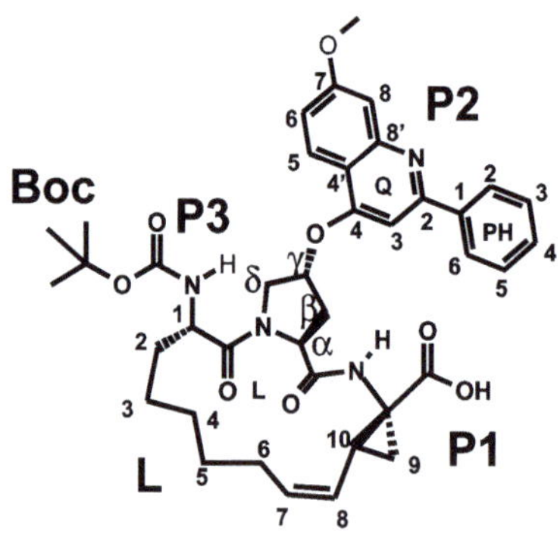

Table 1 Apparent interproton distances (Å) derived from the NMR ROESY and X-ray crystallography for inhibitors **13**[a]

Crosspeak	Free compound **13**	Bound compound **13**
P1–NH–L9	2.7	2.4
P1–NH–L10	3.1	3.6
P1–NH–P2-Hα	2.1	2.5
P2–Hα–P2-Hβ_A	2.4	2.6
P2–Hγ–P2-Hβ_B	3.2	2.6
P2–Hγ–P2-Hβ_A	2.6	2.5
P2–Hγ–P2-Hδ_B	2.7	2.6
P2–Hγ–P2-Hδ_A	2.9	2.8
P2–Hδ_A–P2-Hδ_B	2.3	1.8
P2–Hδ_B–L1	2.3	1.9
P2–Hδ_A–P2-Hβ_A	2.5	2.4
P2–Hδ_B–P2-Hβ_A	3.0	3.6
P3–NH–L1	2.9	3.0
Q–H3–P2-Hβ_B	3.2	3.0
Q–H3–P2-Hδ_B	3.6	4.4
Q–H3–P2-Hγ	2.2	1.8
Q–H5–P2-Hα	3.2	3.6
Q–H5–P2-Hβ_B	4.4	4.9
Q–H5–QH6	2.4	2.4
Q–H6–Q-OMe	3.2	4.0
Q–H8–Q-OMe	2.2	2.6
L7–L6	2.6	3.0
L7–L10	3.2	3.4
L8–L9	2.5	2.7
L8–L6	3.2	3.6
L1–L2	2.4	2.3

[a] Distances for free compound were derived from 200 msec ROESY data (in DMSO-d_6 solvent). Crosspeak volumes were scaled to the Q-H5–Q-H6 correlation which was assigned a distance of 2.4 Å. Scaling also included multiplying the volumes of crosspeaks that involved methyl and *tert*-butyl groups by factors of 0.66 and 0.5, respectively. Distances for bound compound were derived from the X-ray crystal structure, and hydrogens were added to the crystal structure using Accelyrs' InsightII software. Reported distances between the Boc or L linker with other hydrogens were measured to the closest hydrogen. Hydrogens labeled with a subscript B denotes a stereochemistry in which the hydrogen is pointing toward the reader, and vice versa for a hydrogen labeled with a subscript A

3.4 Correlation of ^{13}C T_1 Relaxation with Internal Flexibility and Inhibitor Activity for Small Molecules

As noted in the first account on HCMV protease, it was discovered that there was an inverse correlation between inhibitor potency versus segmental flexibility. The increase in flexibility results in a lower population of free ligands that assume the bioactive conformation. The resultant loss in potency can be explained either as a conformational selection mechanism and an increased entropic cost associated with ligand binding. Given this relationship, a practical means of monitoring flexibility at the atomic-level would be valuable from a drug design point-of-view.

^{13}C T_1 relaxation [15, 69–83] can provide one of the clearest measurements of the dynamic behavior of molecules in solution. However, to be used as a practical tool for medicinal chemistry purposes, data must be acquired and interpreted in an expedient manner given that inhibitor series can evolve rapidly in drug discovery settings. Fortunately, the direct correlation of ^{13}C T_1 relaxation data with molecular flexibility can qualitatively be made for protonated carbons of free ligands, which in general, tumble rapidly and isotropically in solution and satisfy the *extreme narrowing limit* [79–81]. At natural isotopic abundance, T_1 relaxation of ^{13}C nuclei of small molecules is dominated by a dipole-dipole mechanism to their covalently attached hydrogens, at approximately 1.1 Å distance (Fig. 16); and protonated aromatic carbons can also experience contributions via a chemical shift anisotropy mechanism.

Owing to this, ^{13}C T_1 times are highly sensitive to frequencies of motions which occur on the pico- to nano-second timescales. Fortuitously, the internal flexibility of drug-like ligands in the free state, which influences their

$$1/T_1 = nq_{CH}\,[J(\omega_H - \omega_C) + 3J(\omega_C) + 6J(\omega_H + \omega_C)]$$

$J(\omega)$ = spectral density function of the CH vector

$$q_{CH} = (\mu_o/4\pi)^2(\gamma_H\gamma_C h / 2\pi)^2 / 20\mathbf{r}^{\mathbf{6}}_{CH}$$

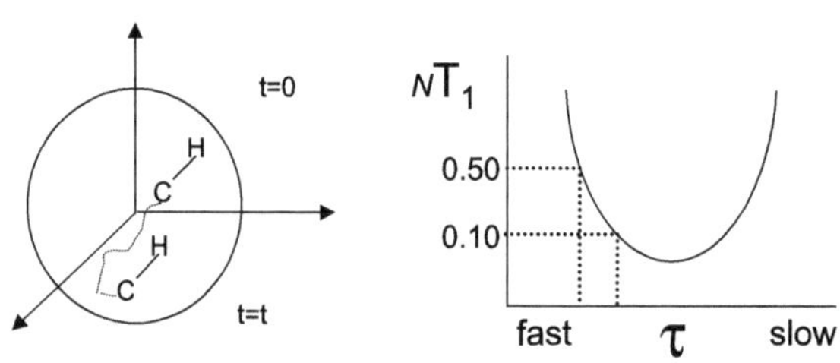

Fig. 16 The relationship between ^{13}C T1 relaxation and spectral density is described mathematically and pictorially

binding affinity to macromolecules via entropic costs or conformational selection, typically occurs within this timescale. Thus, the direct comparison of raw ^{13}C T_1 relaxation times can provide atomic-level details as to the relative flexibility of substituents of a ligand and between related series of compounds that have similar isotropic tumbling properties. Within this context, *longer relaxation times are generally indicative of increased segmental flexibility*. Relaxation data could also be analyzed using a more time-intensive "model-free" approach that has been widely applied for monitoring the dynamics of macromolecules [84]. However, when applied to small molecules, the calculated S2 order parameter of the model-free approach results in an overall inverse correlation with raw ^{13}C T_1 times [73], where S^2 is equal to 1 for a rigid part of a molecule and decreases toward 0 for parts that are less well-ordered. Thus, it is reasonable to probe the internal flexibility of small molecules directly by ^{13}C T_1 measurements which should make it a valuable tool in a pharmaceutical setting.

3.5
^{13}C T_1 to Probe the Influence of Macrocyclization and P1 Cyclization on Flexibility and Potency

^{13}C T_1 times were collected to probe the dynamics effect of macrocyclization by chemically linking the P1 and P3 side-chains. Data were collected on the macrocyclic compound **14** (Fig. 17) and on the related acyclic compound **15**. A comparison of the T_1 times for carbons in the same or similar positions shows that macrocyclization effectively resulted in a reduction in flexibility. This is evident by the observation of shorter T_1 times for the P1 side-chain (cyclopropyl and vinyl) of the macrocyclic compound **14** (Fig. 17, 0.21–0.26 sec) as compared to the acyclic compound **15** (Fig. 17, 0.29–0.32 sec).

Similarly, the P1/P2 backbone of the macrocyclic compound **9** appears to be more rigidified (0.25 sec for the α carbon of P2) as compared to the acyclic compound **3** (0.32 sec for the α-carbon of P2). It is also noteworthy that the macrocycle of compound **14** also succeeds in promoting a rigid P3 backbone as does the acyclic β-branched P3 analogue **15**. Similar ^{13}C T_1 values are observed for the α- and β-carbons of P3 for compound **14** (0.27 and 0.32 sec) and compound **15** (0.27 and 0.31 sec). Given the above data, it is clear that macrocyclic compound **14** has a more rigid backbone and P1 side-chain as compared to the acyclic compound **15**, and it is 10-fold more potent.

^{13}C T_1 data were also used to monitor the dynamics effect of P1 side-chain cyclization. For example, the linear norvaline group experiences significant segmental fluctuations in the free state given the long and incremental increases in T_1 times for the α to δ-carbons of compound **16** (Fig. 18, α 0.34, γ 0.76 and δ 3.45 sec). In contrast, the cyclized P1 side-chain of compound **17** experiences significantly less segmental motion as demonstrated by the rela-

14
$K_i = 0.035\ \mu M$

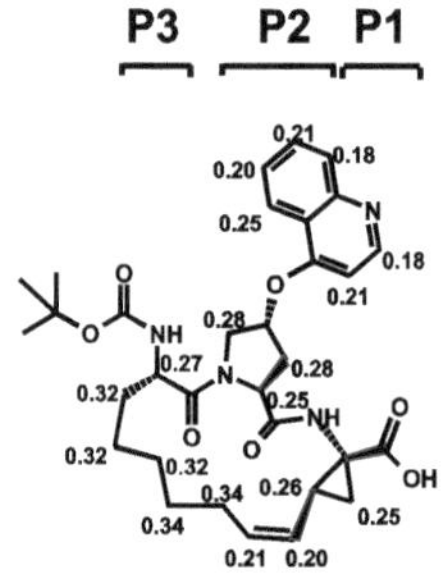

15
$K_i = 0.35\ \mu M$

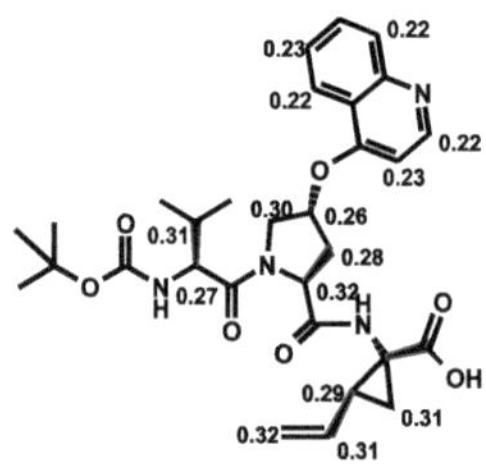

Fig. 17 Inhibitors **14** and **15** are shown with their HCV protease inhibition activity and free-state ^{13}C NT_1 relaxation times. In cases where a carbon has more than one covalently attached hydrogen, NT_1 values are provided where N is the number of attached protons and NT_1 is the product

16
$K_i = 0.61\ \mu M$

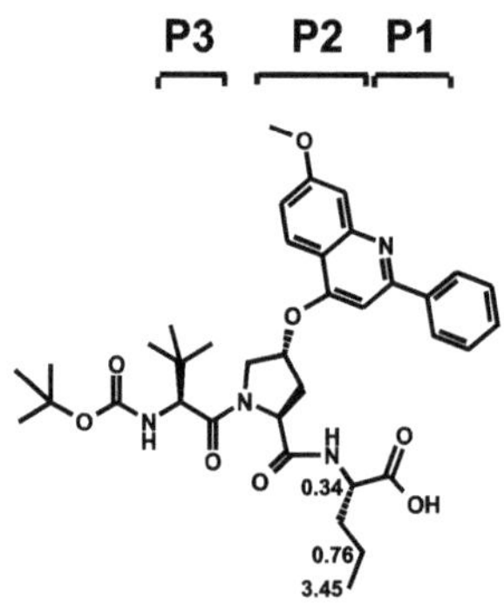

17
$K_i = 0.020\ \mu M$

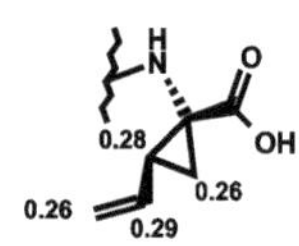

Fig. 18 Inhibitors **16** and **17** are shown with their HCV protease inhibition activity and free-state ^{13}C NT_1 relaxation times

tively lower ^{13}C T_1 times for the vinylcyclopropyl (ranging from 0.26 to 0.29 sec). Although cyclization of the P1 side-chain can contribute to binding affinity in other ways [36], it is clear that it also contributes to a loss in flexibility. Thus, it must be a factor in the observed 30-fold improvement in potency (Fig. 18) when the P1 norvaline (compound **16**, K_i 0.61 μM) is replaced by a P1 vinyl cyclopropyl [59] (compound **17**, K_i 0.020 μM).

3.6
^{13}C T_1 to Probe the Relative Flexibility of P2 Substituents and the Discovery of Long-Range Dynamics Effects

^{13}C T_1 data were also collected on a series of compounds that have different P2 substituents, and which experience a range of potencies. It is clear from Fig. 19 that the more potent compounds (lower K_i values) are also those that exhibit less flexibility as revealed by shorter T_1 times. Perhaps the most remarkable example is the observation that the most potent compound **18** has strikingly shorter T_1 times as compared to the longer times measured for compound **20** which lacks a P2 substituent and measurable activity. In light of this, rigidification to the bioactive conformation must play a role in the > 4000-fold boost in potency observed upon addition of the P2 substituent. However, it must be kept in mind that other binding features must also contribute to the observed boost. Although the binding mode and role of this group is evaluated elsewhere [36], Fig. 15 shows that the large P2 substituent lies over the protease surface and must have some van der Waals contacts. It also partially covers the catalytic triad, suggesting that it may influence the hydrogen bonding network involving the triad with the inhibitors's C-terminal acid (vide infra, Fig. 21A). This latter feature bears a resemblance to the HCMV protease campaign in which it was found that different P2 groups can significantly impact the enzymatic properties (vide supra).

This observation, that multiple parameters can influence structure-activity relationships (SAR), frequently makes it difficult or impossible to deconvolute the relative impact of each binding event. Nonetheless, monitoring the changes in ^{13}C T_1 times helps shed light as to the contributions that are related to changes in ligand dynamics.

More subtle correlations are also found upon comparing the T_1 data for the other P2 substituents. For example, the absence of the third aromatic ring as in compound 18 results in compound **15** which is slightly less potent, and is more flexible (longer T_1 times). Similar trends are also notable with the insertion of a CH2 linker as in compound **19**.

A further comparison of the T_1 data for this series of compounds also leads to an unexpected and interesting observation. P2 groups that have reduced flexibility also result in a reduction in the flexibility of neighboring or distal substituents. For example, shorter ^{13}C T_1 times are observed for the carbons of P1 and P3 of compound **18** as compared to the less potent compounds **15** and **19**, and for compound **20** which lacks a P2 substituent. Overall, these results are consistent with an important correlation

18
$K_i = 0.15\ \mu M$

15
$K_i = 0.35\ \mu M$

19
$K_i = 2.3\ \mu M$

20
$K_i > 600\ \mu M$

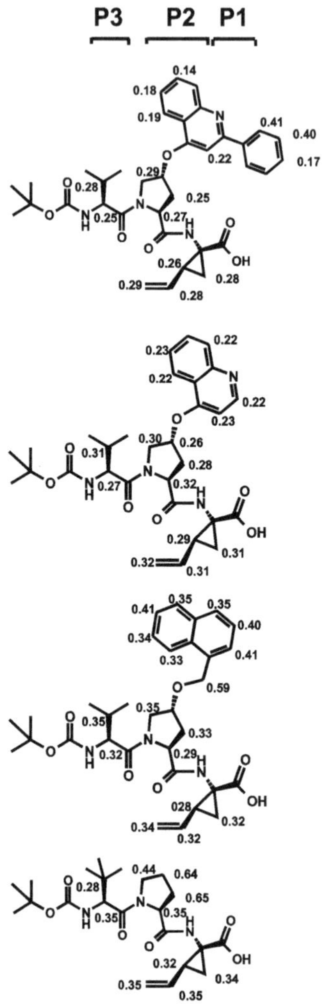

Fig. 19 Inhibitors **15** and **18–20** are shown with their HCV protease inhibition activity and free-state ^{13}C NT1 relaxation times

between short- and long-range, dynamics effects in the free state with inhibitor activity.

3.7
Significant Differences Exist Between the Free and Bound States of Original Peptidic Leads

Given the above trends, it can be imagined that the lead peptide **21**, that has only modest potency, should exhibit important flexibility differences between the free and bound states. The long ^{13}C T_1 times observed for peptide 21 in the free-state is consistent with this (Fig. 20A) [54]. Furthermore, free-state details are also noteworthy. In comparison to the backbone carbons, much longer T_1 times are observed for the side-chains, indicating that they experience greater segmental flexibility.

This is particularly apparent for the norvaline group where long and incremental increases of T_1 times are detected for the α to δ-carbons (Fig. 20A,

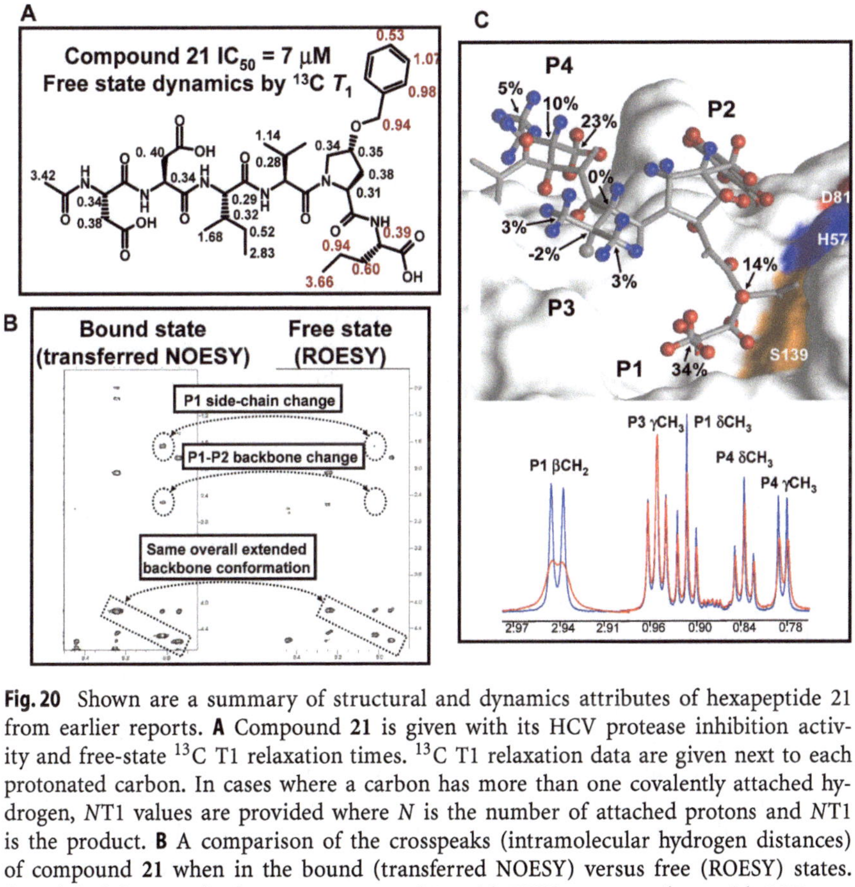

Fig. 20 Shown are a summary of structural and dynamics attributes of hexapeptide 21 from earlier reports. **A** Compound **21** is given with its HCV protease inhibition activity and free-state ^{13}C T1 relaxation times. ^{13}C T1 relaxation data are given next to each protonated carbon. In cases where a carbon has more than one covalently attached hydrogen, *N*T1 values are provided where *N* is the number of attached protons and *N*T1 is the product. **B** A comparison of the crosspeaks (intramolecular hydrogen distances) of compound **21** when in the bound (transferred NOESY) versus free (ROESY) states. **C** Model of the complex between compound **21** with HCV protease, along with DLB perturbation mapping, and transferred ^{13}C T1 data. Shown is a summary of the DLB data which identified the hydrogens that contacted the protease versus those that were solvent exposed. Hydrogens of compound **21** are *colored blue* for resonances in which no broadening perturbations were observed upon binding protease, and hydrogens are *colored red* when significant resonance broadening was observed upon binding. P5 and P6 were determined to be relatively flexible in the bound state and are not shown above. A summary of the transferred ^{13}C T1 data is also displayed as percentage change in ^{13}C T1 before and after the addition of HCV protease

α 0.39, β 0.60, γ 0.94 and δ 3.66 sec). These observations are corroborated by other NMR experiments that compare inter-hydrogen distance information in the bound and free states. The transferred NOESY and ROESY NMR data (Fig. 20B) indicate that both states adopt a predominant extended conformation, whereas important differences are observed for the side-chains. This is also consistent with the novel transferred ^{13}C T_1 experiment that reports ligand sites that become rigidified upon binding. Larger percentage changes in ^{13}C T_1 times upon addition of HCV protease to the NMR tube (Fig. 20C) report those sites that experience significant rigidification upon binding. The potential impact of such information can be substantial. These sites should be re-designed such that they chemically mimic the rigidified bioactive conformation. For example, the P1 carbons in Fig. 20C have large percentage changes, and Fig. 18 illustrates the impact that P1 cyclization can have on lowering the free-state flexibility (lower T_1 times) and increasing activity (i.e. 30-fold boost in potency). One would expect to observe such an improvement in potency when a lower entropic cost is paid upon binding.

The P2 substituent of **21** also exhibits significant flexibility in the free state. Relatively long ^{13}C T_1 times are observed (Fig. 20A) for the benzylic (0.94 sec) and phenyl carbons. The longer relaxation times measured for the ortho (0.98 sec) and meta (1.07 sec) phenyl carbons as compared to the para carbon (0.53 sec) indicates that this aromatic ring undergoes fast rotation or spinning along the benzylic/para-carbon axis. As designed, replacement of the phenyl group with a larger naphthyl group reduced this rotational property; compare the phenyl T_1 times of compound **21** in Fig. 20A with that of the naphthalene of compound **19** in Fig. 19. This design process also required some knowledge of the complex. In this case, the NMR model of the complex involving **21** in Fig. 20C suggested that there should be sufficient space available for large P2 substituents given the surface binding mode. Figure 19 shows examples of other large P2 substituents that were prevalent along our medicinal chemistry campaign.

The P3 valine group of **21** (Fig. 20A) experiences only limited flexibility in the free state as reported by the relatively short ^{13}C T_1 times. Furthermore, the low percentage, transferred ^{13}C T_1 data confirms that only minor changes in flexibility are observed for P3 upon binding. Finally, the DLB and NMR model of the complex reveals that this group has minimal contact with the protease, and is mainly solvent exposed in the bound state (blue-colored hydrogens, Fig. 20C).

Taken together, a proposal for the binding role of the P3 side-chain can be made, particularly in light of its important influence in imparting potency [36] (vide supra, see HCMV protease account). It plays an indirect role in the binding affinity by sterically rigidifying the P3 backbone in the free-state to resemble that of the bound extended conformation [15, 54, 56]. Having taken this into consideration, and given the proximity of the P1 and

P3 side-chains in the bound state (Fig. 20C), the P3 side-chain was rationally replaced by the macrocycle [61, 63].

3.8
HCV Protease Adapts Upon Ligand Binding

Serine proteases have been largely characterized as lock-and-key enzymes [85–87] with a clear exception that the viral HCMV protease assumes induced-fit binding to its substrates and ligands (vide supra) [15]. There is growing evidence that HCV protease also has induced-fit characteristics [88–91] that can impact drug discovery efforts.

Our NMR data reveals that most of the amide NH peaks in the ^{15}N-HSQC NMR map of the HCV protease domain (colored red) change dramatically upon addition of compound **17** (colored blue) (Fig. 21B). Such changes would be expected upon alteration of the global conformation of HCV protease. Data from other reports also suggest that HCV protease undergoes an interdomain rigidification upon addition of the NS4A cofactor or substrate-based ligands [88–95].

We also discovered X-ray crystallographic evidence for this. A superposition of the structure of *apo* HCV protease [96] (pink in Fig. 21A) with one that is bound to compound **17** [57] (green in Fig. 21A) demonstrates that the backbones (displayed as ribbons) can differ in interdomain positions.

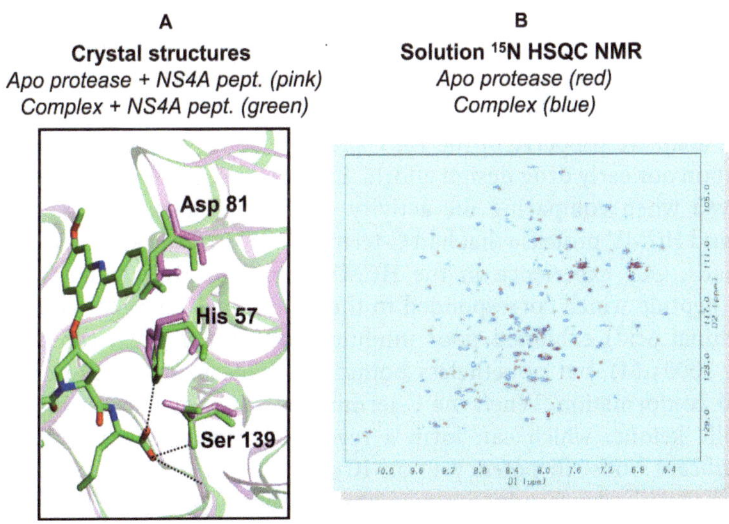

Fig. 21 Comparison of apo versus inhibited HCV protease. **A** Shown is a superposition of X-ray structures of apo protease (*pink ribbon*) and protease when bound to **17** (*green*). **B** Shown is a superposition of ^{15}N-HSQC NMR spectra of apo protease (*red*) and protease when bound to **17** (*blue*)

Local conformational changes are also observed upon binding compound **17**. The residues of the catalytic triad of apo HCV protease are colored pink in Fig. 21A [96], and they shift positions when HCV protease is complexed to compound **17** (colored green in Fig. 21A) [57]. Most notable is that the side-chain of the catalytic Ser 139 rotates away from the carboxylate of the inhibitor. Also, the position of His 57 adjusts slightly to a position which is favorable for forming a hydrogen bond with the carboxylate of the inhibitor, and Asp 81 follows to maintain the hydrogen bond with His 57. Although not shown in Fig. 21A, there appears to be a coordinated conformational change of the side-chains of Arg 155, Asp 168, and Arg 123 upon binding of the inhibitor. This "structural triad" within the active-site appears to play an important role in the binding of compound **17**.

Recently published studies using an HCV subgenomic replicon identified the development of resistance mutations at position 168 in the presence of a compound related to **17** [97] and BILN 2061 [98]. Taken together, these data suggest that HCV protease has adaptive properties that depend on the structural and chemical features of the ligand when bound at the active-site.

The observed plasticity of the triad of HCV protease can be related to several factors. As a viral enzyme that must cleave a variety of sites, HCV protease may have evolutionarily adapted features that help it to modulate the catalytic efficiency and the proper positioning of the triad via influences on the flexibility of the N- and C-terminal subdomains. This can be considered with the context that this protease lacks a disulfide linkage in the vicinity of the catalytic triad and between the subdomains [99, 100], which is a typical feature of extracellular serine proteases. Thus, HCV protease may have established other avenues to manipulate interdomain flexibility and proper active-site alignment (such as the involvement of the NS4A cofactor, zinc ion, and coexistence with the HCV helicase).

The adaptive property of the HCV protease active-site had an important impact on our early drug design efforts. For example, unexpected trends were observed when comparing the activity of N-terminal product peptides of HCV and HCMV protease that had C-terminal acids versus activated carbonyl warheads. Our experience in the HCMV protease program demonstrated that a peptide which corresponded to the N-terminal cleavage product (and C-terminal acid) exhibited poor inhibitory activity (Fig. 22, compound **21** $IC_{50} > 1000\ \mu M$), but nonetheless bound the protease in a well-defined extended conformation. When the C-terminal acid was replaced with an electrophilic ketone, which can form a reversible covalent adduct, there was a significant boost in potency to an IC_{50} of $1.8\ \mu M$ for compound **22**. Expecting that similar trends would be observed for peptides of HCV protease, we were surprised to find that the N-terminal product peptide **23** was more potent than anticipated ($IC_{50} = 17\ \mu M$) and thus became an early lead for further SAR efforts [50]. We immediately sought to replace the C-terminal acid with a classical activated-carbonyl warhead with the expectation that it would

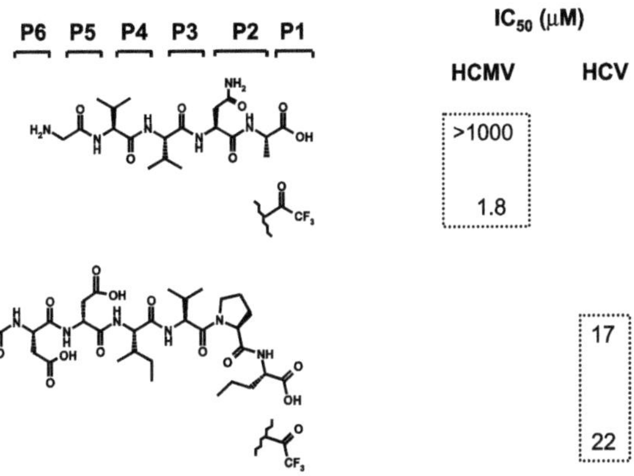

Fig. 22 Inhibitors **21** and **22** of HCMV protease, and inhibitors **23** and **24** of HCV protease are shown

result in an additional boost in activity. Again to our surprise, however, the C-terminal acid and trifluoromethylketone analogues exhibited comparable activity (compounds **23** and **24** in Fig. 22) [52], suggesting that the adaptive properties of HCV protease allow it to bind both types of C-termini.

Unlike the case of HCMV protease where only compounds with an activated carbonyl were active enough to be considered as a lead for optimization (compound **22** is much more active than **21** as shown in Fig. 22) [15, 25], both product and activated carbonyl inhibitors of HCV protease were viable starting points for optimization (compounds **23** and **24**) [50, 52, 53]. The decision to optimize the product peptide, which had a C-terminal acid, stemmed from the fact that the trifluoromethylketone peptide **24** lacked specificity for HCV protease and bound to other proteases with significantly higher affinity (e.g. an $IC_{50} < 0.06\ \mu M$ was measured against human leukocyte elastase). In summary, the unusual adaptive features of the active-site of HCV protease (and sub-domains) likely resulted in the unexpected potency of peptides having a C-terminal acid. As a result, the beneficial properties of this group were exploited throughout our medicinal chemistry campaign that led to the design of the BILN 2061 family of compounds.

3.9 Conclusions

This latter account describes the practical application of ^{13}C T_1 data to monitor the flexibility of free compounds in a drug discovery setting. Overall, the relationship between the free-state bioactive conformation with inhibitory activity was systematically monitored. Furthermore, the strategy that was implemented to monitor ligand dynamics can have general utility and is not restricted to peptide-like compounds. We have recently described a related approach involving non-peptidic inhibitors of HCV polymerase [101]. It would also be interesting to explore other NMR relaxation parameters to obtain further ligand dynamics information as suggested by others [102]. Finally, this account also describes some adaptive features observed for the active-site of HCV protease, along with examples of its impact on drug design.

4 Concluding Remarks

The work described here provides two detailed examples of the exploitability of the adaptive features of ligands and binding pockets. Although other examples have been described in the literature, this work has the distinction of systematically monitoring the structure and dynamics features of ligands using multidisciplinary strategies. It is certain that the well-characterized systems described here are not unique, and that most bi-molecular interactions involve a range of adaptive processes. The simplistic lock-and-key and induced-fit views of ligand binding must evolve to a better understanding of intermediate events and properties. The work described here can serve as examples to monitor and exploit adaptive features.

Acknowledgements The work described herein involved many colleagues at Boehringer Ingelheim, and I would like to acknowledge their valuable contributions. I am also grateful to M. Bös, R. Bethell, and M. Cordingley for their encouragement and support. I also thank N. Aubry, S. Kawai, J. Gillard, P. White, and P. Bonneau for insightful discussions or for critical reading of the manuscript.

References

1. Mocarski ES Jr (1995) In: Fields BN (ed) Virology 2. Lippincott-Raven, Philadelphia, PA, pp 2447–2492
2. Britt WJ, Alford CA (1995) In: Fields BN (ed) Virology 2. Lippincott-Raven, Philadelphia, PA, pp 2493–2525
3. Gibson W, Welch AR, Hall MRT (1994) Perspect Drug Discov Des 2:413
4. Gao M, Matusick-Kumar L, Hurlburt W, DiTusa SF, Newcomb WW, Brown JC, McCann PJ III, Deckman I, Colonno RJ (1994) J Virol 68:3702
5. Preston VG, Coates JAV, Rixon FJ (1983) J Virol 45:1056
6. Matusick-Kumar L, McCann PJ III, Robertson BJ, Newcomb WW, Brown JC, Gao M (1995) J Virol 69:113
7. Casjens S, King J (1975) Annu Rev Biochem 44:555
8. Tong L, Qian C, Massariol M-J, Bonneau PR, Cordingley MG, Lagacé L (1996) Nature 383:272
9. Qiu X, Culp JS, DiLella AG, Hellmig B, Hoog SS, Janson CA, Smith WW, Abdel-Meguid SS (1996) Nature 383:275

10. Shieh H-S, Kurumbail RG, Stevens AM, Stegeman RA, Sturman EJ, Pak JY, Wittwer AJ, Palmier MO, Wiegand RC, Holwerda BC, Stallings WC (1996) Nature 383:279
11. Chen P, Tsuge H, Almassy RJ, Gribskov CL, Katoh S, Vanderpool DL, Margosiak SA, Pinko C, Matthews DA, Kan C-C (1996) Cell 86:835
12. Margosiak SA, Vanderpool DL, Sisson W, Pinko C, Kan C-C (1996) Biochemistry 35:5300
13. Darke PL, Cole JL, Waxman L, Hall DL, Sardana MK, Kuo LC (1996) J Biol Chem 271:7445
14. Bonneau PR, Grand-Maître C, Greenwood DJ, Lagacé L, LaPlante SR, Massariol M-J, Ogilvie WW, O'Meara JA, Kawai SH (1997) Biochemistry 36:12644
15. LaPlante SR, Bonneau P, Aubry N, Cameron DR, Déziel R, Grand-Maître C, Plouffe C, Tong L, Kawai SH (1999) J Am Chem Soc 121:2974
16. LaPlante SR, Aubry N, Bonneau PR, Cameron DR, Lagacé L, Massariol M-J, Montpetit H, Plouffe C, Kawai SH, Fulton BD, Chen ZG, Ni F (1998) Biochemistry 37:9793
17. Schechter J, Berger A (1970) Philos Trans R Soc London, Ser B 257:249
18. Sardana VV, Wolfgang JA, Veloski CA, Long WJ, Legrow K, Wolanski B, Emini EA, LaFemina RL (1994) J Biol Chem 269:14337
19. LaFemina RL, Bakshi K, Long WJ, Pramanik B, Veloski CA, Wolanski BS, Marcy AI, Hazuda DJ (1996) J Virol 70:4819
20. Ni F (1994) Prog NMR Spectrosc 26:517
21. Fersht A (1985) Enzyme Structure and Mechanisms. 2nd edn. Freeman and Co, New York
22. Polgár L (ed) (1989) Mechanisms of Protease Action. Chap. 3. CRC Press Inc, Boca Raton, FL
23. Hedstrom L (2002) Chem Rev 102:4501
24. Mercer DF, Schiller DE, Elliott JF, Douglas DN, Hao C, Rinfret A, Addison WR, Fischer KP, Churchill TA, Lakey JRT, Tyrrell DLJ, Kneteman NM (2001) Nat Med 7:927
25. Ogilvie WW, Bailey M, Poupart M-A, Abraham A, Bhavsar A, Bonneau PR, Bordeleau J, Bousquet Y, Chabot C, Duceppe J-S, Fazal G, Goulet S, Grand-Maître C, Guse I, Halmos T, Lavallée P, Leach M, Malenfant E, O'Meara JA, Plante R, Plouffe C, Poirier M, Soucy F, Yoakim C, Déziel R (1998) J Med Chem 40:4113
26. LaPlante SR, Cameron DR, Aubry N, Bonneau PR, Déziel R, Grand-Maître C, Ogilvie WW, Kawai SH (1998) Angew Chem Int Ed Engl 37:2729
27. Tong L, Qian C, Massariol M-J, Déziel R, Yoakim C, Lagacé L (1998) Nature Struct Biol 5:819
28. Cao Y, Musah RA, Wilcox SK, Goodin DB, McRee DE (1998) Protein Sci 7:72
29. Wolfenden R (1972) Acc Chem Res 5:10
30. Wolfenden R (1976) Annu Rev Biophys 5:271
31. Yoakim C, Ogilvie WW, Cameron DR, Chabot C, Guse I, Haché B, Naud J, O'Meara JA, Plante R, Déziel R (1998) J Med Chem 41:2882
32. Déziel R, Malenfant E (1998) Bioorg Med Chem Lett 8:1437
33. Yoakim C, Ogilvie W, Cameron DR, Chabot C, Grand-Maître C, Guse I, Haché B, Kawai SH, Naud J, O'Meara JA, Plante R, Déziel R (1998) Antiviral Chem Chemother 9:379
34. Borthwick AD, Weingarten G, Haley TM, Tomaszewski M, Wand W, Hu Z, Bédard J, Jin H, Yuen L, Mansour TS (1998) Bioorg Med Chem Lett 8:365
35. Bonneau PR, Hasani F, Plouffe C, Malenfant E, LaPlante SR, Guse I, Ogilvie WW, Plante R, Davidson WC, Hopkins JL, Morelock MG, Cordingley MG, Déziel R (1999) J Am Chem Soc 121:2965
36. LaPlante SR, Llinàs-Brunet M (2005) Curr Med Chem Antiinfective Agents 4:111

37. Choo Q-L, Kuo G, Weiner AJ, Overby LR, Bradley DW, Houghton M (1989) Science 244:359
38. Kuo G, Choo QL, Alter HJ, Gitnick GL, Redeker A G, Purcell RH, Miyamura T, Dienstag JL, Alter MJ, Stevens CE (1989) Science 244:362
39. Di Bisceglie AM (1998) Lancet 351:351
40. World Health Organization (1999) J Virol Hepat 6:35
41. Chander G, Sulkowski MS, Jenckes MW, Torbenson MS, Herlong HF, Bass EB, Gebo KA (2002) Hepatology 36:S135
42. Tan S-L, Pause A, Shi Y, Sonenberg N (2002) Nature Rev Drug Discov 1:867
43. Di Bisceglie AM, McHutchison J, Rice CM (2002) Hepatology 35:224
44. Kolykhalov AA, Mihalik K, Feinstone SM, Rice CJ (2000) Virology 74:2046
45. Lamarre D, Anderson PC, Bailey M, Beaulieu P, Bolger G, Bonneau P, Bös M, Cameron DR, Cartier M, Cordingley MG, Faucher A-M, Goudreau N, Kawai SH, Kukolj G, Lagacé L, LaPlante SR, Narjes H, Poupart M-A, Rancourt J, Sentjens RE, St George R, Simoneau B, Stelnmann G, Thibeault D, Tsantrizos YS, Weldon SM, Yong C-L, Linas-Brunet M (2003) Nature 426:186
46. Hinrichsen H, Benhamou Y, Wedemeyer H, Reiser M, Sentjens RE, Calleja JL, Forns X, Erhardt A, Crönlein J, Chaves R, Yong C-L, Nehmiz G, Steinmann G (2004) Gasteroenterology 127:1347 (Doi:10 1016/j gastro 2004 08 002)
47. Love RA, Parge HE, Wickersham JA, Hostomsky Z, Habuka N, Moomaw EW, Adachi T, Hostomska Z (1996) Cell 87:331
48. Kim JL, Morgenstern KA, Lin C, Fox T, Dwyer MD, Landro JA, Chambers SP, Markland W, Lepre CA, O'Malley ET, Harbeson SL, Rice CM, Murcko MA, Caron PR, Thomson JA (1996) Cell 87:343
49. Kakiuchi N, Hijikata M, Komoda Y, Tanji Y, Hirowatari Y, Shimotohno K (1995) Biochem Biophys Res Comm 210:1059
50. Llinàs-Brunet M, Bailey M, Fazal G, Goulet S, Halmos T, LaPlante S, Maurice R, Poirier M, Poupart M-A, Thibeault D, Wernic D, Lamarre D (1998) Bioorganic and Medicinal Chemistry Letters 8:1713
51. Steinkühler C, Biasiol G, Brunetti M, Urbani A, Koch U, Cortese R, Pessi A, De Francesco R (1998) Biochemistry 37:8899
52. Llinàs-Brunet M, Bailey M, Déziel R, Fazal G, Gorys V, Goulet S, Halmos T, Maurice R, Poirier M, Poupart M-A, Rancourt J, Thibeault D, Wernic D, Lamarre D (1998) Bioorg Med Chem Lett 8:2719
53. Bailey M, Halmos T, Goudreau N, Lescop E, Llinàs-Brunet M (2004) J Med Chem 47:3788
54. LaPlante SR, Cameron DR, Aubry N, Lefebvre S, Kukolj G, Maurice R, Thibeault D, Lamarre D, Llinàs-Brunet M (1999) J Biol Chem 274:18618
55. Llinàs-Brunet M, Bailey M, Fazal G, Ghiro E, Gorys V, Goulet S, Halmos T, Maurice R, Poirier M, Poupart M-A, Rancourt J, Thibeault D, Wernic D, Lamarre D (2000) Bioorg Med Chem Lett 10:2267
56. LaPlante SR, Aubry N, Bonneau P, Kukolj G, Lamarre D, Lefebvre S, Li H, Llinàs-Brunet M, Plouffe C, Cameron D R (2000) Bioorg Med Chem Lett 10:2271
57. Tsantrizos Y, Bolger G, Bonneau P, Cameron DR, Goudreau N, Kukolj G, LaPlante SR, Llinàs-Brunet M, Nar H, Lamarre D (2003) Angew Chem Int Ed 42:1356
58. LaPlante SR, Aubry N, Deziel R, Ni F, Xu P (2000) J Am Chem Soc 122:12530
59. Rancourt J, Cameron DR, Gorys V, Lamarre D, Poirier M, Thibeault D, Llinàs-Brunet M (2004) J Med Chem 47:2511
60. Poupart M-A, Cameron DR, Chabot C, Ghiro E, Goudreau N, Goulet S, Poirier M, Tsantrizos Y (2001) J Org Chem 66:4743

61. Goudreau N, Cameron DR, Bonneau P, Gorys V, Plouffe C, Poirier M, Lamarre D, Llinàs-Brunet M (2004) J Med Chem 47:123
62. Llinàs-Brunet M, Bailey M, Ghiro E, Gorys V, Halmos T, Poirier M, Rancourt J, Goudreau N (2004) J Med Chem 47:6584
63. Goudreau N, Brochu C, Cameron DR, Duceppe J-S, Faucher A-M, Ferland J-M, Grand-Maitre C, Poirier M, Simoneau B, Tsantrizos YS (2004) J Org Chem 69:6185
64. Pause A, Kukolj G, Bailey M, Brault M, Do F, Halmos T, Lagace L, Maurice R, Marquis M, McKercher G, Pellerin C, Pilote L, Thibeault D, Lamarre D (2003) J Biol Chem 278:20374
65. Ali S, Pellerin C, Lamarre D, Kukolj G (2004) J Virol 78:491
66. Llinàs-Brunet M, Bailey M, Bolger G, Brochu C, Faucher AM, Ferland J-M, Garneau M, Ghiro E, Gorys V, Grand-Maitre C, Halmos T, Lapayre-Pachette N, Liard F, Poirier M, Rheame M, Tsantrizos YS, Lamarre D (2004) J Med Chem 47:1605
67. Thibeault D, Bousquet C, Gingras R, Lagace L, Maurice R, White P, Lamarre D (2004) J Virol 78:7352
68. Faucher A-M, Bailey M, Beaulieu PL, Brochu C, Duceppe J-S, Ferland J-M, Ghiro E, Gorys V, Halmos T, Kawai SH, Poirier M, Simoneau B, Tsantrizos YS, Llinàs-Brunet M (2004) Organic Letters 6:2901
69. Kessler H, Bats JW, Griesinger C, Koll S, Will M, Wagner K (1988) J Am Chem Soc 110:1033
70. Dellwo MJ, Wand JJ (1989) J Am Chem Soc 111:4571
71. Palmer AG, Rance M, Wright P (1991) J Am Chem Soc 113:4371
72. Detlefsen DJ, Hill SE, Day SH, Lee MS (1999) Curr Med Chem 6:353
73. Briand J, Kopple KJ (1995) J Biol NMR 6:347
74. Peng JW, Schiffer CA, Xu P, Gunsteren WF, Ernst RR (1996) J Bio NMR 8:453
75. Zhu L, Prendergast FG, Kemple MD (1998) J Biol NMR 12:133
76. Friedrichs MS, Stouch TR, Bruccoleri RE, Mueller L, Constantine KL (1995) J Am Chem Soc 117:10855
77. Ramirez-Alvarado M, Daragen VA, Serrano L, Mayo KH (1998) Protein Sci 7:720
78. Daragen VA, Ilyina EE, Fields GB, Mayo KH (1997) Protein Sci 6:355
79. Abragam A (1961) The Principles of Nuclear Magnetism. Clarendon Press, Oxford
80. Ernst RR, Bodenhausen G, Wokaun A (1987) Principles of Nuclear Magnetic Resonance in One and Two Dimensions. Clarendon Press, Oxford, pp 49–57
81. Levy GC, Lichter RL, Nelson GL (1980) Carbon-13 Nuclear Magnetic Resonance. 2nd edn., Chap. 8. Wiley, New York
82. Peng JW, Wagner G (1992) J Magn Reson 98:308
83. Peng JW, Wagner G (1992) Biochemistry 31:8571
84. Lipari G, Szabo A (1982) J Am Chem Soc 104:4546
85. Fersht A (1985) Enzyme Structure and Mechanisms. 2nd edn. Freeman and Co, New York
86. Polgár L (ed) (1989) Mechanisms of Protease Action. Chap. 3. CRC Press Inc, Boca Raton, FL
87. Hedstrom L (2002) Chem Rev 102:4501
88. Barbato G, Cicero DO, Nardi MC, Steinkühler C, Cortese R, De Francesco R, Bazzo R (1999) J Mol Biol 289:371
89. Love RA, Parge HE, Wickersham JA, Hostomsky Z, Habuka N, Moomaw EW, Adachi T, Margosiak S, Dagostino E, Hostomaska Z (1998) Clinical and Diagnostic Virology 10:151
90. McCoy MA, Senior MM, Gesell JJ, Ramanathan L, Wyss DF (2001) J Mol Biol 305:1099

91. Bianchi E, Pessi A (2002) Biopolymers 66:101
92. Barbato G, Cicero DO, Cordier F, Narjes F, Gerlach B, Sambucini S, Grzesiek S, Matassa VG, De Francesco R, Bazzo R (2000) EMBO J 19:1195
93. Urbani A, Bazzo R, Nardi MC, Cicero DO, De Francesco R, Steinkühler C, Barbato G (1998) J Biol Chem 273:18760
94. Bianchi E, Orrù S, Dal Piaz F, Ingenito R, Casbarra A, Biasiol G, Koch U, Pucci P, Pessi A (1999) Biochemistry 38:13844
95. Archer SJ, Camac DM, Wu ZJ, Farrow NA, Domaille PJ, Wasserman ZR, Bukhtiyarova M, Rizzo C, Jagannathan S, Mersinger LJ, Kettner CA (2002) Chem Biol 9:79
96. Yan Y, Li Y, Munshi S, Sardana V, Cole JL, Sardana M, Steinkühler C, Tomei L, De Francesco R, Kuo L, Chen Z (1998) Protein Sci 7:837
97. Trozzi C, Bartholomew L, Ceccacci A, Biasiol G, Pacini L, Altamura S, Narjes F, Muraglia E, Paonessa G, Koch U, De Francesco R, Steinkühler C, Migliaccio G (2003) J Virology 77:3669
98. Lin C, Lin K, Luong Y-P, Rao RB, Wei Y-Y, Brennan DL, Fulghum JR, Hsiao H-M, Ma S, Maxwell JP, Cottrell KM, Perni RB, Gates CA, Kwong AD (2004) J Biol Chem 279:17508
99. Tong L, Wengler G, Rossman MG (1993) J Mol Biol 230:228
100. Tong L (2002) Chem Rev 102:4609
101. LaPlante SR, Jakalian A, Aubry N, Bousquet Y, Ferland J-M, Gillard J, Lefebvre S, Poirier M, Tsantrizos YS, Kukolj G, Beaulieu PL (2004) Angew Chem Int Ed 43:4306
102. Peng JW (2003) J Am Chem Soc 125:1116 MKW

Author Index Volumes 251–272

Author Index Vols. 26–50 see Vol. 50
Author Index Vols. 51–100 see Vol. 100
Author Index Vols. 101–150 see Vol. 150
Author Index Vols. 151–200 see Vol. 200
Author Index Vols. 201–250 see Vol. 250

The volume numbers are printed in italics

Subject Index

MIX
Papier aus verantwortungsvollen Quellen
Paper from responsible sources
FSC® C105338

If you have any concerns about our products,
you can contact us on
ProductSafety@springernature.com

In case Publisher is established outside the EU,
the EU authorized representative is:
Springer Nature Customer Service Center GmbH
Europaplatz 3, 69115 Heidelberg, Germany

Printed by Libri Plureos GmbH
in Hamburg, Germany